VOLUME EIGHTY SEVEN

ADVANCES IN PARASITOLOGY

Mathematical Models for Neglected Tropical Diseases: Essential Tools for Control and Elimination, Part A

VOLUME EIGHTY SEVEN

ADVANCES IN PARASITOLOGY

Mathematical Models for Neglected Tropical Diseases: Essential Tools for Control and Elimination, Part A

Edited by

ROY M. ANDERSON

Imperial College London, Department of Infectious Disease Epidemiology, Faculty of Medicine, School of Public Health St Mary's Campus, Norfolk Place, London

MARIA GLORIA BASÁÑEZ

Imperial College London, Department of Infectious Disease Epidemiology, Faculty of Medicine, School of Public Health St Mary's Campus, Norfolk Place, London

ELSEVIER

AMSTERDAM • BOSTON • HEIDELBERG • LONDON
NEW YORK • OXFORD • PARIS • SAN DIEGO
SAN FRANCISCO • SINGAPORE • SYDNEY • TOKYO
Academic Press is an imprint of Elsevier

Academic Press is an imprint of Elsevier
125 London Wall, London, EC2Y 5AS, UK
The Boulevard, Langford Lane, Kidlington, Oxford OX5 1GB, UK
225 Wyman Street, Waltham, MA 02451, USA
525 B Street, Suite 1800, San Diego, CA 92101-4495, USA

First edition 2015

ISBN: 978-0-12-803256-5
ISSN: 0065-308X

For information on all Academic Press publications
visit our website at store.elsevier.com

CONTENTS

CONTRIBUTORS

Roy M. Anderson
Imperial College London, Department of Infectious Disease Epidemiology, Faculty of Medicine, School of Public Health St Mary's Campus, Norfolk Place, London, UK

Sarah M. Bartsch
Public Health Computational and Operations Research (PHICOR) and International Vaccine Access Center (IVAC), Johns Hopkins Bloomberg School of Public Health, Baltimore, MD, USA

Dirk Berkvens
Department of Biomedical Sciences, Institute of Tropical Medicine, Antwerp, Belgium

David J. Blok
Department of Public Health, Erasmus MC, University Medical Center Rotterdam, Rotterdam, The Netherlands

Johannes Charlier
Department of Virology, Parasitology and Immunology, Faculty of Veterinary Medicine, Ghent University, Merelbeke, Belgium

Nakul Chitnis
Department of Epidemiology and Public Health, Swiss Tropical and Public Health Institute, Basel, Switzerland; University of Basel, Basel, Switzerland

James A. Cotton
Wellcome Trust Sanger Institute, Wellcome Trust Genome Campus, Cambridge, UK

Thomas Crellen
Department of Infectious Disease Epidemiology, School of Public Health, Faculty of Medicine, Imperial College London, St Mary's Campus, London, UK; Wellcome Trust Sanger Institute, Wellcome Trust Genome Campus, Cambridge, UK

Zulma M. Cucunubá
Department of Infectious Disease Epidemiology, Imperial College London, London, UK; Grupo de Parasitología, Instituto Nacional de Salud, Colombia

Sake J. de Vlas
Department of Public Health, Erasmus MC, University Medical Center Rotterdam, Rotterdam, The Netherlands

Brecht Devleesschauwer
Department of Virology, Parasitology and Immunology, Faculty of Veterinary Medicine, Ghent University, Merelbeke, Belgium; Institute of Health and Society (IRSS), Université catholique de Louvain, Brussels, Belgium

Egil A.J. Fischer
Department of Epidemiology, Crisis Organisation and Diagnostics, Central Veterinary Institute, Part of Wageningen UR, Lelystad, The Netherlands

Manoj Gambhir
Department of Epidemiology and Preventive Medicine, Monash University, Melbourne, VIC, Australia

Katrin M. Gorham
Public Health Computational and Operations Research (PHICOR) and International Vaccine Access Center (IVAC), Johns Hopkins Bloomberg School of Public Health, Baltimore, MD, USA

Sébastien Gourbière
Institut de Modélisation et d'Analyse en Géo-Environnements et Santé (IMAGES), Université de Perpignan Via Domitia, Perpignan, France

Ian M. Hastings
Department of Parasitology, Liverpool School of Tropical Medicine, Liverpool, UK

Matt J. Keeling
Mathematics Institute/WIDER, University of Warwick, Coventry, UK

Poppy H.L. Lamberton
Department of Infectious Disease Epidemiology, School of Public Health, Faculty of Medicine, Imperial College London, St Mary's Campus, London, UK

Bruce Y. Lee
Public Health Computational and Operations Research (PHICOR) and International Vaccine Access Center (IVAC), Johns Hopkins Bloomberg School of Public Health, Baltimore, MD, USA

Bruno Levecke
Department of Virology, Parasitology and Immunology, Faculty of Veterinary Medicine, Ghent University, Merelbeke, Belgium

Edwin Michael
Department of Biological Sciences, University of Notre Dame, Notre Dame, IN, USA

Pierre Nouvellet
Medical Research Council Centre for Outbreak Analysis and Modelling, Department of Infectious Disease Epidemiology, Imperial College London, London, UK

Jan Hendrik Richardus
Department of Public Health, Erasmus MC, University Medical Center Rotterdam, Rotterdam, The Netherlands

Kat S. Rock
Mathematics Institute/WIDER, University of Warwick, Coventry, UK

Brajendra K. Singh
Department of Biological Sciences, University of Notre Dame, Notre Dame, IN, USA

Niko Speybroeck
Institute of Health and Society (IRSS), Université catholique de Louvain, Brussels, Belgium

Wilma A. Stolk
Department of Public Health, Erasmus MC, University Medical Center Rotterdam, The Netherlands

Chris M. Stone
Department of Epidemiology and Public Health, Swiss Tropical and Public Health Institute, Basel, Switzerland; University of Basel, Basel, Switzerland

Steve J. Torr
Warwick Medical School, University of Warwick, Coventry, UK; Department of Vector Biology, Liverpool School of Tropical Medicine, Liverpool, UK

Stefan Van Aelst
Department of Applied Mathematics, Computer Science and Statistics, Faculty of Sciences, Ghent University, Ghent, Belgium; Department of Mathematics, Faculty of Sciences, KU Leuven University, Leuven, Belgium

Jozef Vercruysse
Department of Virology, Parasitology and Immunology, Faculty of Veterinary Medicine, Ghent University, Merelbeke, Belgium

Joanne P. Webster
Department of Infectious Disease Epidemiology, School of Public Health, Faculty of Medicine, Imperial College London, St Mary's Campus, London, UK; Department of Pathology and Pathogen Biology, Centre for Emerging, Endemic and Exotic Diseases (CEEED), Royal Veterinary College, University of London, London, UK

PREFACE

Three decades ago, *Advances in Parasitology* published an article on 'Helminth infections of humans: mathematical models, population dynamics, and control', which attempted to summarize progress at that time in the development of mathematical models for the transmission dynamics of helminths that cause disease in human communities, and their use in evaluating the impact of mass drug administration (MDA) to control infection (Anderson & May, 1985). Expanded interest in the control of the so-called Neglected Tropical Diseases (NTDs) in recent years, spurred on in part by drug donations from the pharmaceutical industry to the World Health Organization to be used in resource-poor settings with endemic infection, and other international initiatives, led us to conclude that it is a sensible time to review again progress on model development and associated applications in the design of control policies, not just for the helminthiases, but for other infections falling under the remit of the NTDs.

The launch of the London Centre for Neglected Tropical Disease Research (LCNTDR) (http://www.londonntd.org/) in January 2013—a collaboration between the London School of Hygiene and Tropical Medicine, Imperial College London, the Natural History Museum and the Royal Veterinary College to further research on how best to control the NTDs—provided an opportunity to discuss the idea for a new review of mathematical models with the present editor of *Advances in Parasitology*, Professor David Rollinson. The launch of the LCNTDR took place on the 30th of January 2013, exactly one year after the London Declaration on Neglected Tropical Diseases (http://unitingtocombatntds.org/resource/london-declaration) and the announcement by the Director-General of the World Health Organization, Dr Margaret Chan, of a roadmap to accelerate work to overcome the impact of NTDs on the 30th of January 2012. The roadmap set targets for the period 2012–2020—with interim milestones to be achieved by 2015—providing an unprecedented impetus for improved control (reduction or elimination of the public health burden) and, where feasible, the national or regional elimination (interruption of transmission) of prioritized NTDs (http://www.who.int/neglected_diseases/NTD_RoadMap_2012_Fullversion.pdf). The release of this volume is, therefore, well timed in the sense that it coincides with the establishment of a consortium of researchers funded by an initiative of the Bill and Melinda Gates

Foundation (B&MGF) to support mathematical modelling studies on these NTDs.

The term 'Neglected Tropical Diseases' (NTDs) was first used in 2005 (http://blog.wellcome.ac.uk/2012/01/10/neglected-tropical-diseases-the-campaign-trail/) to increase both awareness and the global resources available for tackling a group of bacterial, viral, protozoan and helminthic diseases that collectively impose a substantial burden of morbidity and mortality among the poorest populations of our world. In the past, they received much less attention than the so-called 'Big Three' (malaria, HIV/AIDS and tuberculosis) from the global scientific, R&D, funding and health policy communities. The World Health Organization (WHO) has prioritized, for increased control efforts, 17 NTDs that are endemic in 149 countries and affect more than 1.4 billion people, costing developing economies billions of dollars every year (http://www.who.int/neglected_diseases/diseases/en/).

Two main strategies have been endorsed by the WHO for reducing the health impact of NTDs, namely, preventive chemotherapy and intensified disease management. The former (PC) refers to the large scale, regular and prolonged delivery to populations (in the form of MDA) of single-dose medicines efficacious for the treatment of trematode and nematode helminthiases and trachoma, which contribute to averting the morbidity associated with these infections (e.g. the schistosomiases, the intestinal or soil-transmitted helminthiases and the filariases). The latter (IDM) are directed at NTDs for which similar simple tools and treatments are not as yet available (e.g. Buruli ulcer, yaws, leprosy, the leishmaniases, African and American trypanosomiasis, and cestode infections). Complementary measures to support these strategies include control of insect vectors or intermediate hosts, improved water and sanitation infrastructure, coordination with veterinary public health, health education and capacity building.

The Disease Reference Group on human helminthiases (DRG4), convened by the UNICEF/UNDP/World Bank Special Programme for Research and Training in Tropical Diseases (TDR), hosted at the WHO, has published a collection of articles entitled 'A research agenda for the control and elimination of human helminthiases' (http://www.ploscollections.org/static/pntdCollections). This document noted that although it is generally accepted that mathematical models have an important role to play in our understanding of the transmission dynamics of NTDs in general and helminth infections in particular, the potential of models to provide critical insights to inform and support decision making in ongoing control and elimination programmes has not yet been fully realized. Three years later, most authors of the chapters we now present argue, in similar ways, that to fulfil this role, models still need to be further developed, parameterized, and most importantly, validated against epidemiological data. In general, the unifying theme is a call for a closer dialogue between modellers and the wider scientific and stakeholder communities. Progress in this area has been achieved in other infectious disease fields such as malaria, influenza and HIV.

The B&MGF-supported NTD Modelling Consortium (http://www.ntdmodelling.org/) will be crucial in helping to coalesce and energize the NTD modelling community towards helping to achieve the WHO 2020 control and elimination goals. The motivation of the consortium lies in the recognition that many urgent policy issues concerning the control and elimination of NTDs can only be answered through the use of quantitative tools, and that this can only be truly achieved through strong collaborations between modellers, epidemiologists, policy makers and field epidemiologists.

It is in this spirit that we present the first part of this thematic volume of *Advances in Parasitology*. Initially envisaged as a single, stand-alone volume, we found that the very enthusiastic response from contributors warranted the publication of two volumes. This present volume brings together a range of articles looking at transmission dynamics, mathematical model definition and analysis, statistical tools for parameter estimation, molecular epidemiological approaches and health economic perspectives.

Part A starts with a chapter by Manoj Gambhir and colleagues on the role of the Allee effect in the elimination of NTDs, defined in the context of positive density dependence (the positive correlation at low population densities between population size and mean individual fitness—often measured as per capita population growth rate). These authors examine the various density-dependent processes that regulate populations of the mosquito-borne filarial nematode *Wuchereria bancrofti*—the causal agent of lymphatic filariasis—and discuss how these affect two important epidemiological outcome variables that relate to control and elimination programmes, namely, the parasite transmission breakpoint (or extinction threshold) and the reproduction fitness, measured as the effective reproduction ratio. They conclude that although the operation of a single positive density-dependent process can introduce a reasonable chance of achieving elimination, this chance is not appreciably increased by the operation of additional positive density dependencies. Reports by other published work of the possible existence of an Allee effect in trachoma are reviewed.

The paper by David Blok and co-workers focuses on transmission models for leprosy, one of the diseases identified by WHO for intensified case detection and case management. Two compartmental, population models and one individual-based model have been described in the literature, with the latter examining transmission within households and the impact of case finding among contacts of new leprosy patients. The models highlight that among the most relevant factors affecting epidemiological outcomes are contact heterogeneity, heterogeneity in susceptibility and spatial heterogeneity. However, existing models have only been applied to data from a limited number of countries. Parameterization of the models for other areas, particularly those with high incidence, is essential to support current initiatives for the global elimination of leprosy.

Continuing with the theme of IDM diseases, the paper by Kat Rock and colleagues reviews mathematical models of human African trypanosomiasis (sleeping sickness), a vector-borne infection transmitted by tsetse flies. The disease is usually fatal if untreated and transmission occurs in foci across sub-Saharan Africa. Mathematical modelling of African trypanosomiasis began in the 1980s, chiefly based on extensions of the classical Ross-Macdonald malaria model, but progress has been slow and renewed modelling efforts are badly needed. The existing deterministic compartmental models have captured salient epidemiological features and allowed examination of intervention effectiveness (treatment of humans and control of tsetse fly populations) but have, by and large, overestimated infection prevalence and ignored transient dynamics. The authors argue that there is still a need for improved models validated with enhanced data collection, which can provide quantitative predictions to guide control and elimination strategies.

A review of models for the ecology, evolution and control of American trypanosomiasis (Chagas disease) and its triatomine vectors is presented by Pierre Nouvellet and colleagues, who find that the use of transmission dynamics models in Chagas disease, unlike in other vector-borne infections such as malaria and dengue, has been overlooked in the last 100 years of research on *Trypanosoma cruzi* and its vectors. They make a plea for improved modelling, as do the authors of the earlier chapters in this volume.

Introducing the subject of the NTDs that can be addressed with the WHO PC strategy, Bruno Levecke and colleagues discuss the application of hierarchical statistical models in the context of designing surveys for assessing faecal egg counts of soil-transmitted helminths for the purposes of monitoring and evaluation of MDA-based control programmes as well as determination of drug efficacy. As MDA programmes expand worldwide for STH control, the development of new tools to monitor progress and assess drug efficacy is vital.

More advanced in their use for decision support in control and elimination programmes are models for lymphatic filariasis, and Wilma Stolk and co-authors review the status of existing deterministic (EPIFIL) and stochastic (LYMPHASIM) frameworks. Projections from these models lend support to the notion that lymphatic filariasis transmission can be interrupted provided that high coverage of annual MDA is sustained for programme durations probably longer than the 4–6 years initially envisaged, particularly if the initial level of endemicity is high. As in other contributions of this volume, the authors highlight the need for the models to be validated against robust longitudinal epidemiological data to become truly useful. Critical issues identified include the investigation through modelling of the impact of systematic non-compliance, the risk of emerging anthelmintic resistance, and consideration of spatial heterogeneity.

Whilst not ear-marked for global elimination, the WHO roadmap aspires to schistosomiasis being eliminated in selected foci by 2020. This snail-transmitted helminthiasis is possibly one of the NTDs that has seen greatest progress towards understanding the role that genetics and genomics can play in elucidating parasite population structure and how repeated selective pressures due to large-scale MDA treatment programmes may shape parasite populations. Poppy Lamberton and colleagues discuss the potential incorporation of genetic and genomic data—focussing on the recently sequenced *Schistosoma* spp.—into transmission dynamics models to understand the impact of MDA and other control methods on key phenotypes such as those of virulence and/or drug resistance.

Finally, Bruce Lee and co-authors discuss the importance of economic and financing studies to support decision making when resources are limited, as is the case for NTDs in many regions where they are endemic. They suggest that the shortage of resources for NTD control may partly be due to the fact that both the burden of NTDs and the economic value of measures to control them are not well characterized and hence little appreciated. The authors discuss a range of methodologies available to conduct economic evaluations, including studies on cost of illness, cost of intervention, cost and benefit, cost-effectiveness and return-on-investment.

Therefore, this collection of papers provides reviews on a wide range of quantitative approaches for the study of NTD transmission and control of a wide range of pathogens (bacteria to helminths), and highlights that although much progress has been made since 1985 to raise the profile of NTDs, and secure the commitment from the international community for large-scale control programmes, the path to elimination still necessitates much closer collaboration between modellers, field experts and policy makers. It is obvious from the work described in the chapters of this volume, that there is still a pressing need for the models to be improved, calibrated and validated with robust biological, clinical and epidemiological data. We thank the support of the editorial team of *Advances in Parasitology* and hope that our readers will enjoy this volume as much as we have enjoyed preparing it.

Maria-Gloria Basáñez and Roy M. Anderson
February 2015

CHAPTER ONE

The Allee Effect and Elimination of Neglected Tropical Diseases: A Mathematical Modelling Study

Manoj Gambhir*,[1], Brajendra K. Singh[§], Edwin Michael[§]
*Department of Epidemiology and Preventive Medicine, Monash University, Melbourne, VIC, Australia
[§]Department of Biological Sciences, University of Notre Dame, Notre Dame, IN, USA
[1]Corresponding author: E-mail: manoj.gambhir@monash.edu

Contents

Abstract

Elimination and control programmes for neglected tropical diseases (NTDs) are underway around the world, yet they are generally informed by epidemiological modelling only to a rudimentary degree. Chief among the modelling-derived predictors of disease emergence or controllability is the basic reproduction number R_0. The ecological

Advances in Parasitology, Volume 87
ISSN 0065-308X
http://dx.doi.org/10.1016/bs.apar.2014.12.001

systems of several of the NTDs include density-dependent processes — which alter the rate of *e.g.* parasite establishment or fecundity — that complicate the calculation of R_0. Here we show how the forms of the density-dependent functions for a model of the NTD lymphatic filariasis affect the effective reproduction number R_{eff}. We construct infection transmission models containing various density-dependent functions and show how they alter the shape of the R_{eff} profile, affecting two important epidemiological outcome variables that relate to elimination and control programmes: the parasite transmission breakpoint (or extinction threshold) and the reproduction fitness, as measured by R_{eff}. The current drive to control, eliminate or eradicate several parasitic infections would be substantially aided by the existence of ecological Allee effects. For these control programmes, the findings of this paper are encouraging, since a single positive density dependency (DD) can introduce a reasonable chance of achieving elimination; however, there are diminishing returns to additional positive DDs.

1. INTRODUCTION

A renewed urgency on the part of both governmental and nongovernmental organizations to control neglected tropical diseases (NTDs) has led to a debate on whether these diseases can be eliminated or eradicated (World Health Organization, 2012). This debate has highlighted the need for a better understanding of the parameters that govern infection transmission and that can be used to measure the success of proposed intervention strategies on the path to elimination (London Declaration on Neglected Tropical Diseases, 2012). The mature field of mathematical epidemiology has yielded insights on parameters governing many infectious diseases, but it is only drawn-upon sporadically when it comes to large-scale disease control.

Among the predictors of disease emergence or controllability to have come from mathematical work, the basic reproduction number R_0 is perhaps the most widely known; it contains information relating to the epidemic growth rate, the final endemic/epidemic size and the ease or difficulty with which an infection can be controlled. R_0 can also be influenced by a wide variety of environmental and socioeconomic factors; it acts as a threshold quantity for infection transmission since its value should be greater than 1 for infection to become established in a host population (Anderson and May, 1992, Barbour et al., 1996). R_0 is defined as the number of infections a single infectious case gives rise to at the beginning of an epidemic (Diekmann and Heesterbeek, 2000). This definition is closely related to the initial growth rate of an epidemic, which allows R_0 to be estimated by measuring this rate (Donnelly et al., 2003). Many infections — particularly those of macroparasites (e.g. helminths, ticks) — are described by models that may, due to density-dependent population regulation, exhibit a growth rate of zero near the disease-free equilibrium state (Diekmann and Heesterbeek, 2000). These infections, however, are obviously still able to achieve an endemic equilibrium in host populations, as evidenced by the large number of macroparasites that are endemic in human populations. The simple measure, R_0, is therefore inadequate as a summary parameter in this case. The specific type of density-dependent mechanism responsible for this problem is a positive density dependency (DD) or facilitation function (Berec et al., 2007; May, 1977a). Negative DDs reduce and positive DDs increase the growth rate of populations as the numbers of individuals increase in these populations.

1.1 Allee effects and disease elimination

One of the most interesting consequences of the presence of positive DDs is the Allee effect (Courchamp et al., 1999). Ecological Allee effects arise when there exists a population density threshold below which extinction will occur, and positive DDs are thought to be responsible since they require a founder population to colonize or invade (extinction results from a strong Allee effect in which population growth first slows and then goes negative, with decreasing parasite population level; weak Allee effects arise when population growth decreases but does not go negative) (Courchamp et al., 1999). These effects are therefore important from both a parasite elimination and a species conservation point of view. However, since the basic reproduction number alone will not suffice for us to determine whether or not parasite establishment will occur, we need to examine the full functional profile of the effective reproduction number R_{eff} instead of the simple parameter R_0: R_{eff} is equal to R_0 when the parasite population is zero at the very beginning of an epidemic, but its values are modified by the effects of DDs as population densities increase, as infections become established. For models of macroparasites, R_{eff} is defined as the number of parasites *of a particular life stage* that arise from a single parasite of that life stage, during the course of an epidemic (and R_0 applies specifically in the absence of parasites i.e. prior to an epidemic).

In the presence of positive DDs, R_{eff} is a humped function (Berec et al., 2007; Churcher et al., 2005; Basáñez et al., 2009). When the peak value of R_{eff} (R_{peak}) is greater than 1, the effective reproduction number intersects the $R_{eff} = 1$ line twice i.e. there are two equilibria (Figure 1). The higher of these is the stable endemic equilibrium, and the lower is an unstable extinction 'breakpoint' (May, 1977a), which gives rise to the strong Allee effect

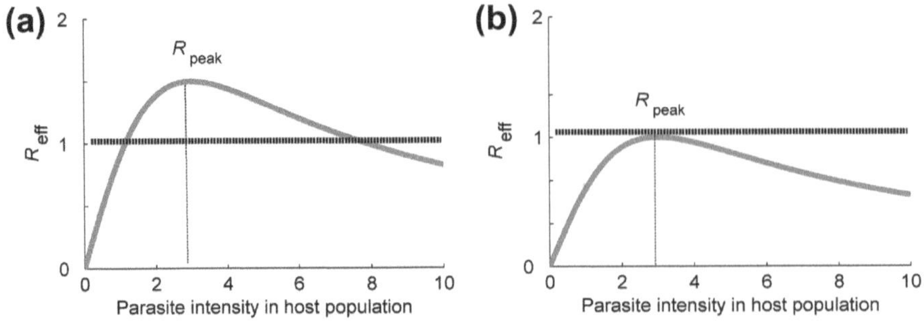

Figure 1 ***The effective reproduction number profile.*** Schematic diagram of the effective reproduction number against the parasite burden in the host population for a system containing at least one positive DD. The humped function reaches its peak value at R_{peak}. (a) Illustrates the case in which the system intersects the $R_{eff} = 1$ line twice, with the lower intersection corresponding to the parasite transmission breakpoint. (b) Illustrates the critical case at which the R_{eff} curve grazes this line at only one point, which corresponds to the maximum possible value of the parasite transmission breakpoint.

(henceforth referred to simply as the Allee effect). As noted above, the Allee effect has been extensively observed and studied in ecology, where the aim is generally to prevent species extinction (Berec et al., 2007; May, 1977a). Here, however, we are interested in circumstances (e.g. intervention situations) that will cause parasite population numbers to drop below this threshold for extinction so that local elimination or even global eradication might be achieved (Gambhir et al., 2010b; Gambhir and Michael, 2008).

While Allee effects are highly applicable to the idea of disease elimination, they have been studied remarkably little by the infectious disease community. The original idea for the Allee effect arose in ecology (Allee, 1949), where it was related to the consequences of density-dependent processes in ecological systems at low population densities; among scientists studying infectious diseases, these ideas have been most amenable to macroparasite researchers, whose systems and models of interest are similar to their counterparts from ecology. Host—parasite systems that contain positive DDs have been widely studied and documented (see e.g. for onchocerciasis (Basáñez et al., 2009; Churcher et al., 2005, 2006; Duerr et al., 2003), schistosomiasis (Spear, 2012), lymphatic filariasis (LF) (Gambhir et al., 2010b; Gambhir and Michael, 2008), malaria (Muriu et al., 2013; Churcher et al., 2010; White et al., 2011)) though few of these studies explicitly focus upon the possibility of extinction of the parasite at low-intensity levels (the paper by White et al. (2011) is a notable exception, perhaps because malaria is a disease for which there is an eradication goal). The paper by Duerr et al. (2005) has come closest to examining systematically the effects of DDs in combination with one another upon the specific goal of driving filarial worms, primarily onchocerciasis and LF, to extinction.

One of the most prominent recent examples of the Allee effect in the infectious disease literature, however, has been an article on the ocular infectious disease trachoma, which is also an NTD (Chidambaram et al., 2005). This article discussed the proposal that there might be an Allee effect operating for trachoma, an idea based on the observation that the community prevalence level in a mass drug administration (MDA) field trial had held steady over a year after the administration of drug to the population (Solomon et al., 2004). This observation stood in contrast to the predictions of modelling and other posttreatment observational studies which found re-emergence of infection and disease to be the most likely outcome (Gambhir et al., 2010a, 2009, 2007; Liu et al., 2014; Burton et al., 2005; West et al., 2005). The proposed mechanisms for an Allee effect for trachoma include reductions in bacterial genotypic diversity leading to more effective immune responses from the host population, and a dose-dependent effect in which lower community prevalence leads to lower average chlamydial load and lower probability of transmission per contact. These phenomena were investigated in detail by fitting a series of mathematical models to baseline and posttreatment data from Ethiopia, and determining that a model exhibiting positive feedback (or positive DD) was preferred as an explanation for the observed data (Liu et al., 2014). However, another article by the same group found that there was no statistically significant change in the effective reproduction number of trachoma 6 and 12 months after each of 3 MDA treatment rounds (Liu et al., 2013). This latter finding would appear to be inconsistent with the presence of an Allee effect, though it should be noted that the community infection prevalence was close to 5% after the third round of MDA, which may not be low enough to reveal the effect of positive DDs.

In this paper, we concentrate upon the effective reproduction number as a means of investigating the impact of Allee effects on the population dynamics of disease elimination. We show how R_{eff} can be formulated for macroparasitic models that include increasing numbers of DDs, focussing on the profiles of this function for a specific parasite transmission model pertaining to the mosquito-borne macroparasitic disease, LF. Our focus upon LF is intended to be illustrative of the kind of analysis that can be accomplished for most NTDs, including those which are more commonly classified as micro- rather than macroparasites. We look specifically at (1) the relationship between transmission breakpoint values and the DDs present and (2) the rate of return of infection following a single high coverage and efficacy simulated round of MDA. While these ideas have been investigated in the literature on microparasitic infections (see e.g. Regoes et al., 2002), and have been mentioned in the context of macroparasites (Basáñez et al., 2009; Churcher et al., 2005, 2006), they have never been systematically investigated and discussed in regard to control interventions. We address this gap by discussing these results in the context of global elimination programmes for macroparasitic diseases.

2. METHODS

We construct a simple transmission model of LF, a major vector-borne macroparasitic infection of humans, expressed as a set of differential equations, where each equation describes the time-development of a specific parasite life stage in the human and mosquito host populations (Anderson and May, 1992; Norman et al., 2000). This is a 'mean-field' model, in the sense that it represents the time evolution of the mean values of the relevant variables across the host population. Disease associated with LF is caused by adult *Wuchereria bancrofti* worms which, in their human host, become stuck in the lymphatics and cause disruption of the immune system. Adult worms sexually reproduce and produce a large quantity of microfilariae (Mf), which can be measured through an e.g. 20-μL fingerprick blood sample. Mosquitoes become infected with Mf when they take a bloodmeal from an infected human host, and the ingested Mf mature into the infective L3 larval stage. Once infected mosquitoes bite uninfected humans, they may pass on these infective L3 larvae into their new host, and the larvae mature into adult worms, beginning the cycle anew (WHO, 2014). We fitted our population model of LF to baseline human infection prevalence data from two Papua New Guinea (PNG) villages (Gambhir et al., 2010b), in which Mf prevalence among human hosts was approximately 50% (Nanaha and Peneng) (see Appendix for method used for model fitting, and Appendix Table 1 for model parameter values). This endemic prevalence is certainly at the higher end of the current global range, but we obtained very similar results to those outlined here, for models fitted to lower endemic settings (such as those in Gambhir et al., 2010b). We used the model thus parametrized to an LF-endemic population to explore systematically the interactive effects of different combinations of positive and negative DDs on the behaviour of the R_{eff} of LF, as follows.

2.1 Multiple positive and negative density dependencies (DD) in an LF transmission model

Equation (1) describes a model of LF transmission as a set of differential equations over human host age and time; this model quantifies the changes in the populations of each of the main parasite life stages within the populations of both the definitive and the intermediate vector hosts and so this is 'macroparasite' model, to use the lexicon of Anderson and May (1992). In this model, the parasite burdens are the state variables upon which we concentrate. Simple immigration–death processes describing the transmission dynamics for each parasite stage are written as (see Appendix for the full model equations and further details)

Human Host

$$\frac{dW}{dt} = \lambda.L.f_1(I).f_2(W) - \mu.W$$

$$\frac{dM}{dt} = \alpha.W.f_3(W) - \delta.M$$

Host Immunity

$$\frac{dI}{dt} = W - \gamma.I$$

Vector

$$\frac{dL}{dt} = \beta.M.f_4(M) - \sigma.L \tag{1}$$

Here, W, the number of adult worms per definitive host, and M, the Mf load per host, refer to parasite life stages within the definitive host population, whereas the L3 infective larval load per mosquito (larvae develop through L1 and L2 stages but only become infective once they reach the L3 stage), L, refers to the parasite life stage within the vector host (Gambhir and Michael, 2008; Michael et al., 2004; Norman et al., 2000). The host immunity variable, I, increases in magnitude at a rate equal to the adult worm burden and has a decay rate of γ, allowing the model to capture the waning of immunity. λ, α, β are the 'immigration' rates of each of the life stages from the 'parent' stage, which is normally the previous life stage of the parasite; μ, δ, σ are the death rates of each of the life stages; and $f_1(I), f_2(W), f_3(W), f_4(M)$ represent the modifying density-dependent functions acting on each of the respective parasite life stage intensities. Details of these functions are given below and in Table 1. The components of the parasite life cycle we consider

Table 1 Functional forms for each of the density-dependent functions included in the fitted lymphatic filariasis model

Density dependency	Expression	Parameters	Type of DD
Larval establishment immunity $f_1(I)$ (Duerr et al., 2005; Gambhir and Michael, 2008)	$\frac{1}{1+c.I}$	c – strength of immunity to larval establishment	Negative
Host immunosuppression $f_2(W)$ (Duerr et al., 2005; Duerr et al., 2003)	$\frac{1+I_C.S_C.W}{1+S_C.W}$	I_C – strength of immunosuppression; S_C – slope of immunosuppression function	Positive (if $I_C > 1$)
Worm mating probability $f_3(W)$ (May, 1977b)	$1-\left(1+\frac{W}{2k}\right)^{-(1+k)}$	k – negative binomial aggregation parameter	Positive
Vector uptake $f_4(M)$ (Gambhir and Michael, 2008)	$1/M\int\pi(a).(1-g(M)).da$ where, $g(M)=\kappa.\left(1-e^{-\frac{r.M}{\kappa}}\right)^i$ where $i=1$ for *Culex* and $i=2$ for *Anopheles* mosquitoes, $\pi(a)$ is the population age distribution.	κ – maximum level of L3 given Mf level; r – initial gradient of uptake function	*Culex*: negative; *Anopheles*: positive

The mathematical expressions are shown along with parameter values used in the model analyzed in this paper; additionally, the nature of the density dependency (DD) (positive or negative) is provided in the rightmost column.

here are illustrated in Figure 2, where we emphasize those parasite stages occurring in the human host and mosquito vector. Also shown on the figure are the density-dependent modifying functions, as described above, acting on the transformation of the parasite from one stage to the next. Note that we only consider density-dependent functions operating on the immigration terms of each stage, though effects on death terms may also be analyzed in the same way. With respect to the DDs included in the model, $f_1(I)$ is a function describing the curtailing effect of host immunity on parasite establishment; $f_2(W)$ is a function representing immunosuppression of the human host induced by the adult parasite population; $f_3(W)$ is the mating probability of adult worms; and $f_4(M)$ is a function describing the survival of Mf, M, ingested from human hosts, into L3 infective larvae, L, in the mosquitoes, referred to here as an uptake function (Gambhir and Michael, 2008; Snow et al., 2006; Snow and Michael, 2002; Stolk et al., 2008) (see Table 1, and note that $f_4(M)$ has two functional forms, one corresponding to *Anopheles* and one *Culex* vector species, which we denote with superscripts i.e. $f_4^{Anopheles}(M)$ and $f_4^{Culex}(M)$). Although there may be other functions governing LF infection, we restrict our attention to these functions, given that they have been shown to be the major density-dependent mechanisms that act to moderate filarial transmission dynamics (Duerr et al., 2005; Gambhir and Michael, 2008; Michael and Bundy, 1998; Stolk et al., 2008). Details of the model parameters, their definitions and values, are given in the

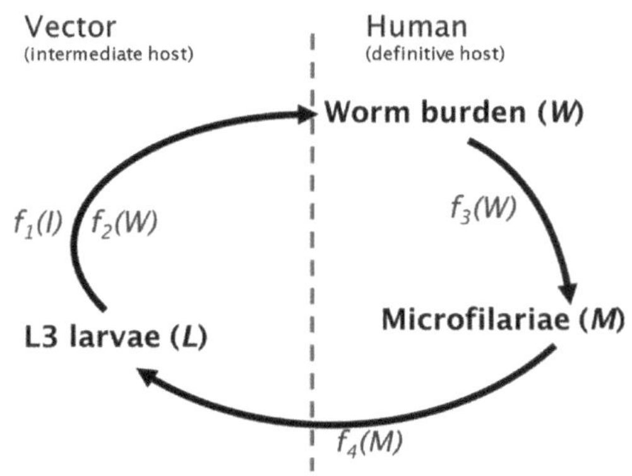

Figure 2 ***The life cycle of the helminth parasite modelled in this paper.*** The simplified life cycle shows the three life stages that are included in the model, the density-dependent functions acting on the transformations between life stages ($f_1(I)$, $f_2(W)$, $f_3(W)$, $f_4(M)$; details of the functions are given in the main text and Table 1), and the host in which the parasite life stages occur.

Appendix, and a thorough analysis of the values obtained through fitting the model to data have been previously published (Gambhir et al., 2010b).

In the absence of DDs, the model becomes (omitting the I equation, since it plays no part in the density-independent model)

$$\frac{dW}{dt} = \lambda.L - \mu.W$$

$$\frac{dM}{dt} = \alpha.W - \delta.M$$

$$\frac{dL}{dt} = \beta.M - \sigma.L \tag{2}$$

Now, we set $\frac{dL}{dt} = 0$ for the instantaneous equilibrium of the L3 larval variable, which gives $L^* = \frac{\beta.M}{\sigma}$. Substituting this expression for the quasi-equilibrium value of L into the equation for W gives

$$\frac{dW}{dt} = \frac{\lambda.\beta.M}{\sigma} - \mu.W \tag{3}$$

Formulating the Jacobian matrix of the (W, M) system and finding its eigenvalues result in the criterion for the positivity of the dominant eigenvalue, which here is $\frac{\lambda.\beta.\alpha}{\mu.\delta.\sigma} > 1$ (see Appendix for details). Since this condition is sufficient for epidemic growth to occur, we can define the basic reproduction number:

$$R_0 = \frac{\lambda.\alpha.\beta}{\mu.\delta.\sigma} \tag{4}$$

This expression can be expanded by substituting in the fine-structure constants from a full model such as that of Norman et al. (2000), Basáñez and Ricardez-Esquinca (2001) and Basáñez and Boussinesq (1999) (i.e. each of the parameters included in the expression for R_0 may themselves be broken down into their constituent components such as annual mosquito-biting rate, probability of parasite transmission per bite etc.; see also the Appendix for further details of the full LF model).

In the presence of the DDs given above in Eqn (1), each of the immigration terms is modified by a multiplicative density-dependent function so that the expression for the reproduction number in Eqn (4) becomes

$$R_{eff} = \frac{\lambda.f_1(I).f_2(W).\alpha.f_3(W).\beta.f_4(M)}{\mu.\delta.\sigma} \tag{5}$$

(In other words, $R_{eff} = R_0.f_1(I).f_2(W).f_3(W).f_4(M)$). The effective reproduction number R_{eff} is a function of the parasite burden and, at the disease-free state $((W,M,L) = 0)$ it reduces to R_0 (for the set of DDs listed in Table 1)

$$R_0 = \frac{\lambda.f_1(0).f_2(0).\alpha.f_3(0).\beta.f_4(0)}{\mu.\delta.\sigma} \tag{6}$$

If at least one of the DDs is positive, we can see that the value of R_0 will be zero in the disease-free state (here $f_2(0) = f_4^{Anopheles}(0) = 0$). Figure 1 illustrated this case.

Negative and positive DDs combine to produce a humped function that may intersect the $R_{eff} = 1$ line twice, depending upon the size of the non-density-dependent ecological parameters (see Figure 1(a)), which determine the overall magnitude of the function. Figure 1(b), however, shows that the maximum possible transmission breakpoint occurs when the non-density-dependent parameters are such that R_{eff} grazes the $R_{eff} = 1$ line at just one point. The functional expression given above (in Eqn (5)) is difficult to calculate analytically but its features can be easily determined numerically. We analyze the model defined by Eqn (1) by adding DDs from Table 1 one at a time to observe their effect on the filarial R_{eff} curve and, specifically, on the magnitudes and locations of transmission breakpoints and the maximum value R_{peak}.

2.2 Age-structured expression of the effective reproduction number

When age-structure is important in the human host population (which it often is, since we usually need to fit our models to parasite prevalence and intensity data that pertain to different age-groups), the above model can be modified from a set of ordinary differential equations in time (t), to a set of partial differential equations (PDEs) in the variable age (a), as well as time. The resulting PDEs correspond to the ODEs in Eqn (1) above, with the addition of age-dependent parameters, of which there is only one, namely the parasite establishment rate λ; we therefore alter this parameter to $\lambda(a)$, now an explicit function of age, since the mosquito human-biting rate may be dependent upon e.g. skin surface area, which increases until adulthood. Next, we insert specific functional forms of these DDs into the model and examine the steady-state system $\left(\frac{\partial}{\partial t} = 0\right)$ to obtain the following equations, whose dynamical behaviour has been explored previously (Gambhir and Michael, 2008; Norman et al., 2000):

$$\frac{dW}{da} = \frac{\lambda.\beta}{\sigma}.f_1(I).f_2(W).M.f_4(M) - \mu.W$$

$$\frac{dM}{da} = \alpha.f_3(W).W - \delta.M$$

$$\frac{dI}{da} = W - \gamma.I \tag{7}$$

Note here that the equilibrium value of the larval load in the mosquitoes $L^* = \beta.M.f_4(M)/\sigma$ has been substituted into the worm burden W equation. Table 1 defines and gives details of the specific functional forms of the DDs used in this paper.

Following Dietz (1982), we integrate these equations over the age of the population when the age distribution ($\pi(a)$) is modelled as a simple exponential decay (i.e. $\pi(a) = e^{-\mu_1 a}/Z$, where μ_1 is the death rate of the human host population and Z is a normalizing constant; overbars indicate age-averages):

$$\overline{W} = \frac{\beta}{\sigma.\mu_1}.\overline{\lambda.f_1(I).f_2(W)}.\overline{M.f_4(M)} - \frac{\mu}{\mu_1}\overline{W}$$

$$\overline{M} = \frac{\alpha}{\mu_1}.\overline{f_3(W).W} - \frac{\delta}{\mu_1}\overline{M} \tag{8}$$

Both of the equations contained in Eqn (3) apply to the system at equilibrium. Multiplying the first by the second – and following a few more simple steps of algebra – results in a functional constraint satisfied at equilibrium:

$$\frac{\alpha.\beta.\overline{\lambda.f_1(I).f_2(W)}.\overline{M.f_4(M)}.\overline{f_3(W).W}}{\sigma.(\delta+\mu_1).(\mu+\mu_1).\overline{M}.\overline{W}} = 1 \tag{9}$$

In the absence of age-structure in the human population, the expression on the left-hand side of the above equation is equal to the expression we previously obtained for the effective reproduction number (Eqn (4)), and we find here that it is equal to 1 when the system is at equilibrium i.e. above and below the value of 1, the state variables will change so as to bring R_{eff} back to the equilibrium value of 1:

$$R_{eff} = \frac{\alpha.\beta.\overline{\lambda.f_1(W).f_2(I)}.\overline{M.f_4(M)}.\overline{f_3(W).W}}{\sigma.(\delta+\mu_1).(\mu+\mu_1).\overline{M}.\overline{W}} \tag{10}$$

We use the above expression to calculate the profile of R_{eff} over the full range of possible parasite intensities and determine how this profile is affected by successively adding or removing DDs.

2.3 Dynamical consequences of the number of DDs on the re-emergence of infection following drug treatment

For four of the models with increasing numbers of DDs (labelled i—iv in Table 2 and Figure 4(a)), we also applied a single simulated round of very high efficacy and coverage MDA (99% worm and Mf kill rate, 100% coverage) in order to depress the infection level to a very low value such that the R_{eff} values differ appreciably between the four models (i—iv) (i.e. <0.1% Mf in Figure 4(a)); we then observe changes in the prevalence of infection over time following this treatment round for each of the models. This extremely high MDA coverage and efficacy is clearly infeasible in a field setting (or with current pharmaceuticals, which have lower efficacies), but we simulate it here purely to bring down the parasite prevalence rapidly to a level at which the R_{eff} value, corresponding to each of four submodels (i—iv), become distinct from one another. In reality, a combination of several rounds of MDA and vector control would be needed to achieve such a reduction in prevalence. This theoretical approach is similar to methods in theoretical and experimental physics, where extremely high-energy collisions (or low temperatures) need to occur to reveal underlying physical laws which are only observable under these extreme conditions.

2.4 A simplified model of R_{eff} for investigating interactions between DDs

Multiple positive and negative DDs may operate at various points in the parasite life cycle, and their individual and combined effects on the

Table 2 Details of the density dependencies (DDs) successively included in the fitted lymphatic filariasis (LF) model

Model number	Density dependencies in LF model	Type of DD
(i)	$f_1(I), f_4^{Culex}(M)$	Negative
(ii)	$f_1(I), f_4^{Culex}(M), f_3(W)$	Negative.positive
(iii)	$f_1(I), f_4^{Anopheles}(M)$	Negative.positive
(iv)	$f_1(I), f_4^{Anopheles}(M), f_3(W)$	Negative.positive.positive
(v)	$f_1(I), f_4^{Anopheles}(M), f_3(W), f_2(W)$	Negative.positive.positive.positive

Each density-dependent function included in models (i)—(v) (see Figure 4) is listed below and details of each function can be found in Table 1.

overall R_{eff} are likely to vary depending on the values of the parameters determining the strength of each individual density-dependent function (Berec et al., 2007). As the LF model described above contains density-dependent functions with parameter values estimated via fitting to community prevalence data, scale variations in the numerical values of fitted parameters may conceal the relative importance of each DD to the overall behaviour of R_{eff}. Another problem pertains to the impact that coexisting non-density-dependent terms may also on the behaviour of R_{eff}. To overcome these problems, and in order to explore the general impact that individual DDs may have on the overall behaviour of R_{eff} for both filariasis and ultimately for macroparasitic infections, we develop here a simplified model of the density-dependent part of R_{eff} − i.e. without the constant (i.e. non-DD) terms comprising R_0 − which we name R_{dd}, to illustrate the effects of multiple positive and negative DDs upon key variables such as the peak value of R_{eff} (i.e. R_{peak}) and the parasite burden at which this value occurs, which we call the maximum worm transmission breakpoint, W^*.

The expression derived for R_{eff} (Eqn (5)) shows that the immigration DDs occur as products on the numerator (death DDs would occur as similar products on the denominator). Here we continue to focus on immigration DDs and, to simplify the mathematics, we use stylized functions to represent the true DDs (such as those shown in Table 1). The stylized forms we use are exponential functions, where

$$\begin{aligned} \textit{negative}\ \ DD &= e^{-AW} \\ \textit{positive}\ \ DD &= \left(1 - e^{-BW}\right) \end{aligned} \tag{11}$$

where A and B are constants denoting the strength of the DD, and W is the worm burden. The negative DDs combine very simply when multiplied together ($e^{A_1 W} e^{A_2 W} \ldots = e^{(A_1 + A_2 \ldots) W}$). The positive DDs are slightly more complicated since $(1 - e^{-B_1 W})(1 - e^{-B_2 W}) = 1 - e^{-B_1 W} - e^{-B_2 W} + e^{-(B_1 + B_2) W}$. However, as detailed in the Appendix, a good approximation to this product is given by the expression,

$$\left(1 - e^{-B_1 W}\right)\left(1 - e^{-B_2 W}\right) \approx \left(1 - e^{-B' W}\right) \tag{12}$$

where $B' = \frac{B_1 B_2 (B_1 + B_2)}{B_1^2 + B_2^2 + B_1 B_2}$. The Appendix shows how positive and negative DDs of this simplified functional form combine to give the humped function that is characteristic of the Allee effect.

Finding the maximum value of this function (as outlined in the Appendix), we show that the parasite density W^* at which the maximum value R_{peak} occurs is

$$W^* = \frac{1}{B}.\ln\left(1 + \frac{B}{A}\right) \tag{13}$$

As previously illustrated in Figure 1, the parasite density associated with R_{peak} corresponds to the maximum transmission breakpoint value for the system. Because the simplified model outlined here only accounts for the density-dependent component of R_{eff} and is therefore not correctly scaled to determine the appropriate lower and upper intersection points of the $R_{eff} = 1$ line, we concentrate upon the *maximum* possible transmission breakpoint value W^*, a quantity independent of the non-density-dependent model parameters.

The corresponding value of R_{peak} here is thus

$$R_{peak} = \left(\frac{A}{(A+B)}\right)^{A/B}\left(\frac{B}{(A+B)}\right) \tag{14}$$

The simple formalism above also allows us to examine the combined effects of several Allee functions or negative DDs on the maximum transmission breakpoint value (W^*), as defined previously.

3. RESULTS

3.1 Effects of DDs on the LF effective reproduction number and equilibria

The LF model was parametrized and fitted to data from two LF-endemic Papua New Guinean communities with Mf prevalence of approximately 50% (Bockarie et al., 2002). An ensemble of 'good' model fits (selected according to the sampling-importance-resampling likelihood-based technique outlined in the Appendix) to the baseline Mf prevalence data for the two PNG villages are shown in Figure 3 (Gambhir et al., 2010b) on the left-hand side, and the corresponding R_{eff} curves — based on fitted parameter values, and calculated by enumerating the age-averaged function in Eqn (10) — are shown in Figure 3 on the right. The highlighted R_{eff} curve in Figure 3(a), corresponding roughly to the median of the curves, was used as a typical profile in the following investigations of the effect of DDs on LF transmission reported here.

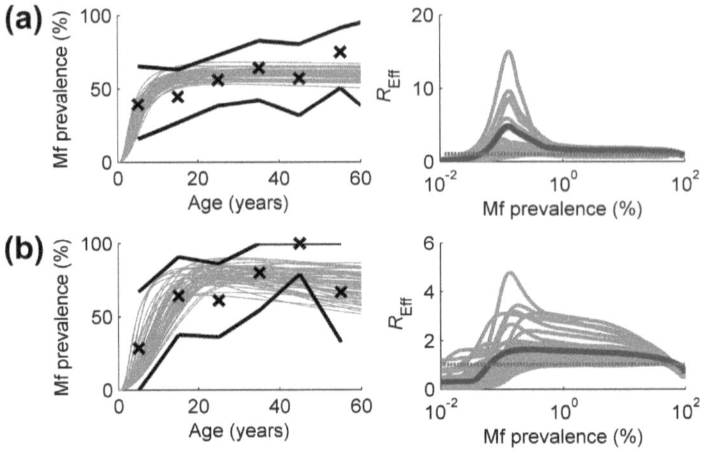

Figure 3 R_{eff} ***curves for an ensemble of model fits.*** The plots on the left of both (a) (top row) and (b) (bottom row) show 100 Mf prevalence curves generated by 100 model parameter sets constituting an ensemble of high-likelihood model fits to the data (crosses and grey lines indicating mid points and 95% confidence intervals of the Mf prevalence data from two PNG villages; see Appendix for an outline of the fitting procedure). On the right of (a) and (b) we illustrate the R_{eff} curves, calculated using Eqn (4), corresponding to each of the parameter sets in the fitted ensemble. The plots of (a) were calculated using the model fitted to Mf prevalence data from the PNG village of Nanaha and those of (b) were from Peneng (Gambhir et al., 2010b). The red (dark grey in print versions) R_{eff} curve shown for each village represents the mean value for each set of curves.

Figure 4(a) depicts the individual and combined effects of each DD for the model corresponding to the highlighted R_{eff} curve in Figure 3(a). Curve (iv) is the actual highlighted curve of Figure 3(a), and contains a full complement of DDs. The DDs present for each of the model-generated curves in the figure are given in Table 2 (see Table 1 for details of the functional forms of the DDs for the LF model, corresponding to each of the functions in Table 2) with each curve labelled on the graph corresponding to each table entry. The x-axis (% Mf prevalence) on the figure is shown in logarithmic scale to highlight the features of the R_{eff} function at low Mf prevalence values.

The results in Figure 4(a) indicate that only when at least one positive DD is present do we obtain the humped curves for R_{eff}, whose maximum points R_{peak} are related to the maximum growth rates of the parasite population. This growth rate is often interpreted as the epidemiological fitness of the parasite and can also be seen as the equivalent to R_0 for systems with positive DDs, in similar fashion to the definition of Regoes et al. (2002). The

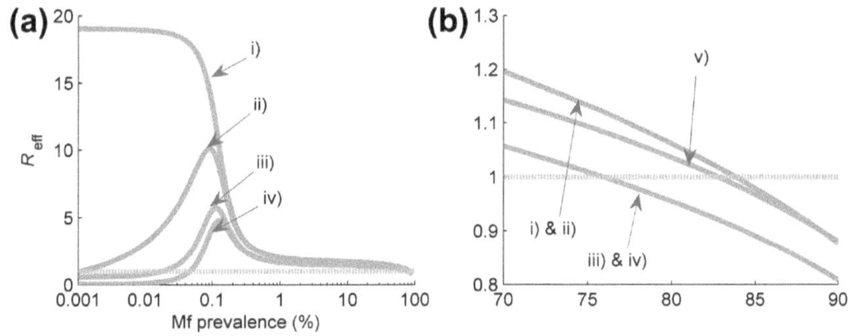

Figure 4 ***The effect of density dependencies on the effective reproduction number.*** Plots illustrating the change in the functional profile of R_{eff} with the successive addition of density-dependent processes. Moving from the top to the bottom of (a), a greater number of density-dependent processes are included and these curves correspond to the models (i)—(v) in Table 2. In addition to immunity against the establishment of adult worms, the models shown in (a) successively include, from the topmost curve: (i) limitation uptake; (ii) limitation uptake, mating probability; (iii) facilitation uptake; (iv) facilitation uptake, mating probability. (b) Zooms in on the right-hand side of (a), where the curves intersect the $R_{eff} = 1$ line at the endemic equilibrium. One further curve becomes distinguishable here: (v) facilitation uptake, mating probability, immunosuppression.

points of intersection with the line $R_{eff} = 1$ give the stable endemic (the rightmost intersection) and the unstable equilibria, with the parasite burden value corresponding to the point at which R_{eff} grazes the $R = 1$ line.

The uppermost curve in Figure 4(a) (labelled (i)) is generated by a model containing only the negatively density-dependent *Culex* uptake function. Curves lying below this correspond to models that contain both positive as well as the negative DDs. Therefore, the two curves lying directly beneath the first were each generated by models that contained one positive DD and one negative DD as follows: curve (ii) the mating probability (positive DD) and *Culex* uptake function (negative DD), and (iii) the anopheline uptake function (positive DD) plus larval establishment immunity (negative DD). These two models show the humped R_{eff} shape characteristic of a combination of negative and positive DDs. The parasite prevalence at which the peak value R_{peak} occurs differs between these two models, attaining a higher value when the anopheline uptake function is included although note that the value of R_{peak} itself is smaller for this case. When we include both mating probability and the positive uptake DD (curve (iv)) the maximum of the curve occurs at a slightly higher parasite prevalence than for (ii) or (iii) and the value of R_{peak} is also lower than either individual case. However, the combination of these two positive DDs (curve (iv)) appears not to be additive in the location or value of R_{peak}; rather, the combined case is very close to that generated by the anopheline uptake function alone (iii). Finally, when we add the immunosuppression positive density-dependent function, we again find that, due to the small parameter values associated with this function, the curves of Figure 4(a) change only negligibly (curve (v) not even distinguishable in Figure 4(a)). However, magnifying the region at which these curves intersect the $R_{eff} = 1$ line, we see in Figure 4(b) that the effect of the immunosuppression DD (curve (v)) can only be observed at a high Mf prevalence.

3.2 The impact of DD on the rate of return of infection following a simulated treatment round

Figure 5(a) shows the time profile of each of the four models (i—iv), following a single very high efficacy and coverage MDA round which severely reduces the Mf prevalence from its baseline value to a very low value, but not one that is below the transmission breakpoint. Figure 5(b) magnifies the time period immediately following treatment and shows that infection in the four models re-emerges at different rates, in accordance with their relative values of R_{eff} at the very low values to which the population Mf prevalence has been depressed. This observation lends credence to the use of R_{eff} in determining the rate of re-emergence of infection immediately following community

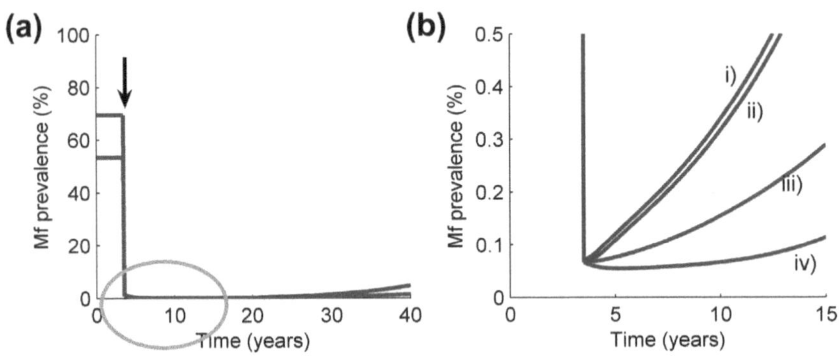

Figure 5 ***Rebound of infection following a single mass drug administration (MDA) round for models containing various numbers of positive density dependency (DD).*** (a) Shows the Mf-reduction effect of a single very high coverage and efficacy MDA treatment round (99% worm and Mf kill rate, 100% coverage) on the four models outlined in Table 2, containing increasing numbers of positive DD. (b) Zooms in on the region enclosed within the ellipse in (a), allowing the rebounding Mf trajectories of each model to be seen.

treatment, at which point the host—vector—parasite system is out-of-, and sometimes far-from-, equilibrium.

3.3 DDs and the Allee effect: the simplified R_{eff} model

Figure 6 illustrates the profile of the simplified R_{eff} model (R_{dd}) for four distinct cases, corresponding to the four cases (i—iv) illustrated for the full effective reproduction number profile R_{eff} which was calculated for LF, as outlined in Table 2. In Figure 6(a), we examine the situation when positive DDs differ in their parameter values while in Figure 6(b), we explore the case when these parameter values are of equal magnitude. The results show that where one of the two positive DDs has a greater magnitude than the other, the four models follow the patterns shown in Figure 4(a), with each of the positive DDs altering the shape of the curve and the position and size of the peak value of the humped shape significantly. However, both such positive DDs acting in combination appear to alter the shape little, i.e. their combined effect on R_{peak} and W^* is sub-additive, indicating that their impact is smaller than the sum of the effects owing to each single DD. When both positive DDs are of equal magnitude, a similar outcome also holds (Figure 6(b)).

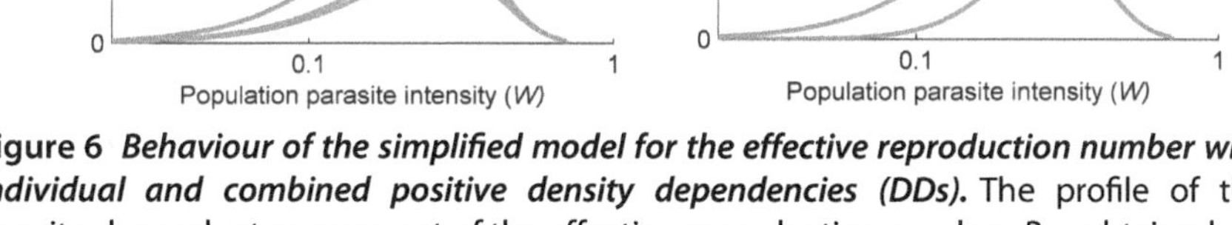

Figure 6 ***Behaviour of the simplified model for the effective reproduction number with individual and combined positive density dependencies (DDs).*** The profile of the density-dependent component of the effective reproduction number, R_{dd}, obtained using the simplified model, for four cases corresponding to the lymphatic filariasis (LF) models outlined in Table 2. The x-axis is drawn on a log scale to allow comparison to the full LF model. DDs contained in the model are: (i) negative ($e^{-A.W}$); (ii) negative and positive ($e^{-A.W}.(1-e^{-B.W})$); (iii) negative and positive ($e^{-A.W}.(1-e^{-C.W})$); (iv) negative and both positives ($e^{-A.W}.(1-e^{-B.W}).(1-e^{-C.W})$), where for figure (a) $A = 0.1$; $B = 0.1$; $C = 0.2$ (units: worm intensity^{-1}) i.e. the positive DDs have different strengths and figure (b) $A = 0.1$; $B = 0.2$; $C = 0.2$ i.e. the positive DDs B and C have the same strength and therefore result in the same R_{dd} curve.

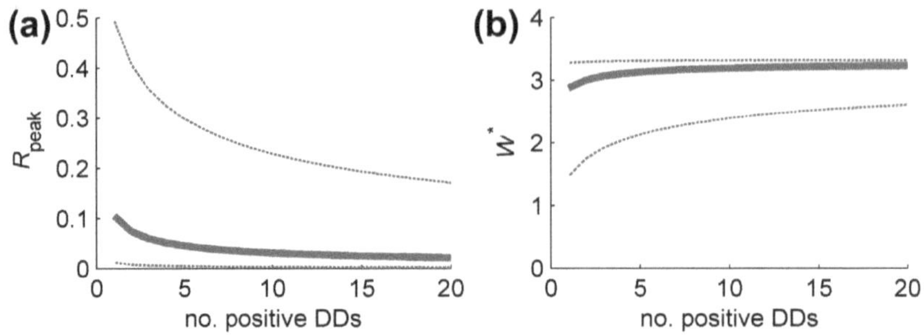

Figure 7 ***Positive and negative density dependencies (DDs) and transmission breakpoints.*** The values of (a) R_{peak} and (b) W^* of Eqns (7) and (8) calculated when the number of negative DDs is fixed to 3, each with a magnitude $A_i = 0.1$ and the number of positive DDs is varied according to the x-axis (thick solid lines). In each graph, the thin solid lines represent the cases for high and low DD magnitudes of $A_i, B_i = 0.01$ and 1.0. The effect of equal magnitude positive DDs are shown here to illustrate how even very large numbers of positive DD tend to alter R_{peak} and W^* very little beyond the initial changes.

Figure 7(a) shows that when Allee functions have density-dependent parameters B_i of equal magnitude, the value of R_{peak} declines rapidly, but that this decline slows with each additional Allee function. However, Figure 7(b) shows that each additional Allee function causes a rise in the value of W^*, with the first Allee function giving a very large jump followed by much smaller increases with each subsequent function. Thus, when Allee functions have an equal magnitude, the first function results in the emergence of a transmission breakpoint, which is almost entirely accounted for by the first positive DD. This observation clearly corresponds to the case in the LF model of successively adding positive DDs (anopheline uptake and the mating probability) to the *Culex* model: the change to R_{peak} and W^* for LF are small when the second of the positive DDs is added. These results highlight the important point that the occurrence of several positive DDs in the transmission dynamics cannot always be expected to be additive in their combined effects on elimination thresholds for parasitic infections.

4. DISCUSSION

In this paper, we have outlined the reasons why macroparasite transmission systems in which there is at least one positive DD, or Allee function, do not have a threshold that is correctly defined as R_0 (following other authors, particularly Regoes et al. (2002)), and why we therefore have to calculate the effective reproduction number, R_{eff}, to define an appropriate epidemic or endemic summary parameter. Since R_0 represents the maximum reproductive potential of a parasite circulating in a host population, the appropriate quantity with which to compare R_0 in the presence of positive DDs is what we have referred to as R_{peak}: the peak value of the full effective reproduction number curve. Despite the relevance of the Allee effect on the possibility of elimination or eradication of parasitic diseases, they have been investigated very little; this is likely due to the assumption that Allee effects either do not exist for important diseases or that the resulting 'breakpoint' parasite levels are so low that they may not appreciably reduce the effort required for elimination.

To illustrate a possible analysis that could be performed for most NTDs, we formulate an expression for the R_{eff} of the macroparasitic disease, LF, for an age-structured transmission model containing a variety of DDs. Our focus upon a macroparasite model is intended to be illustrative, and many of the same calculations could be performed with the susceptible-infected-recovered structure more commonly used for microparasite infections. We show that the overall form and behaviour of the fully specified R_{eff} function — when calculated numerically using the parameters appropriate to LF parasitic ecology in a real endemic community — may vary considerably according to the density-dependent processes that are included in the model. Our analysis of the R_{eff} function for LF transmission confirmed that when there are no positive DDs present (i.e. in the case considered here, when worm mating probability, vector uptake function, and host immunosuppression are excluded), the basic reproduction number R_0 is considerably higher than the peak value R_{peak} calculated when only one positive DD is included. This demonstrates that, for this system, the existence of a single instance of facilitation can considerably curtail the reproductive potential of the parasite.

Since the expression for the LF R_{eff} contains age-averaged density-dependent functions on the numerator and denominator regardless of the life stage in which they occur, the position of the density-dependent process in the life cycle should not affect its contribution to the equilibrium dynamics. However, it is clear that the precise form and parameter values of the function can and do affect the R_{eff} of LF. This can be seen in Figure 4(a), where the mating probability and uptake facilitating DDs have impacts of very different magnitude on the value of R_{peak}. However, the results also highlight that although the value of the LF transmission breakpoint will increase with increasing numbers of facilitating DDs, each successive positive DD appears to alter the overall LF R_{eff} profile only to a diminishing degree.

We developed and used a simplified model of the density-dependent component of R_{eff}, in which non-DD constant terms are excluded and only density-dependent functions are included, in order to explicate the sub-additive impact of positive DDs on LF transmission. Specifically, we used this model to investigate the consequences of combining positive and negative DDs and particularly how such combinations may impact the magnitude of the Allee effect that arises with changing numbers and strengths of positive DD. We find in particular that (1) for a fixed negative DD, increasing numbers of positive DD of the same magnitude increase transmission breakpoint values but decrease R_{peak}; and (2) a single positive DD may contribute to a large part of the transmission breakpoint value but that subsequent positive DDs do not change the transmission breakpoint value by very much.

The presence of many positive DDs might be expected to act either additively or super-additively (Berec et al., 2007) in their impact upon key outcome variables, such as parasite transmission breakpoint levels. For the simple model illustrated here, we showed that the combination of the major positive DDs operating in a macroparasitic infection, affect transmission threshold dynamics sub-additively: there are diminishing returns to each subsequent positive DD i.e. transmission breakpoint rise with the addition of DDs, but by a diminishing amount. The implication of this for macroparasite control programmes is that the appearance of several positive DDs might not drive up the transmission breakpoint level as much as would be liked. On the other hand, these results indicate that the presence of even a single positive DD in the macroparasitic transmission system might be enough to consider it eliminable (Hotez et al., 2007; King, 2009). Increasing numbers of positive DD will certainly raise the barrier to the re-establishment of the parasite in a community but the results here show that multiple Allee effects may not be required to consider the prospect of eliminating a particular macroparasitic disease.

While this result may depend upon the particular functions chosen to represent the DDs of the LF infection ecology — and this is a limitation of the work we present here — the model structure and DDs that we used were typical of helminth models that have been studied over the past 20 years and no special model was constructed for the analysis of this paper. It is likely, therefore, that these results would also apply to models used for other parasitic infections such as schistosomiasis, onchocerciasis and malaria. Recently, some of the concepts discussed in this article have been raised afresh in the context of soil-transmitted helminth control (Truscott et al., 2014; Anderson et al., 2014, 2013, 2012), following a renewed push toward harnessing mathematical predictive tools in the drive toward NTD elimination (London Declaration on Neglected Tropical Diseases, 2012).

The analyses presented in this paper may be used as the basis for future work on models of macro- or microparasitic diseases and in estimating their R_{eff} profiles when data are available. Such data are often collected and may remain unmined for the information they contain on the density-dependent functions operating. As is often the case for dynamical systems, data pertaining specifically to the dynamics of the system can contain the most useful information; such data could most likely be collected immediately following a perturbation of the kind that occurs after MDA, vector control or other highly efficacious community interventions. The R_{eff} profile, as illustrated in this paper, contains much of the information pertaining to parasite elimination, persistence and reproductive fitness in a specific setting and can serve as a focus for parameter estimation (in which R_{peak} or R_0 can be used as parameter selection criteria Spear et al., 2002), or policy studies, in which these thresholds represent targets for the long-term control of infection and disease (Michael et al., 2008). Future work in this area should investigate how one can infer the strength and type of density-dependent functions present through estimations of the profile of R_{eff} from field data, and use such information to quantify the probability of achieving control or elimination of a macroparasitic infection in a given locality.

APPENDIX

Calculating the largest eigenvalue of the system of ODEs at the disease-free equilibrium

The Jacobian matrix J, becomes, in the simplest case outlined in the paper

$$J = \begin{pmatrix} -\mu & \dfrac{\lambda\beta}{\sigma} \\ \alpha & -\delta \end{pmatrix}$$

and the *characteristic polynomial* of this matrix is given by setting the determinant of the Jacobian matrix to zero.

$$\det(J) = \begin{vmatrix} \mu - x & \lambda\beta\sigma \\ \alpha & \delta - x \end{vmatrix} = 0$$

The solutions, x to the resulting quadratic equation are the eigenvalues of the Jacobian matrix, and the largest of these determines the value of R_0. In this case, the characteristic equation becomes

$$(\mu + x)(\delta + x) - \frac{\lambda\beta\alpha}{\sigma} = 0$$

$$\text{so that } x = \frac{-(\mu+\delta) \pm \sqrt{(\mu+\delta)^2 - 4\left(\mu\delta - \frac{\lambda\beta\alpha}{\sigma}\right)}}{2}$$

The dominant eigenvalue is greater than 0 if $\frac{\lambda.\beta.\alpha}{\sigma} > \mu.\delta$ i.e. $\frac{\lambda.\beta.\alpha}{\mu.\delta.\sigma} > 1$, so that

$$R_0 = \frac{\lambda.\beta.\alpha}{\mu.\delta.\sigma}$$

Approximating positive and negative DDs with exponential functions

If we state that all negative DDs can be approximated as negative exponential functions, we see that negative DDs that appear multiplicatively in the expression for R_{eff} will combine so that the exponents of the negative DDs add.

As stated in the main text, positive DDs are approximated as

$$positive\ DD = \left(1 - e^{-BW}\right)$$

where B is the strength of the positive DD. These DDs occur multiplicatively as follows:

$$\left(1 - e^{-B_1 W}\right)\left(1 - e^{-B_2 W}\right) = 1 - e^{-B_1 W} - e^{-B_2 W} + e^{-(B_1+B_2)W}$$

Now, the resulting function can be approximated by a function of the same form as the original, with a modified DD by zeroing the difference between the integral of this approximation and the true function given above. This procedure is performed as a method for minimizing the distances between the curves of the true and approximate DDs; where the curves cross each other the area between the curves is both negative and positive and these areas can partially cancel each other out in the integral below. After performing the calculation, we checked that reasonable approximate DDs had been found. The expression whose zero is to be found is

$$= \int_0^\infty \left(1 - e^{-B'W}\right) - \left(1 - e^{-B_1W} - e^{-B_2W} + e^{-(B_1+B_2)W}\right) dW$$

$$= \left[\frac{e^{-B'W}}{B'} - \frac{e^{-B_1W}}{B_1} - \frac{e^{-B_2W}}{B_2} + \frac{e^{-(B_1+B_2)W}}{B_1 + B_2}\right]_0^\infty$$

This integral is equal to zero (i.e. the areas under the approximation and exact curves are the same) when

$$B' = \frac{B_1 B_2 (B_1 + B_2)}{B_1^2 + B_2^2 + B_1 B_2}$$

When we make this approximation, we are therefore able to multiply any number of positive and negative DDs together and obtain a general solution for the maximum value of the function R_{eff} and the parasite density W^* at which this maximum value occurs. We refer here to the component of the function R_{eff} made up of the density-dependent functions of parasite intensity as $R(W)$:

$$R(W) = e^{-AW}\left(1 - e^{-BW}\right)$$

$$\frac{dR(W)}{dW} = -Ae^{-AW} + (A + B)e^{-(A+B)W}$$

which is zero when

$$W^* = ln\left(1 + \frac{B}{A}\right)^{1/B}$$

And the value of $R(W)$ here is

$$\left(\frac{A}{(A + B)}\right)^{A/B} \left(\frac{B}{(A + B)}\right)$$

Model parameter values and density-dependent functions

The complete LF model, including all of the fine-structure parameters, is given by the following equations:

$$\frac{\partial W}{\partial t} + \frac{\partial W}{\partial a} = ABR.\psi_1.\psi_2 s_2.h(a).L^*.f_1(I).f_2(W) - \mu.W$$

$$\frac{\partial M}{\partial t}+\frac{\partial M}{\partial a}=\alpha.f_3(W,k).W-\delta.M$$

$$\frac{\partial I}{\partial t}+\frac{\partial I}{\partial a}=W-\gamma.I$$

$$\frac{dL}{dt}=\eta.\kappa.d.\int\pi(a).(1-g(M))da-\sigma.L-\eta.\psi_1.L$$

$$L^*=\frac{\eta.\kappa.d.\int\pi(a).(1-g(M))da}{\sigma+\eta.\psi_1}$$

Each of the parameters is defined and its value is presented in Appendix Table 1. These parameter values were obtained by fitting the LF model to human Mf prevalence data from several communities in PNG (Gambhir et al., 2010b; Singh et al., 2013). In the main text, in Eqn (1) and subsequently, the fully expressed model above has been simplified so that the parameter λ represents the set of parameters above $ABR.\psi_1.\psi_2 s_2.h(a)$, and β represents the set of parameters $\eta.\kappa.d.$

Model fitting and uncertainty estimation

We applied a variation of the Bayesian Melding method – specifically using the sampling-importance-resampling algorithm – used previously to quantify the uncertainty associated with deterministic model predictions of, for example, oceanic whale population size and HIV prevalence (Alkema et al., 2008). The original algorithm deals with a deterministic model that relates a set of input parameters and initial conditions θ, to a set of outputs ϕ, though it has been extended recently to take into account stochastic models (Sevcikova et al., 2007). Prior information, based on literature reviews and expert opinion, for both the model inputs and outputs ($p(\theta)$ and $p(\phi)$) are then combined, along with any available data, in the form of likelihood functions for the input and output parameters ($L(\theta)$ and $L(\phi)$). The algorithm we used to quantify uncertainty in the parameters of the present model and hence induced uncertainty in outcomes closely followed the method outlined by Brown et al. (2008):

1. From the prior input parameter distributions, $p(\theta)$, select 100,000 sets of model input parameters.
2. Run the model once for each of the selected parameter sets in order to generate a set of 100,000 model outputs (here Mf prevalence curves).

Appendix Table 1 Description and values of the parameters of the model. The functions/parameters indicated by an asterisk vary over the course of the simulation or over age

Parameter symbol	Definition	Typical values (range of prior distribution (lower, upper))
Model parameters		
ABR	Annual biting rate of mosquitoes per person	Varying*
η	Number of bites per mosquito	[5,10] per month
ψ_1	Proportion of L3 leaving mosquito per bite	[0.12, 0.70]
$\psi_2 s_2$	Proportion of L3 leaving mosquito that enter host × proportion of L3 entering host that develop into adult worms (this product is referred to as the 'establishment rate' in the main text)	[0.00004, 0.004]
μ	Death rate of adult worms	[0.008, 0.018] per month
α	Production rate of Mf per worm	[0.2, 1.5] per month
δ	Death rate of Mf	[0.08, 0.12] per month
γ	Immunity decay rate	0 per month
d	Proportion of mosquitoes which pick up infection when biting an infected host	[0.26, 0.48]
σ	Death rate of mosquitoes	[1.5, 8.5] per month
Model functions		
$k(M)$	Aggregation parameter from negative binomial distribution	k_0+k_1M: [0.0006, 0.0008]+[0, 0.04]M
$h(a)$	Parameter to adjust rate at which individuals of age a are bitten: Linear rise from 0 at age zero to 1 at H years	H: [1, 20] years
L^*	Equilibrium value of the larval density (see Eqn (5))	Varying*
$\pi(a)$	Probability that an individual is of age a	Varying*

3. Calculate the goodness of fit for each of the outputs by computing the likelihood for each ($L(\theta)$), given the prevalence data for each endemic area.
4. Resample, with replacement, 500 times from the original set of 100,000 parameter sets, with the probability of drawing each resample proportional to its likelihood for the data calculated in (3).
5. Run the model to calculate the desired quantities for each of the 500 parameter sets found in (4). These quantities are the Threshold Biting Rate (TBR), transmission breakpoints, R_0 values, and, for the full selection of 500 parameter sets, extinction probabilities. As pointed out by Brown et al. (2008), it is unlikely that the 500 resampled parameter sets will be unique, since those with the highest likelihood will be picked multiple times in the resampling procedure of (4).

ACKNOWLEDGMENTS

The work of MG, BKS and EM was financially supported by NIH grant no. RO1 AI069387-01A1.

REFERENCES

Alkema, L., Raftery, A.E., Brown, T., 2008. Bayesian melding for estimating uncertainty in national HIV prevalence estimates. Sex. Transm. Infect. 84, I11–I16.

Allee, W.C., 1949. Principles of Animal Ecology. Saunders Co., Philadelphia.

Anderson, R., Hollingsworth, T.D., Truscott, J., Brooker, S., 2012. Optimisation of mass chemotherapy to control soil-transmitted helminth infection. Lancet 379, 289–290.

Anderson, R., May, R., 1992. Infectious Diseases of Humans: Dynamics and Control. Oxford University Press.

Anderson, R., Truscott, J., Hollingsworth, T.D., 2014. The coverage and frequency of mass drug administration required to eliminate persistent transmission of soil-transmitted helminths. Philos. Trans. R. Soc. London B Biol. Sci. 369, 20130435.

Anderson, R.M., Truscott, J.E., Pullan, R.L., Brooker, S.J., Hollingsworth, T.D., 2013. How effective is school-based deworming for the community-wide control of soil-transmitted helminths? PLoS Negl. Trop. Dis. 7, e2027.

Barbour, A.D., Heesterbeek, J.A., Luchsinger, C.J., 1996. Thresholds and initial growth rates in a model of parasitic infection. Ann. Appl. Prob. 6, 1045–1074.

Basáñez, M.G., Boussinesq, M., 1999. Population biology of human onchocerciasis. Philos. Trans. R. Soc. London B Biol. Sci. 354, 809–826.

Basáñez, M.G., Ricardez-Esquinca, J., 2001. Models for the population biology and control of human onchocerciasis. Trends Parasitol. 17, 430–438.

Basáñez, M.G., Churcher, T.S., Grillet, M.E., 2009. *Onchocerca–Simulium* interactions and the population and evolutionary biology of *Onchocerca volvulus*. Adv. Parasitol. 68, 263–313.

Berec, L., Angulo, E., Courchamp, F., 2007. Multiple Allee effects and population management. Trends Ecol. Evol. 22, 185–191.

Bockarie, M.J., Tisch, D.J., Kastens, W., Alexander, N.D., Dimber, Z., Bockarie, F., Ibam, E., Alpers, M.P., Kazura, J.W., 2002. Mass treatment to eliminate filariasis in Papua New Guinea. N. Engl. J. Med. 347, 1841–1848.

Brown, T., Salomon, J.A., Alkema, L., Raftery, A.E., Gouws, E., 2008. Progress and challenges in modelling country-level HIV/AIDS epidemics: the UNAIDS Estimation and Projection Package 2007. Sex. Transm. Infect. 84 (Suppl. 1), i5–i10.

Burton, M.J., Holland, M.J., Makalo, P., Aryee, E.A., Alexander, N.D., Sillah, A., Faal, H., West, S.K., Foster, A., Johnson, G.J., Mabey, D.C., Bailey, R.L., 2005. Re-emergence of Chlamydia trachomatis infection after mass antibiotic treatment of a trachoma-endemic Gambian community: a longitudinal study. Lancet 365, 1321–1328.

Chidambaram, J.D., Lee, D.C., Porco, T.C., Lietman, T.M., 2005. Mass antibiotics for trachoma and the Allee effect. Lancet Infect. Dis. 5, 194–196.

Churcher, T.S., Dawes, E.J., Sinden, R.E., Christophides, G.K., Koella, J.C., Basáñez, M.G., 2010. Population biology of malaria within the mosquito: density-dependent processes and potential implications for transmission-blocking interventions. Malar. J. 9, 311.

Churcher, T.S., Ferguson, N.M., Basáñez, M.G., 2005. Density dependence and overdispersion in the transmission of helminth parasites. Parasitology 131, 121–132.

Churcher, T.S., Filipe, J.A., Basáñez, M.G., 2006. Density dependence and the control of helminth parasites. J. Anim. Ecol. 75, 1313–1320.

Courchamp, F., Clutton-Brock, T., Grenfell, B., 1999. Inverse density dependence and the Allee effect. Trends Ecol. Evol. 14, 405–410.

Diekmann, O., Heesterbeek, J.A.P., 2000. Mathematical Epidemiology of Infectious Diseases: Model Building, Analysis and Interpretation. Wiley.

Dietz, K., 1982. The population dynamics of onchocerciasis. In: Anderson, R.M. (Ed.), Population Dynamics of Infectious Diseases. Chapman and Hall, London.

Donnelly, C.A., Ghani, A.C., Leung, G.M., Hedley, A.J., Fraser, C., Riley, S., Abu-Raddad, L.J., Ho, L.M., Thach, T.Q., Chau, P., Chan, K.P., Lam, T.H., Tse, L.Y., Tsang, T., Liu, S.H., Kong, J.H.B., Lau, E.M.C., Ferguson, N.M., Anderson, R.M., 2003. Epidemiological determinants of spread of causal agent of severe acute respiratory syndrome in Hong Kong. Lancet 361, 1761–1766.

Duerr, H.P., Dietz, K., Eichner, M., 2005. Determinants of the eradicability of filarial infections: a conceptual approach. Trends Parasitol. 21, 88–96.

Duerr, H.P., Dietz, K., Schulz-Key, H., Buttner, D.W., Eichner, M., 2003. Density-dependent parasite establishment suggests infection-associated immunosuppression as an important mechanism for parasite density regulation in onchocerciasis. Trans. R. Soc. Trop. Med. Hyg. 97, 242–250.

Gambhir, M., Basáñez, M.G., Blake, I.M., Grassly, N.C., 2010a. Modelling trachoma for control programmes. Adv. Exp. Med. Biol. 673, 141–156.

Gambhir, M., Basáñez, M.G., Burton, M.J., Solomon, A.W., Bailey, R.L., Holland, M.J., Blake, I.M., Donnelly, C.A., Jabr, I., Mabey, D.C., Grassly, N.C., 2009. The development of an age-structured model for trachoma transmission dynamics, pathogenesis and control. PLoS Negl. Trop. Dis. 3, e462.

Gambhir, M., Basáñez, M.G., Turner, F., Kumaresan, J., Grassly, N.C., 2007. Trachoma: transmission, infection, and control. Lancet Infect. Dis. 7, 420–427.

Gambhir, M., Bockarie, M., Tisch, D., Kazura, J., Remais, J., Spear, R., Michael, E., 2010b. Geographic and ecologic heterogeneity in elimination thresholds for the major vector-borne helminthic disease, lymphatic filariasis. BMC Biol. 8, 22.

Gambhir, M., Michael, E., 2008. Complex ecological dynamics and eradicability of the vector borne macroparasitic disease, lymphatic filariasis. PLoS One 3, e2874.

Hotez, P.J., Molyneux, D.H., Fenwick, A., Kumaresan, J., Sachs, S.E., Sachs, J.D., Savioli, L., 2007. Control of neglected tropical diseases. N. Engl. J. Med. 357, 1018–1027.

King, C.H., 2009. Toward the elimination of schistosomiasis. N. Engl. J. Med. 360, 106–109.

Liu, F., Porco, T.C., Mkocha, H.A., Munoz, B., Ray, K.J., Bailey, R.L., Lietman, T.M., West, S.K., 2014. The efficacy of oral azithromycin in clearing ocular chlamydia: mathematical modeling from a community-randomized trachoma trial. Epidemics 6, 10–17.

Liu, F., Porco, T.C., Ray, K.J., Bailey, R.L., Mkocha, H., Munoz, B., Quinn, T.C., Lietman, T.M., West, S.K., 2013. Assessment of transmission in trachoma programs over time suggests no short-term loss of immunity. PLoS Negl. Trop. Dis. 7, e2303.

London Declaration on Neglected Tropical Diseases, 2012. Uniting to Combat Neglected Tropical Diseases. Ending the Neglect and Reaching 2020 Goals. Available. https://www.gov.uk/government/uploads/system/uploads/attachment_data/file/67443/NTD_20Event_20-_20London_20Declaration_20on_20NTDs.pdf.

May, R.M., 1977a. Thresholds and breakpoints in ecosystems with a multiplicity of stable states. Nature 269, 471–477.

May, R.M., 1977b. Togetherness among schistosomes - effects on dynamics of infection. Math. Biosci. 35, 301–343.

Michael, E., Bundy, D.A., 1998. Herd immunity to filarial infection is a function of vector biting rate. Proc. Biol. Sci. 265, 855–860.

Michael, E., Malecela-Lazaro, M.N., Simonsen, P.E., Pedersen, E.M., Barker, G., Kumar, A., Kazura, J.W., 2004. Mathematical modelling and the control of lymphatic filariasis. Lancet Infect. Dis. 4, 223–234.

Michael, E., Malecela, M.N., Zervos, M., Kazura, J.W., 2008. Global eradication of lymphatic filariasis: the value of chronic disease control in parasite elimination programmes. PLoS One 3, e2936.

Muriu, S.M., Coulson, T., Mbogo, C.M., Godfray, H.C., 2013. Larval density dependence in *Anopheles gambiae* s.s., the major African vector of malaria. J. Anim. Ecol. 82, 166–174.

Norman, R.A., Chan, M.S., Srividya, A., Pani, S.P., Ramaiah, K.D., Vanamail, P., Michael, E., Das, P.K., Bundy, D.A., 2000. EPIFIL: the development of an age-structured model for describing the transmission dynamics and control of lymphatic filariasis. Epidemiol. Infect 124, 529–541.

Regoes, R.R., Ebert, D., Bonhoeffer, S., 2002. Dose-dependent infection rates of parasites produce the Allee effect in epidemiology. Proc. Biol. Sci. 269, 271–279.

Sevcikova, H., Raftery, A.E., Waddell, P.A., 2007. Assessing uncertainty in urban simulations using Bayesian melding. Transp. Res. Part B Methodol. 41, 652–669.

Singh, B.K., Bockarie, M.J., Gambhir, M., Siba, P.M., Tisch, D.J., Kazura, J., Michael, E., 2013. Sequential modelling of the effects of mass drug treatments on anopheline-mediated lymphatic filariasis infection in Papua New Guinea. PLoS One 8, e67004.

Snow, L.C., Bockarie, M.J., Michael, E., 2006. Transmission dynamics of lymphatic filariasis: vector-specific density dependence in the development of *Wuchereria bancrofti* infective larvae in mosquitoes. Med. Vet. Entomol. 20, 261–272.

Snow, L.C., Michael, E., 2002. Transmission dynamics of lymphatic filariasis: density-dependence in the uptake of *Wuchereria bancrofti* microfilariae by vector mosquitoes. Med. Vet. Entomol. 16, 409–423.

Solomon, A.W., Holland, M.J., Alexander, N.D., Massae, P.A., Aguirre, A., Natividad-Sancho, A., Molina, S., Safari, S., Shao, J.F., Courtright, P., Peeling, R.W., West, S.K., Bailey, R.L., Foster, A., Mabey, D.C., 2004. Mass treatment with single-dose azithromycin for trachoma. N. Engl. J. Med. 351, 1962–1971.

Spear, R.C., 2012. Internal versus external determinants of *Schistosoma japonicum* transmission in irrigated agricultural villages. J. R. Soc. Interface 9, 272–282.

Spear, R.C., Hubbard, A., Liang, S., Seto, E., 2002. Disease transmission models for public health decision making: toward an approach for designing intervention strategies for *Schistosomiasis japonica*. Environ. Health Perspect. 110, 907–915.

Stolk, W.A., De vlas, S.J., Borsboom, G.J., Habbema, J.D., 2008. LYMFASIM, a simulation model for predicting the impact of lymphatic filariasis control: quantification for African villages. Parasitology 135, 1583–1598.

Truscott, J.E., Hollingsworth, T.D., Brooker, S.J., Anderson, R.M., 2014. Can chemotherapy alone eliminate the transmission of soil-transmitted helminths? Parasit. Vectors 7, 266.
West, E.S., Munoz, B., Mkocha, H., Holland, M.J., Aguirre, A., Solomon, A.W., Bailey, R., Foster, A., Mabey, D., West, S.K., 2005. Mass treatment and the effect on the load of *Chlamydia trachomatis* infection in a trachoma-hyperendemic community. Invest Ophthalmol. Vis. Sci. 46, 83–87.
White, M.T., Griffin, J.T., Churcher, T.S., Ferguson, N.M., Basáñez, M.G., Ghani, A.C., 2011. Modelling the impact of vector control interventions on *Anopheles gambiae* population dynamics. Parasit. Vectors 4, 153.
WHO, 2014. Lymphatic Filariasis [Online]. Available: http://www.who.int/mediacentre/factsheets/fs102/en/ (accessed 05.11.14.).
World Health Organization, 2012. Accelerating Work to Overcome the Global Impact of Neglected Tropical Diseases: A Roadmap for Implementation. World Health Organization, Geneva.

CHAPTER TWO

Mathematical Modelling of Leprosy and Its Control

David J. Blok*, Sake J. de Vlas*, Egil A.J. Fischer[$], Jan Hendrik Richardus*,[1]
*Department of Public Health, Erasmus MC, University Medical Center Rotterdam, Rotterdam, The Netherlands
[$]Department of Epidemiology, Crisis Organisation and Diagnostics, Central Veterinary Institute, Part of Wageningen UR, Lelystad, The Netherlands
[1]Corresponding author: E-mail: j.richardus@erasmusmc.nl

Contents

Abstract

Leprosy or Hansen's disease is an infectious disease caused by the bacterium *Mycobacterium leprae*. The annual number of new leprosy cases registered worldwide has remained stable over the past years at over 200,000. Early case finding and multidrug therapy have not been able interrupt transmission completely. Elimination requires innovation in control and sustained commitment. Mathematical models can be used to predict the course of leprosy incidence and the effect of intervention strategies. Two compartmental models and one individual-based model have been described in the literature. Both compartmental models investigate the course of leprosy in populations and the long-term impact of control strategies. The individual-based model focusses on transmission within households and the impact of case finding among contacts of new leprosy patients. Major improvement of these models should result from a better understanding of individual differences in exposure to infection and

Advances in Parasitology, Volume 87
ISSN 0065-308X
http://dx.doi.org/10.1016/bs.apar.2014.12.002

developing leprosy after exposure. Most relevant are contact heterogeneity, heterogeneity in susceptibility and spatial heterogeneity. Furthermore, the existing models have only been applied to a limited number of countries. Parameterization of the models for other areas, in particular those with high incidence, is essential to support current initiatives for the global elimination of leprosy. Many challenges remain in understanding and dealing with leprosy. The support of mathematical models for understanding leprosy epidemiology and supporting policy decision making remains vital.

1. INTRODUCTION

1.1 Disease

Leprosy or Hansen's disease is an infectious disease caused by the bacterium *Mycobacterium leprae*. Most people are able to clear the bacterium before disease occurs, or are resistant against infection (Fine, 1982). For those developing disease, leprosy affects the skin, the peripheral nerves, the mucosa of the upper respiratory tract and the eyes. The different clinical signs of leprosy depend on the response of the immune system of the patient. When the cellular immune response is strong enough to keep the infection localized, the tuberculoid form will develop. If the cellular response is insufficient or not present, the bacterium can spread systemically and cause lepromatous leprosy. Lepromatous leprosy has many more bacilli in lesions than the tuberculoid form. For treatment purposes, cases are classified into paucibacillary (PB) and multibacillary (MB) leprosy, based on the extent of the disease in terms of bacterial load and number of skin patches (WHO, 1998).

The infection can cause nerve function impairments, leading to secondary complications, such as infection of untreated wounds and ulcers on palms and soles. Nerve function impairment can develop gradually, or during periods of inflammation, called reactions. Chronic disability and social stigma cause substantial suffering to those affected by leprosy. The median incubation time is 3.5 years for PB leprosy and 10.0 years for MB leprosy (Fine, 1982; Meima et al., 2004). The fact that very young children are found with symptomatic leprosy, and that some veterans develop leprosy over 20 years after returning from endemic areas (Noordeen, 1985) shows the wide variation in the incubation period.

1.2 Transmission, Treatment and Control

Although *M. leprae* can remain viable for some time outside the human body (Desikan, 1977), it is commonly accepted that the main route of infection is through direct transmission from an infectious to a susceptible person.

Patients can shed many bacilli through their nose, and nasal carriage of healthy persons indicates that direct respiratory transmission through aerosols is the most likely route of transmission (Hatta et al., 1995), although skin-to-skin transmission is also considered to be possible (Noordeen, 1985). Both routes require close and direct contact. Due to their higher number of bacilli and poorer immune response, patients with MB leprosy are thought to be the only infectious individuals, or at least the most infectious individuals (Fine, 1982). The detection of leprosy is based on clinical signs: skin lesions, loss of sensitivity of skin lesions and thickened nerves, thus established after physical examination. The basis for leprosy control is treatment with multi-drug therapy (MDT), a combination of two or three antibiotics, including rifampicin, according to the type of leprosy, PB or MB (WHO, 1994).

In 1991, the 44th World Health Assembly adopted the objective of eliminating leprosy globally as a public health problem by the year 2000 (WHO, 1991). Leprosy elimination was thereby defined as reducing the prevalence rate to less than 1 case per 10,000 population. Although this was achieved at the global level by the end of 2000, in many countries a sizable leprosy problem still persists. The current leprosy control strategy is formulated by the World Health Organization (WHO) as the 'Enhanced global strategy for further reducing the disease burden due to leprosy 2011–2015' (WHO, 2009). The strategy aims to reduce the global rate of new cases with grade-2 (i.e. visible) disabilities per 100,000 population by at least 35% by the end of 2015, compared with the baseline at the end of 2010. The approach underlines the importance of early detection and quality of care in an integrated service setting. The WHO expects this strategy to reduce the transmission of the disease in the community and thus lower the occurrence of new cases. Recently, WHO has formulated 'roadmap targets' to overcome the global impact of 17 neglected tropical diseases, including leprosy. These targets are set for the period 2015–2020 and for leprosy are defined as (1) global interruption of transmission by 2020 and (2) reduction of grade-2 disabilities in newly detected cases to below 1/million population at global level by 2020 (WHO, 2012).

2. THE CURRENT EPIDEMIOLOGICAL SITUATION AND CHALLENGES

In the year 2012 a total of 232,857 new leprosy cases were registered in the world and less than 20 countries reported >1,000 new cases, indicating that leprosy is gradually becoming limited to a few countries

(WHO, 2013). Three endemic countries (India, Brazil and Indonesia) account for nearly 80% of all new cases in the world. This global annual number of newly detected leprosy cases has been fairly stable over the past 7 years, indicating that transmission of *M. leprae* is ongoing (Figure 1). The WHO provides annual statistics reported by 115 countries from different WHO regions on leprosy, with information on new cases detected, number of cases with grade-2 disability, number of children and women and treatment completion rates of patients with MB leprosy. An estimated three million people live with disability due to leprosy (Britton and Lockwood, 2004) and it is expected that up to one million people will continue to suffer from disability in the next decades (Meima et al., 2008).

It has long been argued that elimination of leprosy cannot be achieved by a strategy based on MDT alone and that new tools and technologies are needed to attain this goal (Richardus and Habbema, 2007; Smith and Richardus, 2008; Rodrigues and Lockwood, 2011). Intensified, population-based approaches to case detection are no longer cost-effective and a new approach is now indicated that is appropriate to the current epidemiological situation. New cases are relatively rare even in endemic countries, health care resources are scarce with many competing health care demands and leprosy control activities are difficult to sustain within integrated programmes. The main risk of exposure to leprosy is in close contacts of new, untreated cases and the risk of exposure to leprosy in the general community is very low. An increasing proportion of new cases will be from household contacts (Richardus et al., 2005). In the past years, progress has been made in the areas of chemoprophylaxis and immunoprophylaxis (vaccination) to prevent leprosy and these interventions have focussed primarily on contacts of leprosy patients (Moet et al., 2008; Duthie et al., 2012).

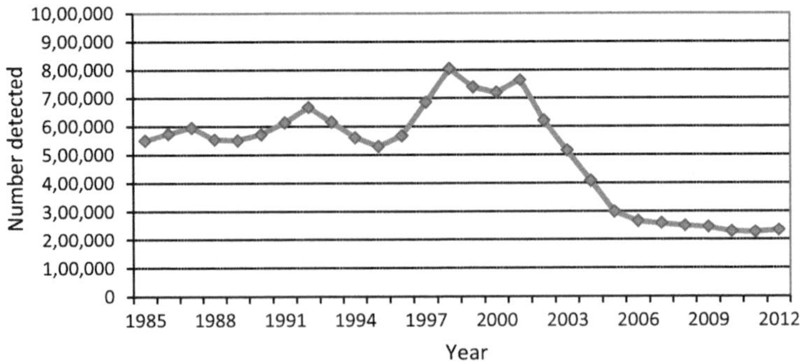

Figure 1 Global leprosy new case detection 1985–2012, based on the figures reported annually by the WHO.

3. HETEROGENEITY IN LEPROSY

Heterogeneity is due to differences between individuals in exposure to infection with *M. leprae* and in developing leprosy after exposure. Relevant forms of heterogeneity in the population are contact heterogeneity, heterogeneity in susceptibility and spatial heterogeneity. These forms of heterogeneity are not mutually exclusive.

Infection with a directly-transmitted bacterial infection, such as *M. leprae*, needs contact between an infectious host and a susceptible host. By heterogeneity in the contact structure of a population, individuals have different risks of coming into contact with infectious individuals. Thus contact heterogeneity plays a major role in the infection dynamics of directly-transmitted diseases (Wallinga et al., 1999). In several studies of leprosy, this risk based upon contact status has been investigated. In Bangladesh it was shown that close contacts of leprosy patients, such as household members, are at a higher risk of developing leprosy themselves (Moet et al., 2004). This has been shown for different countries and continents (Bakker et al., 2005; Fine et al., 1997; Rao et al., 1975; van Beers et al., 1999). The role of close contacts in the epidemic differs between areas. In low incidence areas, the relative risk of contacts is higher than in high incidence areas (Richardus et al., 2005). In some high incidence situations, almost half of the population is a close contact of leprosy patients (Bakker et al., 2004).

Even if all exposure to *M. leprae* would be the same, some people react differently to infections than others. In addition, not all people that are exposed to *M. leprae* develop leprosy. It is not clear whether these individuals clear the bacilli efficiently or are resistant to infection (Fine, 1982; Meima et al., 2004; Noordeen, 1985). It is thought that only a fraction (5—20%) of the population is susceptible to development of leprosy after exposure. Differences in susceptibility can be genetic or caused by environmental factors that alter the health status of a person. Genetic studies found an association of both susceptibility to leprosy (Fitness et al., 2002; Mira et al., 2004; Zhang et al., 2009) and the type of leprosy — tuberculoid or lepromatous — with genetic factors (Mira et al., 2003). In an epidemiological study, Bakker et al. (Bakker et al., 2005) found that approximately 50% of the susceptibility was explained by inheritance. Also, Moet et al. (Moet et al., 2006) found an association between leprosy prevalence and being a relative of a patient. It is, however, difficult to separate relationship from contact status, such as being a household member (Moet et al., 2006). Susceptibility to leprosy is also related to a common environment and the risk of family members might be caused by the fact that all household members share the same environment, including wealth. Poverty, and in particular recent food shortage, have been shown to be a risk factor for leprosy on a population level (Feenstra et al., 2012).

Finally, spatial heterogeneity means that the occurrence of an infectious disease is not evenly distributed over space, which can have several underlying reasons. Leprosy is found to be unevenly distributed in villages (Bakker et al., 2005; van Beers et al., 1999), although this was not observed consistently (Fischer et al., 2008b), and also at higher aggregated area levels, such as districts (Montenegro et al., 2004; Opromolla et al., 2006; Sterne et al., 1995; Fischer et al., 2008a). The uneven spatial distribution of leprosy can be the result of contact heterogeneity, especially clustering at a low level, e.g. village level. If neighbours have intense contact, neighbours will have a higher risk of infection (Moet et al., 2006). This is expected to result in spatial clustering of cases in villages. However, other underlying spatial factors might determine the clustered occurrence of leprosy. It is, for example, associated with impoverished areas (Montenegro et al., 2004; Opromolla et al., 2006). Geographic features include a decreased leprosy incidence with the distance of households to a river or lakeshore in Malawi, and the risk increased with the distance to a main road (Sterne et al., 1995). These features might, however, change from country to country, as for example, in the Nilphamari district of Bangladesh with many water bodies and rivers, where no relationship with the distance to water was found (Fischer et al., 2008a). Leprosy is often described as a rural disease (Fine, 1982; Sterne et al., 1995). However, clustering in urban areas has been reported in Brazil and around urban areas in Bangladesh (Montenegro et al., 2004; Opromolla et al., 2006).

4. LEPROSY MODELS

Three mathematical models for leprosy transmission and control have been described. Two of the three models are compartmental models (Lechat et al., 1985; Meima et al., 1999) and one is a microsimulation or individual-based model (Fischer et al., 2010). Another model in the literature combines a simple leprosy model with a tuberculosis model (Lietman et al., 1997). The purpose of the latter model was to explore whether immunity acquired from tuberculosis infection could have contributed to the disappearance of leprosy in Western Europe. Below, we only focus on models with the primary aim to predict the course of leprosy incidence or to evaluate control strategies. Table 1 presents an overview of these models.

4.1 Lechat's Leprosy Model

Lechat et al. developed the first mathematical model for leprosy in the 1970s and 1980s (Lechat et al., 1974, 1985, 1987, 1990; Lechat, 1992). The model enabled investigation of the course of leprosy in populations under different assumptions and the impact of long-term leprosy control strategies, such as dapsone monotherapy, MDT treatment and Bacillus Calmette-Guérin (BCG)-like vaccines (Lechat et al., 1977, 1985, 1987). The structure of the model is presented in Figure 2. In the Lechat's model,

Table 1 Comparison of leprosy models

	Lechat's models	SIMLEP	SIMCOLEP
Type	**Compartmental**	**Compartmental**	**Individual-based**
Demography			
Birth, ageing, death	Yes	Yes	Yes
Household structures	—	—	Yes
Disease			
Natural immunity	—	Yes	Yes
Asymptomatic infection	Yes	Yes	Yes
Symptomatic infection	Yes	Yes	Yes
Difference in infectiousness	Yes	Yes	Yes
Heterogeneity in susceptibility	—	—	Yes
Contact heterogeneity	—	—	Yes
Transmission			
General population	Yes	Yes	Yes
Within-households	—	—	Yes
Treatment/interventions			
Passive case finding	—	Yes	Yes
Active case finding	—	Yes	Yes
Early diagnosis	Yes	Yes	Yes
Dapsone monotherapy	Yes	Yes	Yes
MDT treatment	Yes	Yes	Yes
Chemoprophylaxis	—	—	Yes
Vaccination (e.g. BCG or leprosy-specific)	Yes	Yes	Yes

BCG, bacillus calmette-guérin; MDT, multidrug therapy.

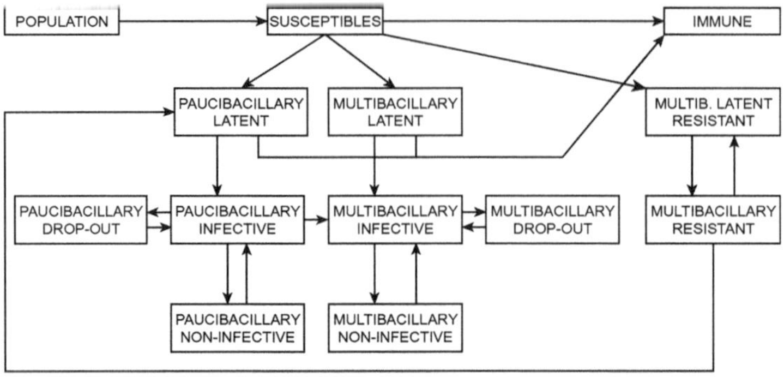

Figure 2 Structure of Lechat's model (Lechat, 1992).

the whole population was considered susceptible for leprosy. New infections were modelled as a function of the number of infected persons in the population. An infected individual first entered the latent stage, followed by the disease stage. Treatment would affect the course of the disease and infectiousness of individuals. These models also made a first attempt to model heterogeneity in infectiousness. Patients in different stages of the disease or under treatment would also have different capacity to transmit leprosy (Lechat, 1981). Lechat's models helped considerably to clarify the thinking about leprosy control. However, there was room for substantial refinement of this model.

In 1999, Meima et al. (1999) developed a new modelling framework, SIMLEP, which builds on the approach of Lechat's models (Lechat et al., 1974, 1985). SIMLEP allowed for more variations of model assumptions to investigate uncertainties in leprosy epidemiology. Assumptions regarding natural immunity, the incubation period and asymptomatic infection could be investigated. Moreover, delays of awareness and treatment were incorporated. However, this model did not include the disease dynamics in households or heterogeneity in susceptibility, both of which were required to evaluate the effects of interventions targeted at household members, such as early diagnosis and chemoprophylaxis. For this reason SIMCOLEP, an individual-based model, was developed. This model was able to take into account transmission in households and test for different assumptions on heterogeneity in susceptibility to leprosy (Fischer et al., 2010).

4.2 The SIMLEP Model

The SIMLEP framework was developed to investigate the many uncertainties in leprosy epidemiology and in response to the need for simulation models to make predictions of future trends (Meima et al., 1999; Anonymous, 1992). The purpose of this model was to take into account variations in the assumptions regarding natural immunity, the incubation period and asymptomatic infection, and delays in awareness and treatment of leprosy. In addition, it permitted testing of different mechanisms describing leprosy transmission by making assumptions about the level of contagiousness per type of infection.

SIMLEP is a compartmental model that describes the process of leprosy transmission, disease and control in a population (see Figure 3). The framework specifies assumptions about demography, leprosy and interventions. Demography is only represented by birth and death processes. In the model, leprosy can only be acquired by susceptible individuals. The model enabled the testing of various assumptions on natural immunity by introducing a nonsusceptible compartment. Upon infection, an individual moved into the stage of asymptomatic infection, which is an infection without the manifestation of symptoms. Asymptomatic infections could heal spontaneously. Untreated patients could enter a symptomatic infection stage, which was subdivided into three types: (1) self-healing symptomatic, (2) downgrading symptomatic and (3) strongly contagious symptomatic. Downgrading symptomatic infections are not strongly contagious, but will downgrade to the strongly contagious symptomatic stage over time. People with a strongly contagious symptomatic infection were assumed to be always highly infectious. The infectiousness of individuals with an asymptomatic infection, a self-healing symptomatic infection or downgrading symptomatic infection could be varied (i.e. noninfectious or infectious) to test assumptions. The transmission was determined as a function of infectious individuals in the population taking into account the differences in infectiousness depending on the stage in which a patient is.

Leprosy control in the model comprises vaccination, diagnosis and chemotherapy. Vaccination refers to BCG vaccination which is assumed to have a protective effect by reducing leprosy susceptibility (Fine and Smith, 1996). Vaccination takes place only at birth in the model. Diagnosis includes the delay of awareness and treatment (or reporting delay). Early case detection reduces the delay between onset of symptomatic leprosy and the start of treatment. However, cases must first become aware of the disease before they can seek care. Delays in detection could be varied. Chemotherapy reflects all possible treatment, including dapsone monotherapy and MDT (Meima et al., 1999). Chemotherapy stops infectiousness of patients immediately after treatment. After treatment a fraction of the people

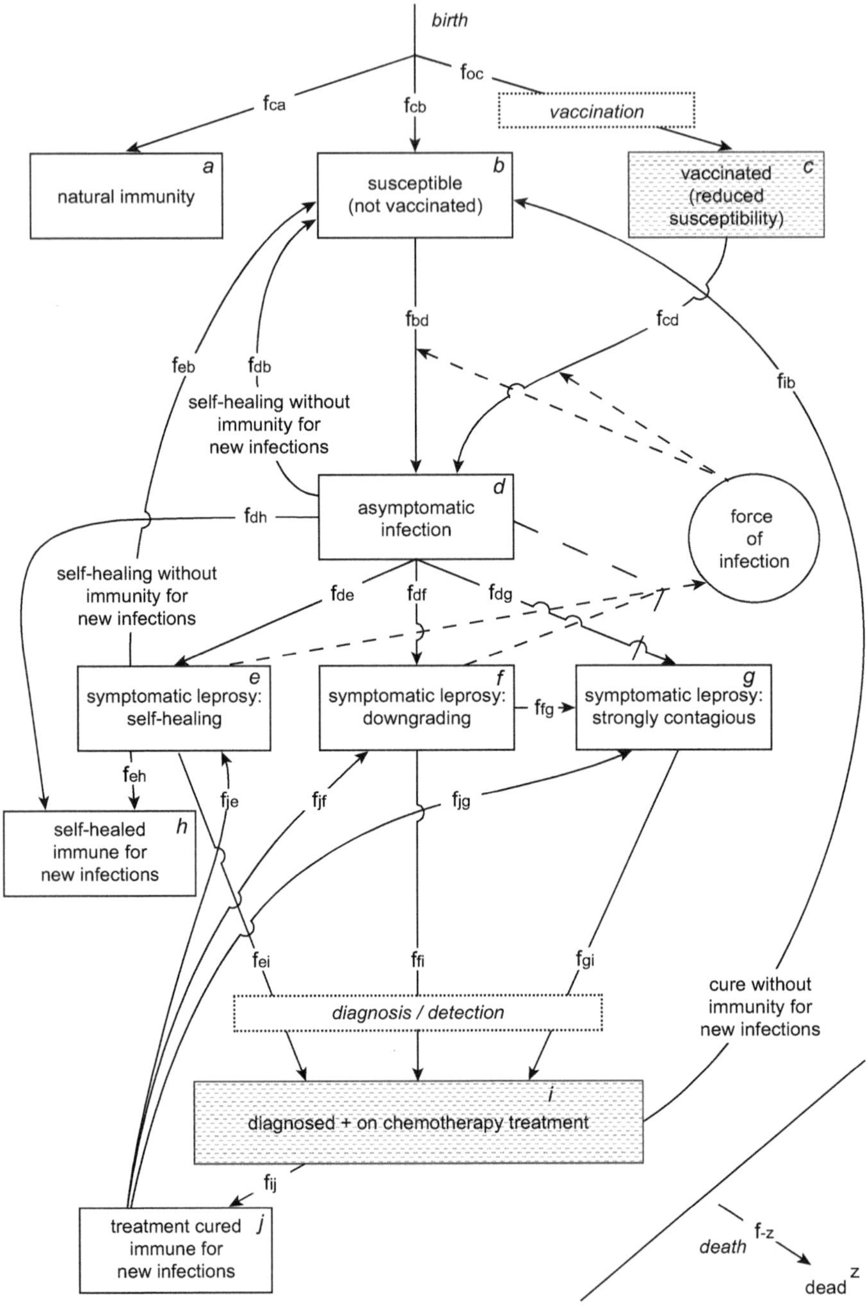

Figure 3 Model structure of SIMLEP. *For details we refer to Meima et al. (1999).*

will go to the compartment 'cured and immune' and the other fraction will go to the compartment 'susceptible' again.

4.2.1 Applications

SIMLEP was used to investigate the disappearance of leprosy from Norway, for which it was found that a model with heterogeneity in age of exposure, heterogeneity in susceptibility and a long tail to the distribution of the incubation period gave the best fit to the data (Meima et al., 2002). Using the SIMLEP modelling framework to predict future trends shows that a failure to maintain early case detection would be devastating, and that elimination of leprosy can only be a long-term goal. A second application of SIMLEP investigated the impact of BCG vaccination at birth and early diagnosis in India. Both interventions showed a decrease in the level of incidence (Gupte et al., 2000).

4.3 The SIMCOLEP Model

SIMCOLEP is a micro-simulation or a stochastic individual-based model, which models leprosy transmission in a population (Fischer et al., 2010). The step from compartmental modelling to individual-based modelling is driven by the need to take into account household structure and heterogeneity in susceptibility. An individual-based model is particularly useful to model networks, household structures and individual heterogeneities.

SIMCOLEP simulates the life histories of individuals and the natural history of infection with *M. leprae* (Fischer et al., 2010). The state of an individual changes during events that are scheduled in time. The timing of events is determined by probability distributions taking into account the current state and history of an individual. The model is divided into two modules: a population module and a disease module (see Figure 4).

The population module describes processes that are not related to the disease or infection, including birth, death and household processes. Birth and death are independent of disease and are determined such that the population size follows a designated growth curve.

SIMCOLEP is the first leprosy model to include household formation, dissolution and changes. This enables the evaluation of interventions targeted at household members and takes into account the transmission between households by movement of infected people. Each individual in the model is part of either his/her own household (single person) or a multi-person household. The model assumes that birth only takes place in households of married couples. At birth individuals are placed in a

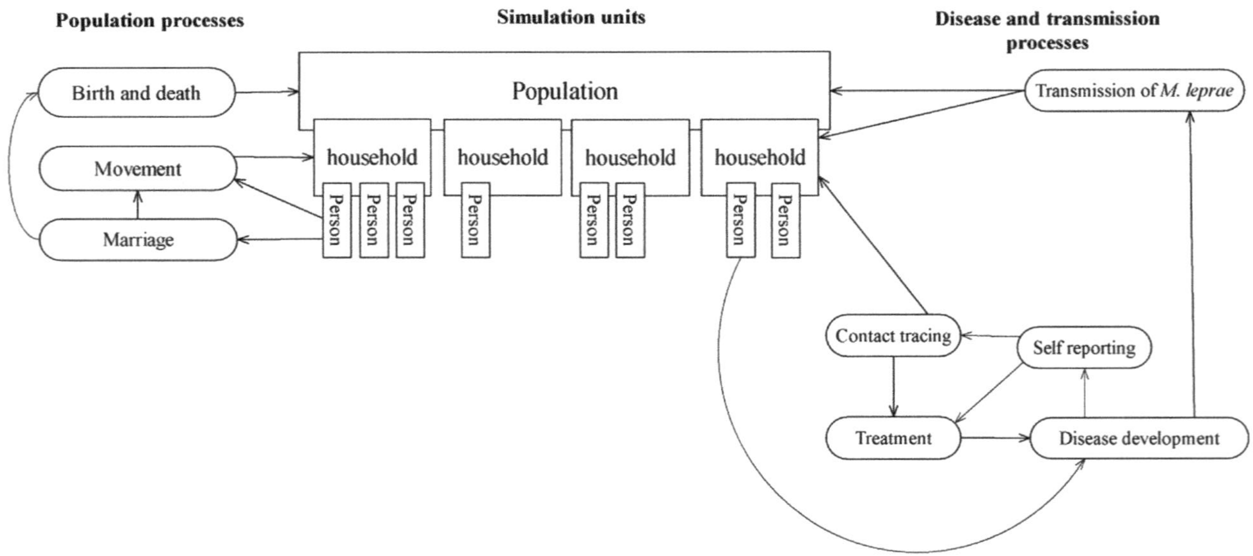

Figure 4 Model structure of SIMCOLEP (Fischer et al., 2010).

household. New households are formed after marriage or when a child leaves its parents to start its own household. Movements between households can occur during adolescence, after marriage or after the death of a spouse. A household dissolves after death, or when the surviving spouse will join the household of their children.

The disease module simulates processes of disease, infection, leprosy control and interventions. The natural history of disease is modelled following SIMLEP (Meima et al., 1999). Transmission occurs when an infectious individual has contact with a susceptible individual in the general population with mass action, which means that the number of infectious contacts is independent of the size of the population. Besides transmission in the general population, SIMCOLEP also adds within-household transmission. Within-household transmission only takes place during contacts with household members with a pseudo-mass action model, thus assuming an increase of the number of contacts with an increasing household size.

Leprosy control in SIMCOLEP includes passive detection and treatment. Passive detection is represented by a detection delay, which can be varied over time. At the moment of passive case detection the individual is diagnosed with either a self-healing (PB) or chronic (MB) infection. Household members of a detected case can be subjected to contact tracing, which can additionally be followed up annually. The model allows testing assumptions on detection. In the default setting 10% of symptomatic cases are missed and asymptomatic cases can also be diagnosed as 'no disease'. The model further allows simulation of interventions, including dapsone monotherapy, MDT and BCG coverage. BCG vaccination is assumed to have a protective effect (Schuring et al., 2009). Besides mimicking the current control programs, the model allows for testing new interventions: active case detection, early diagnosis and administering chemoprophylaxis to contacts. Active case detection is similar to contact tracing. The probability of finding a case is determined by the detection probability of the infection state (Fischer et al., 2010).

Due to the flexibility of individual-based simulations, SIMCOLEP could also be used to test various scenarios of heterogeneity in susceptibility in the population, as it allows for the testing of six mechanisms describing heterogeneity in susceptibility (Fischer et al., 2010). The simplest mechanism assumes a random distribution of susceptible individuals in the population. In the second mechanism, inhabitants can be susceptible in 25% of the households due to a common household factor, such as poverty. However, to allow for variation within the household, not all members of a susceptible household are susceptible. The third and fourth mechanisms are genetic. Mendelian inheritance of one gene determines leprosy susceptibility, and a second gene determines the type of infection (self-healing vs chronic). The 'dominant' mechanism considers both genes to be dominant and the 'recessive' mechanism considers both genes to be recessive. The fifth and sixth mechanisms are a combination of the genetic and household mechanisms: household and dominant, and household and recessive. In these scenarios it is assumed that leprosy susceptibility is caused by genetic factors, but will only present itself due to living in a susceptible household, such that each factor accounts for half of the susceptibility based on Bakker et al. (2005).

4.3.1 Applications

SIMCOLEP was used to investigate which mechanism for heterogeneity of leprosy susceptibility can explain the observed clustering in household contacts of leprosy patients in northwest Bangladesh (Fischer et al., 2010). Results of this study could not rule out any mechanism to explain clustering in household contacts of leprosy. SIMCOLEP was also used to evaluate different intervention strategies in the same region (Fischer et al., 2011). Seven potential intervention scenarios were tested for the future control of leprosy: (1) baseline scenario, which represents the current practice; (2) no contact tracing; (3) administering chemoprophylaxis (single dose of rifampicin) to each individual in contact with a leprosy patient; (4) early diagnosis of subclinical leprosy; (5) BCG vaccination to all newly born infants in the area; (6) combination of BCG and chemoprophylaxis; (7) combination of BCG and early diagnosis of subclinical leprosy. Early diagnosis showed the largest effect on reducing new cases in the population followed by chemoprophylaxis.

5. FUTURE CHALLENGES

Many uncertainties remain with respect to leprosy. A variety of host immunogenic factors influences both an individual's susceptibility to infection with *M. leprae* and the pathologic course of the disease; research in this area is ongoing (Adams et al., 2012; Alter et al., 2011). In particular, questions remain regarding mechanisms of natural immunity and susceptibility to the MB and PB forms of leprosy, which show marked variation in distribution in different parts of the world. SIMCOLEP has explored the likelihood of the contribution of different mechanisms determining susceptibility to leprosy, but could not rule out any of the proposed mechanisms

(Fischer et al., 2010). These studies also showed that the expected effect of interventions differs for each of these mechanisms (Fischer et al., 2011). Better understanding of these mechanisms is therefore important, because the choice of susceptibility mechanism determines the outcome of model predictions.

There are also uncertainties about the transmission of *M. leprae* and whether environmental reservoirs and animal hosts play a role (Turankar et al., 2014). For human-to-human transmission, it is still unclear when an infected person becomes infectious, how long a person stays infectious, and what the role is of healthy carriers and subclinical infections among household contacts of leprosy patients (Araujo et al., 2012). Models can play an important role in explaining these uncertainties by allowing the testing of various assumptions with regard to the transmission of *M. leprae*. Such modelling exercises depend on the availability of suitable data sets, which are scarce. Large prospective cohort studies following patients and their contacts in different areas of the world with different endemic levels of leprosy would be extremely valuable to address these issues.

Although the overall trend in new cases is declining worldwide, there are still key policy challenges and questions that need to be addressed. An important challenge is to determine which interventions at population level have the highest impact on future incidence of disease through the interruption of transmission. Focus should be also on the effect of interventions targeting contacts of leprosy patients, including contact tracing, chemoprophylaxis, immunoprophylaxis (e.g. BCG vaccination or a specific BCG-like leprosy vaccine) and early diagnosis of leprosy by means of diagnostic tests for infection or tests that predict clinical disease. A question that naturally follows is how and when elimination of leprosy can be achieved. Depending on the definition used, elimination is defined as reaching a prevalence rate of less than 1 case per 10,000 population or reducing the incidence of leprosy to zero. Perhaps an even greater challenge is to investigate whether we can move from elimination to eradication of leprosy, defined as the complete worldwide interruption of transmission of *M. leprae*. Key policy questions that follow from elimination or eradication targets are how to evaluate post-elimination monitoring. Mathematical models and, in particular, individual-based models may help to address these questions. Furthermore, continuation of provision of health care to people affected by leprosy through lasting impairment and disability should not be disregarded.

Recent applications of existing leprosy models have only focussed on current and past endemic regions in India, Norway and Bangladesh

(Gupte et al., 2000; Meima et al., 2002; Fischer et al., 2011). Worldwide, nearly 80% of all new cases of leprosy are found in India, Brazil and Indonesia (WHO, 2013). There is still a challenge to apply these models to these specific countries and endemic regions within those countries for the purposes of answering key policy questions. Parameterization of these models with data from those areas is an important task.

6. CONCLUSION

Although three different mathematical models have been developed for leprosy, mathematical models in leprosy have not been applied extensively. This is in part due to the limited size of the leprosy problem in terms of numbers and health burden compared to many other infectious diseases such as HIV/AIDS, tuberculosis and malaria. Even within the group of neglected tropical diseases, the contribution of leprosy is modest. Few scientists have taken lasting interest in leprosy and funding for research is limited. Also, leprosy control has for many years been based on case finding and the provision of MDT to patients, which have effectively reduced the prevalence of leprosy. There has not been much emphasis by policy-making bodies such as the WHO on developing new and innovative disease elimination strategies, because of undue optimism that leprosy is no longer a public health problem and will disappear quietly in the near future. The reality is, however, different and the global number of new cases has plateaued at around 220,000 per year. Many challenges remain in understanding and dealing with the disease. The support of mathematical models for understanding leprosy epidemiology and supporting policy decision making remains vital.

REFERENCES

Adams, L.B., Pena, M.T., Sharma, R., Hagge, D.A., Schurr, E., Truman, R.W., 2012. Insights from animal models on the immunogenetics of leprosy: a review. Mem. Inst. Oswaldo Cruz 107 (Suppl. 1), 197–208.

Alter, A., Grant, A., Abel, L., Alcais, A., Schurr, E., 2011. Leprosy as a genetic disease. Mamm. Genome 22, 19–31.

Anonymous, 1992. International meeting on epidemiology of leprosy in relation to control held in Jakarta, Indonesia, 17-21 June 1991. Lepr. Rev. 63, 1s–126s.

Araujo, S., Lobato, J., Reis Ede, M., Souza, D.O., Goncalves, M.A., Costa, A.V., Goulart, L.R., Goulart, I.M., 2012. Unveiling healthy carriers and subclinical infections among household contacts of leprosy patients who play potential roles in the disease chain of transmission. Mem. Inst. Oswaldo Cruz 107 (Suppl. 1), 55–59.

Bakker, M.I., Hatta, M., Kwenang, A., Faber, W.R., Van Beers, S.M., Klatser, P.R., Oskam, L., 2004. Population survey to determine risk factors for *Mycobacterium leprae* transmission and infection. Int. J. Epidemiol. 33, 1329–1336.

Bakker, M.I., May, L., Hatta, M., Kwenang, A., Klatser, P.R., Oskam, L., Houwing-Duistermaat, J.J., 2005. Genetic, household and spatial clustering of leprosy on an island in Indonesia: a population-based study. BMC Med. Genet. 6, 40.

Britton, W.J., Lockwood, D.N., 2004. Leprosy. Lancet 363, 1209–1219.

van Beers, S.M., Hatta, M., Klatser, P.R., 1999. Patient contact is the major determinant in incident leprosy: implications for future control. Int. J. Lepr. Other Mycobact. Dis. 67, 119–128.

Desikan, K.V., 1977. Viability of *Mycobacterium leprae* outside the human body. Lepr. Rev. 48, 231–235.

Duthie, M.S., Saunderson, P., Reed, S.G., 2012. The potential for vaccination in leprosy elimination: new tools for targeted interventions. Mem. Inst. Oswaldo Cruz 107 (Suppl. 1), 190–196.

Feenstra, S.G., Nahar, Q., Pahan, D., Oskam, L., Richardus, J.H., 2012. Recent food shortage is associated with leprosy disease in Bangladesh: a case-control study. PLoS Negl. Trop. Dis. 5, e1029.

Fine, P.E., 1982. Leprosy: the epidemiology of a slow bacterium. Epidemiol. Rev. 4, 161–188.

Fine, P.E., Smith, P.G., 1996. Vaccination against leprosy—the view from 1996. Lepr. Rev. 67, 249–252.

Fine, P.E., Sterne, J.A., Ponnighaus, J.M., Bliss, L., Saui, J., Chihana, A., Munthali, M., Warndorff, D.K., 1997. Household and dwelling contact as risk factors for leprosy in northern Malawi. Am. J. Epidemiol. 146, 91–102.

Fischer, E., Pahan, D., Chowdhury, S., Richardus, J., 2008a. The spatial distribution of leprosy cases during 15 years of a leprosy control program in Bangladesh: an observational study. BMC Infect. Dis. 8, 126.

Fischer, E., Vlas, D.S., Meima, A., Habbema, D., Richardus, J., 2010. Different mechanisms for heterogeneity in leprosy susceptibility can explain disease clustering within households. PLoS One 5.

Fischer, E.A., De Vlas, S.J., Habbema, J.D., Richardus, J.H., 2011. The long-term effect of current and new interventions on the new case detection of leprosy: a modeling study. PLoS Negl. Trop. Dis. 5, e1330.

Fischer, E.a. J., Pahan, D., Chowdhury, S.K., Oskam, L., Richardus, J.H., 2008b. The spatial distribution of leprosy in four villages in Bangladesh: an observational study. BMC Infect. Dis. 8.

Fitness, J., Tosh, K., Hill, A.V., 2002. Genetics of susceptibility to leprosy. Genes. Immun. 3, 441–453.

Gupte, M.D., Kishore Kumar, B., Elangovan, A., Arokiasamy, J., 2000. Modelling epidemiology of leprosy. Indian J. Lepr. 72, 305–316.

Hatta, M., Van Beers, S.M., Madjid, B., Djumadi, A., De Wit, M.Y., Klatser, P.R., 1995. Distribution and persistence of *Mycobacterium leprae* nasal carriage among a population in which leprosy is endemic in Indonesia. Trans. R. Soc. Trop. Med. Hyg. 89, 381–385.

Lechat, M.F., 1981. The torments and blessings of the leprosy epidemiometric model. Lepr. Rev. 52, 187–196.

Lechat, M.F., 1992. Epidemiometric modeling in leprosy based on Indian data. Lepr. Rev. 63, S31–S39.

Lechat, M.F., Declercq, E.E., Mission, C.B., Vellut, C.M., 1990. Selection of MDT strategies through epidemiometric modeling. Int. J. Lepr. 58, 296–301.

Lechat, M.F., Misson, C.B., Bouckaert, A., Vellut, C., 1977. An epidemiometric model of leprosy: a computer simulation of various control methods with increasing coverage. Int. J. Lepr. 45, 1–8.

Lechat, M.F., Misson, C.B., Lambert, A., 1985. Simulation of vaccination and resistance in leprosy using an epidemiometric model. Int. J. Lepr. 53, 461–467.
Lechat, M.F., Misson, C.B., Vanderveken, M., 1987. A computer simulation of the effect of multidrug therapy on the incidence of leprosy. Ann. Soc. BELG Med. Trop. 67, 59–65.
Lechat, M.F., Misson, J.Y., Vellut, C.M., 1974. An epidemetric model of leprosy. Bull. WHO 51, 361–373.
Lietman, T., Porco, T., Blower, S., 1997. Leprosy and tuberculosis: the epidemiological consequences of cross-immunity. Am. J. Public Health 87, 1923–1927.
Meima, A., Gupte, M.D., Van Oortmarssen, G.J., Habbema, J.D.F., 1999. SIMLEP: a simulation model for leprosy transmission and control. Int. J. Lepr. Other Mycobact. Dis. 67, 215–236.
Meima, A., Irgens, L.M., Van Oortmarssen, G.J., Richardus, J.H., Habbema, J.D., 2002. Disappearance of leprosy from Norway: an exploration of critical factors using an epidemiological modelling approach. Int. J. Epidemiol. 31, 991–1000.
Meima, A., Smith, W.C., Van Oortmarssen, G.J., Richardus, J.H., Habbema, J.D., 2004. The future incidence of leprosy: a scenario analysis. Bull. World Health Organ. 82, 373–380.
Meima, A., Van Veen, N.H., Richardus, J.H., 2008. Future prevalence of WHO grade 2 impairment in relation to incidence trends in leprosy: an exploration. Trop. Med. Int. Health 13, 241–246.
Mira, M.T., Alcais, A., Di Pietrantonio, T., Thuc, N.V., Phuong, M.C., Abel, L., Schurr, E., 2003. Segregation of HLA/TNF region is linked to leprosy clinical spectrum in families displaying mixed leprosy subtypes. Genes. Immun. 4, 67–73.
Mira, M.T., Alcais, A., Nguyen, V.T., Moraes, M.O., Di Flumeri, C., Vu, H.T., Mai, C.P., Nguyen, T.H., Nguyen, N.B., Pham, X.K., Sarno, E.N., Alter, A., Montpetit, A., Moraes, M.E., Moraes, J.R., Dore, C., Gallant, C.J., Lepage, P., Verner, A., Van De Vosse, E., Hudson, T.J., Abel, L., Schurr, E., 2004. Susceptibility to leprosy is associated with PARK2 and PACRG. Nature 427, 636–640.
Moet, F.J., Meima, A., Oskam, L., Richardus, J.H., 2004. Risk factors for the development of clinical leprosy among contacts, and their relevance for targeted interventions. Lepr. Rev. 75, 310–326.
Moet, F.J., Pahan, D., Oskam, L., Richardus, J.H., 2008. Effectiveness of single dose rifampicin in preventing leprosy in close contacts of patients with newly diagnosed leprosy: cluster randomised controlled trial. BMJ 336, 761–764.
Moet, F.J., Pahan, D., Schuring, R.P., Oskam, L., Richardus, J.H., 2006. Physical distance, genetic relationship, age, and leprosy classification are independent risk factors for leprosy in contacts of patients with leprosy. J. Infect. Dis. 193, 346–353.
Montenegro, A.C., Werneck, G.L., Kerr-Pontes, L.R., Barreto, M.L., Feldmeier, H., 2004. Spatial analysis of the distribution of leprosy in the State of Ceara, Northeast Brazil. Mem. Inst. Oswaldo Cruz 99, 683–686.
Noordeen, S.K., 1985. The epidemiology of leprosy. In: Hastings, R.C. (Ed.), Leprosy. Churchill Livingstone, Edinburgh.
Opromolla, P.A., Dalben, I., Cardim, M., 2006. Geostatistical analysis of leprosy cases in the State of Sao Paulo, 1991-2002. Rev. Saude Publica 40, 907–913.
Rao, P.S., Karat, A.B., Kaliaperumal, V.G., Karat, S., 1975. Transmission of leprosy within households. Int. J. Lepr. Other Mycobact. Dis. 43, 45–54.
Richardus, J.H., Habbema, J.D.F., 2007. The impact of leprosy control on the transmission of *M. leprae*: is elimination being attained? Lepr. Rev. 78, 330–337.
Richardus, J.H., Meima, A., Van Marrewijk, C.J., Croft, R.P., Smith, T.C., 2005. Close contacts with leprosy in newly diagnosed leprosy patients in a high and low endemic area: comparison between Bangladesh and Thailand. Int. J. Lepr. Other Mycobact. Dis. 73, 249–257.

Rodrigues, L.C., Lockwood, D.N., 2011. Leprosy now: epidemiology, progress, challenges, and research gaps. Lancet Infect. Dis. 11, 464–470.

Schuring, R.P., Richardus, J.H., Pahan, D., Oskam, L., 2009. Protective effect of the combination BCG vaccination and rifampicin prophylaxis in leprosy prevention. Vaccine 27, 7125–7128.

Smith, C., Richardus, J.H., 2008. Leprosy strategy is about control, not eradication. Lancet 371, 969–970.

Sterne, J.A., Ponnighaus, J.M., Fine, P.E., Malema, S.S., 1995. Geographic determinants of leprosy in Karonga district, northern Malawi. Int. J. Epidemiol. 24, 1211–1222.

Turankar, R.P., Lavania, M., Chaitanya, V.S., Sengupta, U., Darlong, J., Darlong, F., Siva Sai, K.S., Jadhav, R.S., 2014. Single nucleotide polymorphism-based molecular typing of M. leprae from multicase families of leprosy patients and their surroundings to understand the transmission of leprosy. Clin. Microbiol. Infect. 20, O142–O149.

Wallinga, J., Edmunds, W.J., Kretzschmar, M., 1999. Perspective: human contact patterns and the spread of airborne infectious diseases. Trends Microbiol. 7, 372–377.

WHO, 1991. Elimination of Leprosy: Resolution of the 44th World Health Assembly (Resolution No. WHA 44.9). WHO, Geneva.

WHO, 1994. Chemotherapy of leprosy. World Health Organ. Tech. Rep. Ser. 847, 1–24.

WHO, 1998. WHO expert committee on leprosy. World Health Organ. Tech. Rep. Ser. 874, 1–43.

WHO, 2009. Enhanced Global Strategy for Further Reducing the Disease Burden Due to Leprosy (Plan Period: 2011–2015) [Online]. World Health Organization, Regional Office for South-East Asia, New Delhi. Available: http://www.searo.who.int/entity/leprosy/documents/SEA_GLP_2009_3/en/index.html (accessed 17.05.13.).

WHO, 2012. Accelerating Work to Overcome the Global Impact of Neglected Tropical Diseases – a Roadmap for Implementation. World Health Organization, Geneva, Switzerland.

WHO, 2013. Global leprosy: update on the 2012 situation. Wkly. Epidemiol. Rec. 88, 365–380.

Zhang, F.R., Huang, W., Chen, S.M., Sun, L.D., Liu, H., Li, Y., Cui, Y., Yan, X.X., Yang, H.T., Yang, R.D., Chu, T.S., Zhang, C., Zhang, L., Han, J.W., Yu, G.Q., Quan, C., Yu, Y.X., Zhang, Z., Shi, B.Q., Zhang, L.H., Cheng, H., Wang, C.Y., Lin, Y., Zheng, H.F., Fu, X.A., Zuo, X.B., Wang, Q., Long, H., Sun, Y.P., Cheng, Y.L., Tian, H.Q., Zhou, F.S., Liu, H.X., Lu, W.S., He, S.M., Du, W.L., Shen, M., Jin, Q.Y., Wang, Y., Low, H.Q., Erwin, T., Yang, N.H., Li, J.Y., Zhao, X., Jiao, Y.L., Mao, L.G., Yin, G., Jiang, Z.X., Wang, X.D., Yu, J.P., Hu, Z.H., Gong, C.H., Liu, Y.Q., Liu, R.Y., Wang, D.M., Wei, D., Liu, J.X., Cao, W.K., Cao, H.Z., Li, Y.P., Yan, W.G., Wei, S.Y., Wang, K.J., Hibberd, M.L., Yang, S., Zhang, X.J., Liu, J.J., 2009. Genomewide association study of leprosy. N. Engl. J. Med. 361, 2609–2618.

CHAPTER THREE

Mathematical Models of Human African Trypanosomiasis Epidemiology

Kat S. Rock*, Chris M. Stone[§,¶], Ian M. Hastings[||], Matt J. Keeling*, Steve J. Torr[†,‡], Nakul Chitnis[§,¶,1]

*Mathematics Institute/WIDER, University of Warwick, Coventry, UK
[§]Department of Epidemiology and Public Health, Swiss Tropical and Public Health Institute, Basel, Switzerland
[¶]University of Basel, Basel, Switzerland
[||]Department of Parasitology, Liverpool School of Tropical Medicine, Liverpool, UK
[†]Department of Vector Biology, Liverpool School of Tropical Medicine, Liverpool, UK
[‡]Warwick Medical School, University of Warwick, Coventry, UK
[1]Corresponding author: E-mail: nakul.chitnis@unibas.ch

Contents

Advances in Parasitology, Volume 87
ISSN 0065-308X
http://dx.doi.org/10.1016/bs.apar.2014.12.003

Abstract

Human African trypanosomiasis (HAT), commonly called sleeping sickness, is caused by *Trypanosoma* spp. and transmitted by tsetse flies (*Glossina* spp.). HAT is usually fatal if untreated and transmission occurs in foci across sub-Saharan Africa. Mathematical modelling of HAT began in the 1980s with extensions of the Ross–Macdonald malaria model and has since consisted, with a few exceptions, of similar deterministic compartmental models. These models have captured the main features of HAT epidemiology and provided insight on the effectiveness of the two main control interventions (treatment of humans and tsetse fly control) in eliminating transmission. However, most existing models have overestimated prevalence of infection and ignored transient dynamics. There is a need for properly validated models, evolving with improved data collection, that can provide quantitative predictions to help guide control and elimination strategies for HAT.

1. INTRODUCTION

Tsetse flies (*Glossina* spp.) transmit pathogens of the species *Trypanosoma* which can cause African trypanosomiasis in humans and livestock, across approximately 10 million square kilometres of sub-Saharan Africa. Human African trypanosomiasis (HAT), commonly called sleeping sickness, comprises two diseases caused by either *Trypanosoma brucei gambiense* (Gambian HAT) or *Trypanosoma brucei rhodesiense* (Rhodesian HAT). Animal African trypanosomiasis (AAT), also known as nagana, is predominantly caused by *Trypanosoma congolense* and *Trypanosoma vivax*.

Generally, HAT is fatal if untreated; similarly AAT is fatal to some livestock species, particularly more productive breeds exotic to Africa, and has a significant impact on the productivity of indigenous animals. Although this group of tsetse-borne pathogens is found only in sub-Saharan Africa, its impact on health and productivity is comparable to that of diseases with a global distribution. Since 1995, the average number of cases of HAT recorded per year has ranged between 37,977 (1998) and 6,750 (2011) and estimates of the global burden of HAT has ranged between 1.82 million disability adjusted life years (DALYs) in 2000 and 560,000 DALYs in 2010 (Hackett et al., 2014). Like many other neglected tropical diseases, the true burden of HAT is probably higher. Animal trypanosomiasis kills more than 1 million cattle each year (Shaw et al., 2013) and the annual economic loss due to the combined impact of human and animal trypanosomiases has been estimated at $2–4.5 billion.

There have been a number of recent reviews of HAT in general (Brun et al., 2010; Maudlin et al., 2004; Welburn and Maudlin, 2012), and on particular matters such as chemotherapy (Brun et al., 2011), drug resistance (Barrett et al., 2011), interactions between trypanosomes and tsetse (Dyer et al., 2013; Peacock et al., 2012; Walshe et al., 2009), behaviour of tsetse (Gibson and Torr, 1999; Torr and Solano, 2010), epidemiology (Fèvre et al., 2006; Leak, 1999) and vector control (Torr et al., 2005). We therefore only summarize some of the important aspects of trypanosomiasis that are particularly relevant to the development and use of epidemiological models, focussing on the basic disease progression in humans and animals, the demographics of tsetse flies, trypanosome biology and efforts at controlling the infection.

1.1 Human African trypanosomiasis

The two forms of HAT have distinctive epidemiological, geographical, clinical and therapeutic characteristics as summarized below. Currently, the annual number of recorded cases is less than 10,000 per year, but about 70 million people are estimated to be at risk over an area of 1.5 million km^2 (Simarro et al., 2012).

1.1.1 Gambian HAT

More than 95% of all cases of HAT are caused by *T. b. gambiense* transmitted by the *Palpalis*-group species of tsetse, particularly subspecies of *Glossina fuscipes* and *Glossina palpalis*. This group of tsetse, commonly called 'riverine tsetse', infest relatively humid habitats fringing the rivers, lake shores and wetlands of West and Central Africa.

The disease progresses over several years from the initial symptoms of fever, headaches and lymphadenopathy (known as stage I disease) through neuropsychiatric disorders (known as stage II disease) and sleep disturbance (hence the common name of sleeping sickness) and in most cases death.

In contrast to most other important vectors (such as mosquitoes, blackflies, sandflies), both sexes of tsetse rely exclusively on blood for all their nutritional needs. Tsetse become infected after feeding on an infected host. The trypanosomes undergo a complex process of maturation in the fly and after a period of about 20–40 days the infective forms appear in the salivary glands of the fly which, thereafter remains potentially infective to any humans it bites. Flies vary in their susceptibility to initial infection with *Trypanosoma brucei*. Most flies are inherently refractory to infection (Welburn and Maudlin, 1999) and even those that are not refractory are most susceptible during their first blood meal. Unsurprisingly, the proportion of infectious tsetse is generally less than 1%. Gambian HAT is generally regarded as an anthroponosis with the parasite being transmitted between humans and tsetse only.

1.1.2 Rhodesian HAT

The remaining (5%) cases of HAT are caused by *T. b. rhodesiense* transmitted by the *Morsitans*-group of tsetse, especially subspecies of *Glossina morsitans* and *Glossina pallidipes* found in the savannah woodlands of east and southern Africa. The transmission cycle for Rhodesian HAT is broadly similar to that of Gambian HAT except that Rhodesian HAT is a zoonosis with wild and domestic ungulates (e.g. warthog, bushbuck, buffalo, cattle) acting as important reservoir hosts. Indeed, many of the foci for Rhodesian HAT are wilderness areas where wild hosts and tsetse are abundant. Rhodesian HAT is a more acute disease than the Gambian form with the late stage neuropsychiatric disorders occurring within weeks and death in months.

1.2 Animal African trypanosomiasis

The focus of this review is HAT but the disease is often co-endemic with animal trypanosomiasis. Accordingly, national and global strategies to control these diseases are closely linked, and consideration of the animal disease provides insights into the epidemiology of HAT. The transmission cycle of AAT is broadly similar to that of HAT except that the hosts are livestock and wild ungulates. As with Rhodesian HAT, wild hosts can serve as a reservoir of disease for livestock especially in areas where livestock are close to wild hosts, for example, farming areas in Kenya and Tanzania adjacent to the Serengeti, Ngorongoro and Masai Mara National Parks. The causative agents of animal trypanosomiasis have less complex and generally faster rates of development within the fly than the subspecies of *T. brucei*; tsetse biting hosts infected with *T. vivax* and *T. congolense* become infective within around 10 and 20 days, respectively, and the probability that a mature infection results following feeding on an infected host is higher (Peacock et al., 2012). Tsetse are also more susceptible to infection from *T. vivax* throughout their life since the 'teneral effect', in which newly emerged flies are more susceptible to infection than older ones, is unique to *T. brucei*. The upshot of these differences is that in areas where HAT and AAT are co-endemic, the prevalence of *T. vivax* and *T. congolense* in tsetse populations is higher (typically 1–5%) than that for *T. brucei* (about 0.1%).

1.3 Tsetse fly bionomics

1.3.1 Births and deaths

Tsetse have an unusual form of reproduction, termed adenotrophic viviparity, in which the larva develops within the female. From the age of about 6 days, adult females produce a single egg which matures in the uterus for about 7–12 days, the duration being dependent on temperature (Hargrove, 1995). A single mature third-stage larva is deposited by the female on loose soil and it burrows into the ground and pupates. The time to deposition for subsequent larvae is shorter than for the first offspring and the time between offspring decreases with increasing temperature. A mature adult fly emerges 20–40 days after deposition and this pupal duration also decreases with increasing temperature (Hargrove, 2003; Phelps and Burrows, 1969b,c). This resultant low rate of reproduction is only sustainable because tsetse are relatively long lived. Ignoring deaths of larvae and pupae, adult female tsetse flies must live at least 25 days to ensure that they each produce two adult progeny. Such a low reproductive rate means that tsetse populations can only persist if the mean adult daily female mortality is less than 3.5% (Hargrove, 1988) which provides a crude target figure for control measures. Additionally, there can be substantial mortality in the pupal stage, due to predation (Rogers and Randolph, 1990), parasitism (Hargrove, 1999;

Heaversedge, 1969) and extremes of temperature, the latter being important in setting the absolute limits to the distribution of tsetse flies and consequently HAT transmission.

While many mathematical models of vector-borne diseases assume a constant death rate, laboratory (Curtis and Jordan, 1970; Jordan and Curtis, 1968, 1972) and field (Hargrove et al., 2011a) evidence indicates significant changes in tsetse mortality with age. In one field study, female mortality was about 10% per day in newly emerged flies, fell to about 2% by age 10 days, increasing slowly thereafter. The pattern was similar for males but mortality increased much more rapidly with age. The changes relate to low-fat levels and poorly developed flight musculature in newly emerged flies, resulting in the double difficulty of needing to find and feed on a host rapidly but with a limited flight capacity (Hargrove, 1975). Accordingly, many young tsetse flies either die of starvation, or by attempting to feed off high-risk hosts such as humans (Hargrove, 1975). Those that do feed successfully build up energy reserves and flight muscles, and subsequent mortality declines. Increased mortality in older flies is associated with increased wing wear, resulting in diminished flight capacity. Adult mortality also increases with temperature and may be up to six times as high in the hot part of the year as in the cool part (Hargrove et al., 2004; Hargrove and Williams, 1998).

1.3.2 Abundance and distribution

Tsetse populations are able to persist at remarkably low densities. Obtaining reliable estimates of absolute densities of tsetse for sparse populations is difficult, but it seems likely that they can survive at densities of 1 tsetse/km^2 or less (Glasgow, 1963; Vale and Torr, 2005), while the maximum density seems to be in the order of 10,000 tsetse/km^2 (Hargrove and Williams, 1998; Phelps and Vale, 1978; Vale et al., 1988). Traps used to sample tsetse are relatively inefficient; a single trap probably catches between 0.1% and 2% of the population per day (Barclay and Hargrove, 2005). The combination of tsetse persisting at low densities and inefficient sampling devices means that trypanosomiasis may be detected in the apparent absence of tsetse. This phenomenon is sometimes cited as evidence for mechanical transmission of trypanosomes by other biting flies such as horseflies and stable flies. However, it is more likely that in sub-Saharan Africa, low densities of tsetse are transmitting trypanosomes (Leak, 1999).

Tsetse are highly mobile, moving up to 1 km per day (Hargrove, 2000). Savannah tsetse in relatively homogeneous woodland (Hargrove, 1981) or riverine tsetse in extensive wetlands (Rogers, 1977) can be modelled adequately by assuming diffusive movement of tsetse through the habitat. For the more typical habitats of riverine tsetse, however, flies are confined to the vegetation fringing rivers and lakes and hence displacement is largely along river margins and lake shores. The important consequence of the high mobility of tsetse is that populations of tsetse are seldom isolated and hence migration of tsetse into and out of an HAT focus is the norm.

1.3.3 Feeding interval

Males and females are obligate blood feeders obtaining meals at 2–5 day intervals, as gauged from mark-recapture studies (Glasgow, 1961; Jackson, 1933, 1954; Rogers, 1977; Rogers and Randolph, 1986) or through analysis of wild-caught tsetse (Hargrove and Packer, 1993; Langley and Wall, 1990; Randolph and Rogers, 1978; Randolph, Rogers, Dransfield and Brightwell, 1991). The blood meal provides all of the fly's requirements for energy, water and growth, including production of larvae. This contrasts with most other vectors (e.g. mosquitoes, blackflies, sandflies) where only the female feeds on blood and both sexes obtain some of their nutritional needs from feeding on plant sugars. Hence for tsetse there is an imperative to find and feed on a host regularly to avoid starvation.

The defensive behaviour of hosts and predatory insects (e.g. Asilidae (robber or assassin flies) and Bembicidae (a family of wasps)) in the host vicinity also pose a risk to feeding tsetse. Models of the costs and benefits of feeding have been used to analyse how tsetse balance the risk of starvation against feeding-related mortality. One analysis suggested that tsetse have an extended nonfeeding phase of 3–4 days followed by a high-feeding rate thereafter, it being assumed that tsetse can locate and feed on hosts efficiently (Randolph et al., 1992). In contrast, a second analysis suggested that tsetse do not display a marked change in feeding behaviour but rather become more responsive as their nutritional reserves decline (Hargrove and Williams, 1995). Consequently, the probability of feeding on a host is modulated by the fly's reserves: newly emerged tsetse and those that have not fed for several days are more likely to feed than mature flies or those that have fed recently (Torr and Hargrove, 1998; Vale, 1974). The balance between risk of starvation and feeding-related mortality probably varies with size of the fly and habitat. Smaller flies are less mobile and hence less efficient at locating hosts. Host location is also hampered by dense vegetation such as occurs in riverine habitats. As a consequence, a risk-averse feeding strategy is less marked in smaller tsetse (e.g. males, smaller species of tsetse) and riverine species (Vale et al., 2014).

1.3.4 Host selection

Tsetse use a combination of olfactory and visual cues to locate their hosts (Gibson and Torr, 1999; Torr and Solano, 2010). Clausen et al. (Clausen et al., 1998) provide summary results from analyses of 29,245 tsetse collected from 63 separate studies and covering 11 species of tsetse. In general, savannah tsetse feed on Suidae and Bovidae, particularly warthog and buffalo in wilderness areas and cattle where they are present. Experimental analysis of the attraction and feeding of savannah tsetse on various potential hosts demonstrated that a host's mass and its inherent rate of defensive behaviour determine its importance in the diet of tsetse (Hargrove et al., 1995; Torr, 2009; Vale, 1977). In part, the avoidance of feeding on humans by *Morsitans*-group tsetse is modulated by hunger. Hungry flies are more likely to feed on humans. Humans are rarely bitten by savannah tsetse; odours and visual stimuli produced by humans are repellent and those tsetse that do bite humans tend to be young and/or in an advanced stage of starvation (Hargrove, 1976; Torr et al., 2012; Vale, 1974).

Riverine tsetse have a broader range of hosts which can include primates, Suidae, Bovidae and reptiles, particularly the Nile monitor lizard (Clausen et al., 1998). Human odours and visual stimuli do not appear to be repellent to riverine tsetse (Omolo et al., 2009; Rayaisse et al., 2010). As a consequence, humans can form an important part of the diet. For instance, pooled analysis of blood meals from three species of riverine tsetse showed that 8% (454/5,544) were from primates compared to only 1% (88/8,660) for pooled data from four savannah species of tsetse (Clausen et al., 1998). The proportion of reptiles in the diet of some species of riverine tsetse is important since species of trypanosome pathogenic to humans and livestock cannot persist in reptiles (Njagu et al., 1999). Analysis of blood meals from *G. f. fuscipes* has shown that the proportion of meals from reptiles can be greater than 90% in tsetse caught from the shore of Lake Victoria where monitor lizards are abundant and about 40% elsewhere (Clausen et al., 1998).

Studies of tsetse (*G. pallidipes*) feeding on groups of cattle suggest that most (over 90%) meals are taken from a single host (in one feed) (Torr et al., 2007b) but analyses of the sources of blood meals suggest that mixed meals (from multiple feeds in one gonotrophic cycle) are not rare. For instance, a study in Mali found that 30% ($n = 76$) of *Glossina tachinoides* and 17% of *Glossina palpalis gambiensis* ($n = 203$) contained blood from both humans and cattle (Hoppenheit et al., 2013). Such mixed meals may arise from feeding on cattle and humans on successive days. The presence of blood from different hosts in a fly suggests that individual tsetse do not necessarily feed repeatedly on the same host species despite experimental evidence that *G. tachinoides* might 'learn' to feed preferentially on particular species (Bouyer et al., 2007).

Experimental studies of the behavioural responses of tsetse to hosts, and the marked preponderance or absence of potential host species in the diet of tsetse populations suggest that tsetse are selecting particular host species. However, we are aware of only one study that has simultaneously quantified the source of blood meals and the density of potential hosts simultaneously (the 'forage ratio' in mosquito literature): in the Serengeti National Park, Auty et al. used a combination of polymerase chain reaction-based methods to identify sources of blood meals in tsetse and game count transects to quantify densities of hosts (Auty et al., Unpublished data). Their results confirmed that tsetse preferentially select warthog and buffalo which are also important hosts for *T. b. rhodesiense*.

There is some evidence that selection of an individual host is modulated by its infection status. Studies in Kenya showed that feeding rates were higher on cattle infected with *T. vivax* and *T. congolense* (Baylis, Mbwabi and Stevenson, 1994; Baylis and Mbwabi, 1995; Baylis and Nambiro, 1993; Moloo et al., 2000). The higher rates may be due to infected animals attracting more tsetse and/or higher proportions of attracted tsetse feeding successfully; cattle infected with trypanosomes may display lower rates of defensive behaviour, for instance. Alternatively, these animals may be more likely to be infected because they are more attractive to flies.

1.4 Trypanosome biology

1.4.1 Dynamics of trypanosomes in humans and animal hosts

The latent period between initial inoculation of *T. brucei* and a mammalian host becoming infectious is estimated to be about 7—14 days (Ashcroft et al., 1959). A systematic review of observed durations of subsequent infection for *T. b. gambiense* suggests that the mean ranges between several months and years, and is strongly right-skewed with some cases persisting for many years without morbidity or death (Checchi et al., 2008a; Jamonneau et al., 2004). Analysis of 298 cases of Gambian HAT from northern Uganda suggests mean durations of 17 months for stage I and 16 months for stage II (Checchi et al., 2008b). Severe illness and/or hospitalization during stage II may effectively remove the human host from the population of hosts by preventing contact with tsetse flies, in which case the mean duration of infectiousness would be about 17 months.

Recent studies of HAT patients in Côte d'Ivoire suggest that trypanotolerance, known to occur in wild hosts and some breeds of cattle, may also occur in humans (Jamonneau et al., 2012). Rather than the classical progression through stage I, stage II and death, some patients who showed clear parasitological and/or serological indications of infection with *T. b. gambiense* never developed neurological symptoms and ultimately appeared to self-cure, despite never receiving treatment. These infections were long-lasting (5–15 years) and hence this phenomenon could provide an important source of *T. b. gambiense*-infected individuals for onwards transmission.

Duration of infection with *T. b. rhodesiense* in humans is much shorter, being of the order of about 6 months, with the distinction between first and second stages of disease less clear (Apted, 1970).

Trypanosoma brucei infections in wild hosts and some livestock species and breeds may persist for longer periods (Welburn et al., 2008) without severe adverse effects on animal mortality or morbidity (Mulla and Rickman, 1988). Wild animal hosts which maintain chronic infections, however, have a lower parasitaemia (Ashcroft et al., 1959; Moloo et al., 1999) and hence may be less infectious to tsetse (Moloo et al., 1992a).

Infections in cattle tend to result in an acute phase that occurs within weeks of infection that coincides with the first peak of parasitaemia followed by a chronic phase with a variably cryptic parasitaemia that can last for months to years (Murray, 1989). Duration, magnitude and frequency of parasitaemia depend on breed of host, species and strain of trypanosome and the size of the inoculum.

Relatively few studies have experimentally inoculated and followed trypanosome infections in wildlife species for more than 60 days. However, the data available suggest that compared to cattle counterparts, parasitaemia is lower, its onset is later and the observed pathology is less significant (Dwinger et al., 1986; Grootenhuis et al., 1990; Rurangirwa et al., 1986).

Following infection with *T. b. rhodesiense* via tsetse bite, parasites were detectable in warthog blood very sporadically for up to 5–25 days postinfection (Ashcroft et al., 1959). Infections persisting for several months have been measured in eland, impala, Thomson's gazelle and reedbuck (Ashcroft et al., 1959; Corson, 1939). However, results from these studies must be treated with caution since the animals were often inoculated rather than infected by the bite of a tsetse fly, and animals kept in captivity may respond differently to those in the wild.

The trypanosomes pathogenic to humans and livestock evade the host's immune system by a process of antigenic variation in which the variant specific glycoprotein coating the cell changes during the course of an infection (Barry and McCulloch, January 2001). Consequently, no vaccine against HAT has been developed and the prospects of one being developed are poor. A second consequence of antigenic variation is that the parasitaemia varies markedly and unpredictably with time (Morrison et al., 2005) which may affect the likelihood of a susceptible tsetse being infected as it feeds.

1.4.2 Transmission of trypanosomes between hosts and vectors

The probability of a fly being infected depends not only on the levels of parasitaemia in the host but also the age, nutritional condition, sex and species of the fly itself.

The 'teneral effect'. Newly emerged unfed tsetse ('teneral' flies) that are typically less than 3 days old are much more susceptible to infection with subspecies of *T. brucei* at their first blood meal than older flies (Haines, 2013). Indeed, many models of HAT assume that infection can only occur in the first meal and thereafter flies are refractory ((Rogers, 1988) and references therein). The 'teneral effect' appears to be modulated by the nutritional status of a fly as evidenced by laboratory studies in which older refractory flies became susceptible to infection if they were starved (Akoda et al., 2009). The epidemiological significance of this effect of starvation is uncertain, however. Flies that are in advanced stages of starvation are also at a greater risk of mortality, and older flies that become infected will need to survive for a further 20 days or more before they are themselves infectious.

Effect of species, sex and strain. Susceptibility of tsetse to infection is low according to laboratory studies of *G. pallidipes* and *Glossina morsitans centralis* feeding on cattle infected with *Trypanosoma brucei brucei.* One experiment, using a Tanzanian strain of *T. b. brucei*, showed that 3.7% of male and 1.8% of female *G. pallidipes* developed a mature (salivary gland) infection compared to 26.6% and 10.2% of *G. m. centralis.* In a second experiment using a Nigerian strain of *T. b. brucei*, none of the *G. pallidipes* developed a mature infection whereas 13.3% of male *G. m. centralis* and 5.6% of females did (Moloo et al., 1992b). Studies of different strains of *T. b. rhodesiense* have shown that mean duration between initial uptake of trypanosomes and a mature infection varied between 18 and 23 days, and even within strains variation was high with a standard error between 2 and 8 days (Dale et al., 1995). The probability of a mature infection developing in a fly appears to be negatively correlated with the rate of maturation, i.e., the longer midgut infections take to mature, the lower the probability of maturing (Dale et al., 1995) and the lower the subsequent pathogenecity (Welburn et al., 1995). Similar interstrain differences and generally low rates of maturation (0–10%) (Ravel et al., 2006) are also seen with *T. b. gambiense*. As with experimental analyses of trypanosomes in mammalian hosts, laboratory-based studies of trypanosomes with tsetse should be treated with caution because dynamics in wild strains may be different.

1.5 Control of African trypanosomiasis

The management of animal and human trypanosomiasis through various combinations of chemotherapies and vector control has been reviewed multiple times (Brun et al., 2010; Maudlin, 2006; Maudlin et al., 2004; Welburn and Maudlin, 2012) and hence we only summarize the important points.

1.5.1 Gambian HAT

Gambian HAT is generally assumed to infect humans only so its control has largely relied on the detection and treatment of cases. Starting in the 1920s and continuing through to the late 1950s, large-scale campaigns of active case detection and treatment were conducted by the colonial authorities in the French and Belgian territories of West Africa (e.g. present day Burkina Faso, Cameroon and the Democratic Republic of Congo (DRC)). Following the strategy developed by Jamot in the 1920s in Cameron, mobile teams screened millions of people (Steverding, 2008) for HAT.

Programmes of mass treatment with pentamidine were introduced as an early form of prophylactic chemotherapy, from the mid 1940s onwards with millions of people along the Congo River being regularly treated during the 1950s. These tactics reduced the annual number of recorded cases from over 50,000 per year in the 1930s to less than 10,000 per year by the early 1960s (Simarro et al., 2008).

Efforts against Gambian HAT decreased in the immediate post-colonial period of the 1960s. The relatively low number of cases made HAT less important for newly independent countries with more pressing demands on their health services. Over the next 30 years, decline in the capacity of health services, socioeconomic disturbances and war in many countries most at risk of HAT (e.g. Angola, DRC, Sudan, South Sudan, Uganda) contributed to a further decline in global efforts against Gambian HAT. Consequently, the global number of cases reported annually steadily increased to a 1998 peak of 37,385. A revival of the global effort against HAT, led by the World Health Organization and strongly supported by donations of chemotherapeutic drugs and diagnostics, has helped disease endemic nations to reduce the annual number of cases to the same levels as those reported in the early 1960s. The achievements of the past 20 years have been realized almost exclusively through active case detection and treatment. Screening of the population is based on the use of the Card Agglutination Test for Trypanosomiasis followed by staging of positive cases through examination of the cerebrospinal fluid which involves a lumbar puncture. Currently, the drugs used to treat Gambian HAT are pentamidine for stage I and nifurtimox–eflornithine combination therapy for stage II. New drugs and diagnostics are in development. These include rapid diagnostic tests based on immunological and molecular methods (Deborggraeve and Büscher, 2012; Mitashi et al., 2012) and drugs such as fexinidazole and benzoxaboroles (WHO, 2013). While vector control has not been an important part of efforts against Gambian sleeping sickness, the recent development of cost-effective methods to control riverine tsetse suggests that tsetse control will form an important part of efforts to eliminate Gambian HAT in the future (Solano et al., 2013).

1.5.2 Rhodesian HAT

Rhodesian HAT is a zoonosis so it cannot be controlled solely through detection and treatment of human cases, which are usually considered a spill over from a far larger number of chronic and less pathogenic cases occurring in populations of reservoir hosts. Treatment of human cases is with either suramin for stage I disease or melarsoprol for stage II disease.

Most Rhodesian HAT foci are associated with wilderness areas where wild animals such as warthog, bushbuck and buffalo are the natural hosts of *T. b. rhodesiense* and various species of savannah tsetse are the vector. The difficulty of eliminating trypanosomes from wild hosts means that vector control is the only control option available. Currently, tsetse control methods used in wilderness areas include aerial application of nonpersistent insecticides (Kgori et al., 2006) or use of artificial baits to lure and kill tsetse (Vale et al., 1988; Dransfield et al., 2009; Willemse, 1991). Savannah tsetse are highly responsive to host odours and insecticide-treated targets baited with a blend of artificial host odours and deployed at densities of 4 targets/km^2 can eliminate populations of tsetse (Vale and Torr, 2004). In southeast Uganda, *T. b. rhodesiense* is transmitted by a species of riverine tsetse (*G. f. fuscipes*) and cattle rather than wild hosts are the reservoir host. In this particular setting, therefore, mass treatment of cattle with trypanocide and insecticide is a highly cost-effective option for control (Hargrove et al., 2012; Torr et al., 2007a).

1.5.3 Animal African trypanosomiasis

In contrast to HAT, cheap, effective and safe prophylactic drugs exist for animal African trypanosomiasis as well as curative ones. In addition to being cheap (about $1 per dose), the drugs are easily administered through a single intramuscular injection and are widely available. As a consequence, chemotherapy is the mainstay of efforts against animal trypanosomiasis with an estimated 35 million doses of trypanocide being administered each year. This approach has become particularly important following the decline in government-supported provision of veterinary services. However, chemotherapy is not without its problems. In particular, AAT cannot be eliminated without close to full coverage of the host population (Hargrove et al., 2012), which is particularly problematic where livestock mix with wild hosts, and drug resistance is increasing (Burri et al., 2004). In addition, the productivity of livestock maintained in a tsetse-infested area, through regular treatment with trypanocide, is reduced compared to livestock kept in tsetse-free areas (Shaw et al., 2013). Tsetse control also plays an important role in the control of AAT. In areas where densities of cattle are relatively high (greater than 10 animals/km^2), the use of insecticide-treated cattle is particularly cost-effective since the formulations effective against tsetse can also control tick-borne diseases of livestock (Eisler et al., 2003). Livestock are, however, not evenly distributed and in areas where their densities are low then other methods (insecticide-treated targets, aerial spraying, ground spraying) are used.

The mobility of savannah tsetse (up to 1 km/day) means that vector control must generally be applied at a scale (over 500 km^2) that cannot be achieved by individual livestock keepers (Hargrove, 2000; Torr et al., 2005). In the colonial period, government-supported tsetse control departments implemented large-scale control operations over areas of up to 10,000 km^2 each year (Allsopp et al., 2004) but the capacity to conduct operations on this scale of operation has declined since the 1960s. Consequently, sustainable and large-scale vector control operations are currently rare and identifying mechanisms where livestock keepers can implement tsetse control on a large scale without government or donor support is proving elusive.

1.5.4 Drug and insecticide resistance

A continuing worry with any drug- or insecticide-based intervention is that resistance may arise and undermine the control measures. Complacency is dangerous, but there are good grounds for believing insecticide resistance

(IR) is likely to be less of a problem in controlling HAT than for many other diseases (malaria being the obvious co-endemic example). Tsetse are highly sensitive to insecticides compared to other insects, so the belief is that substantial increases in IR levels would have to occur before they became operationally significant. Tsetse flies can respond to selection pressures for resistance, but since they are relatively long lived, with long generation times and low fecundity, they cannot respond as rapidly as short generation time and highly fecund species such as anopheline mosquitoes. Insecticide use for *T. b. gambiense* control is also highly focussed: the main tsetse vectors are highly mobile riverine species and control is focussed on individual, relatively small, historically stable HAT foci. These control measures will generate relative low selection pressures within the large, mobile contiguous populations of these species. Vector control mainly uses deltamethrin (a pyrethroid) but the exophilic lifestyle of the riverine tsetse species mean they rarely, if at all, enter huts where they could encounter pyrethroids used on bed nets or as residual sprays on wall. Also, unlike mosquitoes, tsetse demography makes them unlikely to encounter insecticides used in agriculture. There have been sporadic attempts to investigate resistance in the field (Turner and Golder, 1986) and by modelling (Maudlin and Green, 1981) (notably predicting that it was unlikely to occur). The current belief in the tsetse community is therefore that IR is unlikely to be a significant operational threat in the foreseeable future.

Drug resistance in Gambian HAT is more plausible than IR in tsetse: the explicit aim of control efforts is to treat as many human infections as possible so that, given the belief that humans are the only significant non-tsetse host, this would imply very high selective pressure for resistance. Several drugs are used for stage I infections and are all administered under close clinical scrutiny. This means that factors such as poor patient compliance and underdosing, believed to contribute to drug resistance in other pathogens, are less likely in treatment of Gambian HAT. Close clinical scrutiny also means that patients who do not respond adequately would be identified but, to date, no large-scale drug failures have been noted. Drug resistance, or at least reduced sensitivity, has been suspected to melarsoprol but only after around 50 years of use. The zoonotic nature of *T. b. rhodesiense* means that human treatment will leave a vast untreated reservoir of trypanosomes in other mammalian hosts so selection for resistance driven by human drug use is likely to be low. It is, of course, vital to ensure cattle are not treated with the same drug used to treat humans or this could drive widespread resistance. Finally, we note that people under treatment for HAT are usually very sick and likely to be confined to huts, or even treatment centres. They are therefore inaccessible to tsetse flies so that even if resistance does sporadically arise, it is unlikely to be transmitted back into the wider trypanosome population.

1.6 What has modelling ever done for HAT?

In contrast to the large number and widespread use of mathematical models to guide efforts against diseases as diverse as onchocerciasis (Basáñez and Ricárdez-Esquinca, 2001; Habbema et al., 1996), malaria (The malERA Consultative Group on Modeling, 2011) and influenza (Nsoesie et al., 2014), models of HAT are few and their impact on practice and policy limited. Can models help in the war against trypanosomiasis? If so, why have they not been developed and used more widely?

We believe that there is a pressing need for models to answer the following questions to scientifically underpin the development of global and national strategies to eliminate HAT (Aksoy, 2011; World Health Organization, 2012): a) what percentage of a human population needs to be screened and treated to eliminate a focus of Gambian HAT transmission?; b) at what frequency should such a screen and treat program be repeated?; c) what reduction in the density and longevity of tsetse flies is required to eliminate transmission of HAT?; d) how long must interventions be applied to eliminate a focus?; e) why is HAT not more widespread, and will movement of reservoir hosts, tsetse or infected humans cause it to spread?; f) how does HAT persist at the low levels of prevalence that have been measured in humans and tsetse flies? These questions require mechanistic models that capture the known behaviour of humans, tsetse and trypanosomes as well as matching the current epidemiological situation.

These questions were of less applied importance when national and global strategies were focussed on eradicating the vector and disease from countries or so-called 'fly belts' (Jordan, 1986, 1985). This was achieved in a few cases (Botswana (Kgori et al., 2006), Southeastern Zimbabwe (Jordan, 1986), South Africa (Du Toit, 1954; Kappmeier et al., 1998), Northern Nigeria (Davies, 1971), Zanzibar (Vreysen et al., 2000)), but the general experience has been that attempts to completely and permanently eliminate tsetse and trypanosomiasis were unsuccessful. However, as we shall see, the few models of HAT that do exist suggest that elimination of HAT will not require killing the last fly or finding and treating the last case. Rather, reducing the mean infectious period of humans through case detection and treatment, and reducing the longevity and density of tsetse flies through vector control may eliminate transmission.

Following the longer history of malaria modelling, mathematical modelling of AAT began with Milligan and Baker (1988) and HAT began with Rogers (1988). Since then, multiple models of human and animal African trypanosomiasis have been built and analysed to investigate various aspects of trypanosome biology and epidemiology. Five of the most commonly considered aspects are: dynamics of infection in multiple mammalian host species, variations in tsetse fly susceptibility over its life span, stages of disease progression in mammalian hosts, spatial heterogeneity in transmission and tsetse population dynamics. Many models and analysis have then compared the effectiveness of the two widely used control interventions: treatment of infection humans and tsetse fly control. Recent models have also considered novel interventions such as targeting infection in tsetse flies.

We first outline the history of HAT modelling, including the malaria models that formed the foundations of the first (and most subsequent) HAT models. We then review the five features of HAT biology outlined above, describing the models and modelling approaches used and some of their key analysis and conclusions. We also derive and simulate extended models that combine some assumptions of published models to determine the implications of these assumptions. Finally, we summarize the progress made in modelling the effects of control interventions.

Most of the analysis of HAT models and the effects of interventions have been on the derivation of threshold conditions for persistence and general statements on the necessary conditions to eliminate HAT transmission. These models have substantially improved our understanding of HAT epidemiology, transmission and control but as described in the Outlook, §6, many features of HAT biology remain open to modelling. Additionally, given the current global call for HAT elimination (World Health Organization, 2012), there is a need for properly validated models that can provide quantitative predictions of the impact and cost-effectiveness of control interventions in reducing transmission and provide estimates on the time to potential elimination of control strategies.

2. NOTATION

Previous models used a variety of symbols for the same state variables and parameters. Here we adopt a standardized notation for all equations and figures to remain consistent and allow comparison across different models. The notation for state variables is described in Table 1 and for the parameters in Tables 2 and 3. We use upper case Latin letters to denote state variables or parameters

Table 1 Description of state variables of African trypanosomiases models in this review article. The subscript *i* represents hosts of different types and should be not confused with the state variables, i_{hi} and i_v, which represent proportions of infectious hosts or vectors

Variable	Description
S_{hi}	Number of susceptible hosts of type i (who can become infected when bitten by infectious tsetse flies).
E_{hi}	Number of exposed hosts of type i (who are infected with trypanosomes but are not yet infectious to tsetse flies).
I_{hi}	Number of infectious hosts of type i.
R_{hi}	Number of removed hosts of type i (who are not infectious to tsetse flies and who may not get infected when bitten by infectious tsetse flies).
N_{hi}	Total number of hosts of type i. In some models, N_{hi} may be a parameter.
S_v	Number of susceptible tsetse flies.
E_v	Number of exposed tsetse flies.
G_v	Number of tsetse flies who are not infected but are not susceptible to infection.
I_v	Number of infectious tsetse flies.
N_v	Total number of tsetse flies. In some models, N_v may be a parameter.
s_{hi}	Proportion of susceptible hosts of type i (who can become infected when bitten by infectious tsetse flies). $s_{hi} = S_{hi}/N_{hi}$.
e_{hi}	Proportion of exposed hosts of type i (who are infected with trypanosomes but are not yet infectious to tsetse flies. $e_{hi} = E_{hi}/N_{hi}$.
i_{hi}	Proportion of infectious hosts of type i. $i_{hi} = I_{hi}/N_{hi}$.
r_{hi}	Proportion of removed hosts of type i (who are not infectious to tsetse flies and who may not get infected when bitten by infectious tsetse flies). $r_{hi} = R_{hi}/N_{hi}$.
s_v	Proportion of susceptible tsetse flies. $s_v = S_v/N_v$.
e_v	Proportion of exposed tsetse flies. $e_v = E_v/N_v$.
g_v	Proportion of tsetse flies which are not infected and are not susceptible to infection. $g_v = G_v/N_v$.
i_v	Proportion of infectious tsetse flies. $i_v = I_v/N_v$.

Table 2 Description and dimension of parameters of African trypanosomiases models in this review article. In some cases where parameter values are the same for different types of hosts, the subscript *i* is dropped. For specialized models with other additional parameters, these parameters are described separately

Parameter	Description
N	Number of different types of hosts that tsetse flies bite. These hosts may or may not be susceptible to trypanosomiases. Dimensionless.
μ_{hi}	Per capita death rate of hosts of type i for $1 \leq i \leq n$. Dimension: Time^{-1}.
μ_v	Per capita death rate of tsetse flies. Dimension: Time^{-1}.
B_{hi}	Total (net) birth rate of hosts of type i for $1 \leq i \leq n$. Dimension: Animals $\times$ Time^{-1}
B_v	Total (net) birth rate of tsetse flies. Dimension: Animals $\times$ Time^{-1}.
α	Biting rate of tsetse flies on all hosts. Dimension: Time^{-1}.
f_i	Proportion of bites of tsetse flies on hosts of type i for $1 \leq i \leq n$ given equal availability of all hosts. f_i measures the biting preference of tsetse flies on hosts of type i. $\sum_{i=1}^{n} f_i = 1$. Dimensionless.
δ_{hi}	Per capita disease-induced death rate of hosts of type i for $1 \leq i \leq n$. Dimension: Time^{-1}.
σ_{hi}	Per capita rate of progression of a host of type i for $1 \leq i \leq n$ from the exposed (latent) stage to the infectious stage. Dimension: Time^{-1}.
σ_v	Per capita rate of progression of tsetse flies from the exposed (latent) stage to the infectious stage (assuming an exponential distribution for the latency period). Dimension: Time^{-1}.
T_{hi}	Fixed duration of the intrinsic incubation period of host i. Dimension: Time.
T_v	Fixed duration of the extrinsic incubation period. Dimension: Time.
p_{hiv}	Probability of transmission of infection from an infectious tsetse fly to a susceptible host of type i (for $1 \leq i \leq n$) per bite. Dimensionless.
p_{vhi}	Probability of transmission of infection from an infectious host of type i (for $1 \leq i \leq n$) to a susceptible tsetse fly per bite. Dimensionless.
φ_{hi}	Per capita recovery rate for hosts of type i (for $1 \leq i \leq n$) from the infectious state to the recovered state. $1/\varphi_{hi}$ is the average duration of the infectious period. Dimension: Time^{-1}.
γ_{hi}	Per capita rate of loss of immunity for hosts of type i (for $1 \leq i \leq n$). $1/\gamma_{hi}$ is the average duration of the immune period. Dimension: Time^{-1}.

Table 3 Description of derived parameters of human African trypanosomiases models in this review article

Parameter	Description
N_{hi}	Stable population size of hosts of type i in the absence of disease when the per capita host birth rate is equal to the host death rate. $N_{hi} = B_{hi}/\mu_{hi}$. However, in some models, N_{hi} may be a state variable. Dimension: Animals.
N_v	Stable tsetse fly population size when the per capita fly birth rate is equal to the fly mortality rate. $N_v = B_v/\mu_v$. However, in some models, N_v may be a state variable. Dimension: Animals.
λ_{hi}	Force of infection on (per capita infection rate of) hosts of type i. Dimension: Time^{-1}.
λ_v	Force of infection on (per capita infection rate of) tsetse flies. Dimension: Time^{-1}.
α_i	Biting rate of tsetse flies on hosts of type i for $1 \le i \le n$. $\alpha_i = \alpha f_i$. Dimension: Time^{-1}.
m_i	Ratio of tsetse flies to number of hosts of type i. $m_i = N_v/N_{hi}$. Dimensionless.

Table 4 State variables representing numbers of hosts or flies in previous models of HAT. The subscript *i* denotes the type of host. Unused state variables are left blank

This review	S_{hi}	E_{hi}	I_{hi}	R_{hi}	N_{hi}	S_v	E_v	G_v	I_v	N_v
Artzrouni and Gouteux, (1996a)[a]	H_s	H_i	H_a	H_r		V_s	V_i		V_a	
Funk et al. (2013)[b]			I_a		N_a		C_v^c	G_v^c	I_v	N_v

[a]Model only has human hosts, i denotes 'incubating' not host type.
[b]a denotes host type where $1 \le a \le n$, instead of using i.
[c]These are only used for an extension to the model.

Table 5 State variables representing proportions of hosts or flies in previous models of HAT. The subscript *i* denotes the type of host. Unused state variables are left blank

This review	s_{hi}	e_{hi}	i_{hi}	r_{hi}	s_v	e_v	g_v	i_v
Aron and May (1982)			x			z		y
Baker et al. (1990)			y					$\hat{y}$
Milligan and Baker (1988)	x	h	y	z[a]	$\hat{t}$	$\hat{h}$	$\hat{x}$	$\hat{y}$
Rogers (1988)		w_i	x_i	z_i		f		y

[a]Cattle were assumed to be immune due to chemoprophylaxis. Wild animals had no immune class.

representing numbers of animals; lower case Latin letters to denote state variables representing proportions of animals or dimensionless parameters; and lower case Greek letters to denote parameters representing rates.

The corresponding notation used in previous modelling studies, described in this review, is shown in Tables 4 and 5 for state variables and in Table 6 for parameters.

Table 6 Parameters in previous models of HAT. The subscript *i* denotes the type of host. Unused parameters are left blank. We use the birth rate parameters B_v and B_{hi} to model a constant birth and emergence rate of tsetse and hosts; however, the models considered in this table do not explicitly use these parameters and assume that the birth rate is equal to the product of the mortality rate and the constant population size

This review	μ_{hi}	μ_v	α	f_i	δ_{hi}	σ_{hi}	σ_v	T_{hi}	T_v	p_{hiv}	p_{vhi}	φ_{hi}	γ_{hi}	m_i
Aron and May (1982)		μ	a						τ	b		r		m
Artzrouni and Gouteux (1996a)[a]	m_h	m_v	τ_1^{b}		r_2	q_h	q_v			τ_3	τ_2	r_1		
Baker et al. (1990)		b	β				[f]			f	1^{g}	μ		V^{-1}
Funk et al. (2013)	μ_a	μ_v	τ	f_a			α_v^{e}			b_a	b_v	γ_a		
Milligan and Baker (1988)	μ	b	β	p, q	α			d	τ	f_1	$f_{2,3}$	r^{c}	γ	v
Rogers (1988)[d]		u	a_i^{b}			i_i			T	b_i	c	r_i	v_i	m_i

[a]These parameters are scaled by bite rate^{-1} so that time is measured in 3 day periods rather than 1 day.
[b]This is actually αf_1 i.e. the human biting rate.
[c]Cattle move from the infectious class to the recovered class while wild animals return from the infectious class to the susceptible class.
[d]Rogers does not explicitly provide explicit parameter descriptions for T_{hi} but uses a circumflex to denote the value of state variables T_v or T_{hi} days in the past.
[e]This is only used for an extension to the model.
[f]Baker et al. do not explicitly model exposed flies but simply use a probability, P of the survival of infected flies to become infectious.
[g]Baker et al. assume that the transmission probability from infectious hosts to teneral vectors of *Rickettsia*-like organism-infected flies is 100% and no other flies can get infected.

3. MODELS OF AFRICAN TRYPANOSOMIASIS

3.1 History of trypanosomiasis modelling

Mathematical epidemiology of malaria has a rich history, beginning with Ronald Ross' dynamical model in the early twentieth century and George Macdonald's application of this model to epidemiological and entomological data in the mid-twentieth century (Macdonald, 1950, 1952; Ross, 1911, 1916). A series of models based on this analysis, commonly known as the Ross–Macdonald model, have formed the basis of much of the mathematical modelling of vector-borne diseases to date (Reiner et al., 2013; Smith et al., 2012). In a 1982 review of malaria modelling, Aron and May (Aron & May, 1982) presented multiple versions of the Ross–Macdonald model, beginning with a description and analysis of a system of ordinary differential equations (ODEs) of the proportion of infected humans and mosquitoes and their interactions, assuming that the total population size of humans and mosquitoes was fixed. Although Aron and May's review focusses on malaria, we summarize it here because it is a valuable description of the Ross–Macdonald approach that has formed the basis of most HAT models to date. Figure 1 shows a schematic of the dynamics of the corresponding Ross–Macdonald model based on the number of susceptible and infectious hosts and vectors and the system of equations is given in Eqn (1).

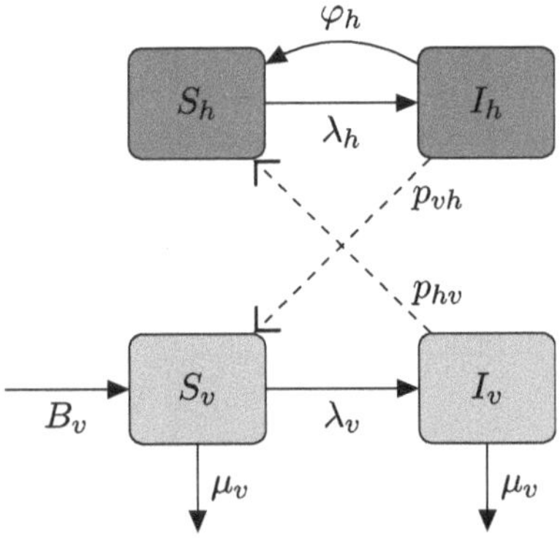

Figure 1 Compartmental diagram of the standard Ross–Macdonald model showing the number of hosts and vectors and transitions between compartments. Susceptible hosts (S_h) become infected from the bites of infected vectors (I_v) before returning to the susceptible state at a constant per capita rate. Susceptible vectors (S_v) emerge at a constant rate, B; become infected when they bite infected humans and remain infected for life. All vectors face a constant per capita mortality rate, μ_v. Solid lines show movement of individuals from one class to another, dashed lines show the paths of infection.

$$\begin{aligned} &\text{Hosts} \quad \begin{cases} \dfrac{dS_h}{dt} = \varphi_h I_h - \lambda_h S_h \\ \dfrac{dI_h}{dt} = \lambda_h S_h - \varphi_h I_h \end{cases} \\ &\text{Vectors} \quad \begin{cases} \dfrac{dS_v}{dt} = B_v - \lambda_v S_v - \mu_v S_v \\ \dfrac{dI_v}{dt} = \lambda_v S_v - \mu_v I_v \end{cases} \end{aligned} \tag{1}$$

where the constant total population of hosts is $N_h = S_h + I_h$; the force of infection on hosts, λ_h, is:

$$\lambda_h = \frac{\alpha p_{hv} I_v}{N_h}$$

and the force of infection on vectors, λ_v, is:

$$\lambda_v = \frac{\alpha p_{vh} I_h}{N_h}.$$

The total vector population remains at a constant size $N_v = S_v + I_v$ if $B_v = N_v \mu_v$.

A simplified two-dimensional system of equations for the proportion of infected humans and vectors is shown in Eqn (2); this is based on the fact that host and vector population sizes remain constant so eliminating the need for two variables. We note here that Aron and May assumed that the probability of transmission of infection per bite of a susceptible vector on an infected human was one, $p_{vh} = 1$, and did not explicitly include a parameter for this probability. This probability was added to subsequent versions of the Ross–Macdonald model, starting from Anderson and May (Anderson and May, 1991); it is often labelled as c, but we retain the label p_{vh} for consistency of notation.

$$\begin{aligned} &\text{Hosts} \quad \frac{di_h}{dt} = m\alpha p_{hv} i_v (1 - i_h) - \varphi_h i_h \\ &\text{Vectors} \quad \frac{di_v}{dt} = \alpha p_{vh} i_h (1 - i_v) - \mu_v i_v \end{aligned} \tag{2}$$

One important simplifying assumption of this Ross–Macdonald model was that it ignored the latent period between infection with the parasite and becoming infectious and able to transmit the parasite. In humans, this latent period is small compared to the average life span so can be ignored. However, in mosquitoes, this period is almost as long as the average life span so most infected mosquitoes die before becoming infectious. Aron and May described a system of delay differential equations (DDEs) that included this latent period in mosquitoes. A schematic of the dynamics of the corresponding model based on the number of susceptible and infectious hosts, and susceptible, exposed and infectious vectors is shown in Figure 2 and the system of equations is given in (3).

$$
\begin{aligned}
&\text{Hosts} \quad
\begin{cases}
\dfrac{\mathrm{d}S_h(t)}{\mathrm{d}t} = \varphi_h I_h(t) - \lambda_h(t) S_h(t) \\[2ex]
\dfrac{\mathrm{d}I_h(t)}{\mathrm{d}t} = \lambda_h(t) S_h(t) - \varphi_h I_h(t)
\end{cases} \\[2ex]
&\text{Vectors} \quad
\begin{cases}
\dfrac{\mathrm{d}S_v(t)}{\mathrm{d}t} = B_v - \lambda_v(t) S_v(t) - \mu_v S_v(t) \\[2ex]
\dfrac{\mathrm{d}E_v(t)}{\mathrm{d}t} = \lambda_v(t) S_v(t) - \lambda_v(t - T_v) S_v(t - T_v) e^{-\mu_v T_v} - \mu_v E_v(t) \\[2ex]
\dfrac{\mathrm{d}I_v(t)}{\mathrm{d}t} = \lambda_v(t - T_v) S_v(t - T_v) e^{-\mu_v T_v} - \mu_v I_v(t)
\end{cases}
\end{aligned}
\tag{3}
$$

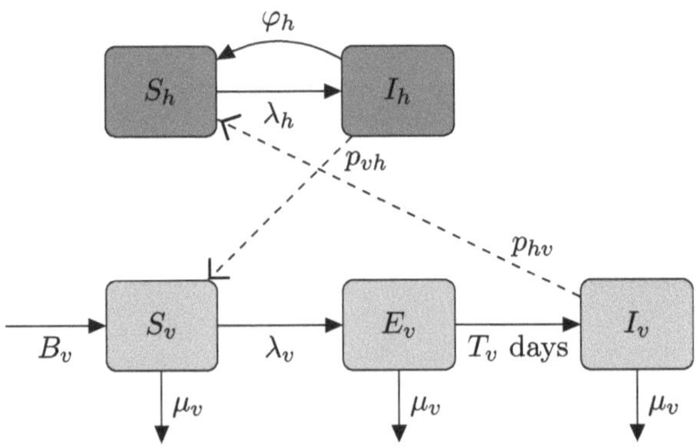

Figure 2 Compartmental diagram of the delay differential equation Ross–Macdonald model showing the number of hosts and vectors. Susceptible hosts become infected from the bites of infected vectors before returning to the susceptible state at a constant per capita rate. Susceptible vectors emerge at a constant rate; become exposed to malaria when they bite infected humans; become infected after a fixed time, T_v and remain infected for life. All vectors face a constant per capita mortality rate.

where the total (constant) population of hosts is $N_h = S_h(t) + I_h(t)$; the total (constant) population of vectors is $N_v = S_v(t) + E_v(t) + I_v(t)$ (assuming $B_v = N_v\mu_v$); the force of infection on hosts, $\lambda_h(t)$, is:

$$\lambda_h(t) = \frac{\alpha p_{hv} I_v(t)}{N_h}$$

and the force of infection on vectors, $\lambda_v(t)$, is:

$$\lambda_v(t) = \frac{\alpha p_{vh} I_h(t)}{N_h}$$

Similar to the reduction from Eqns (1) to (2), Aron and May reduced the five-dimensional system Eqn (3) to a three-dimensional DDE model for the proportion of infectious humans and exposed and infectious vectors, as shown in (4).

$$\text{Hosts} \quad \frac{di_h(t)}{dt} = m\alpha p_{hv} i_v(t)(1 - i_h(t)) - \varphi_h i_h(t)$$

$$\frac{de_v(t)}{dt} = \alpha p_{vh} i_h(t)(1 - e_v(t) - i_v(t))$$

$$\text{Vectors} \quad -\alpha p_{vh} i_h(t - T_v)(1 - e_v(t - T_v) - i_v(t - T_v))e^{-\mu_v T_v} - \mu_v e_v(t) \tag{4}$$

$$\frac{di_v(t)}{dt} = \alpha p_{vh} i_h(t - T_v)(1 - e_v(t - T_v) - i_v(t - T_v))e^{-\mu_v T_v} - \mu_v i_v(t)$$

The basic reproduction number, R_0, is defined in mathematical epidemiology as the expected number of secondary infectious caused by one infectious individual in a fully susceptible population through the duration of the infectious period. R_0 provides a threshold condition for epidemic models: when $R_0 < 1$, the trivial disease-free equilibrium point is asymptotically stable and introduced cases do not lead to epidemics or the establishment of the disease in the population. When $R_0 > 1$, the disease-free equilibrium point is unstable and introduced cases can lead to epidemics or to a new endemic steady state. Hence calculation of R_0 is a critical step in understanding persistence or elimination of infection.

For this model, Aron and May define the basic reproduction number as the number of new infections in humans from one infected human through a generation of infected vectors, as shown in Eqn (5).

$$R_0 = \begin{pmatrix}\text{Number of}\\ \text{vector bites}\\ \text{per human}\\ \text{per time}\end{pmatrix}\begin{pmatrix}\text{Probability of}\\ \text{transmission}\\ \text{from human}\\ \text{to vector}\end{pmatrix}\begin{pmatrix}\text{Duration of}\\ \text{infection}\\ \text{in humans}\end{pmatrix}$$
$$\times(\text{Probability of vector surviving latent period})$$
$$\times\begin{pmatrix}\text{Number of}\\ \text{bites on humans}\\ \text{per vector}\\ \text{per time}\end{pmatrix}\begin{pmatrix}\text{Probability of}\\ \text{transmission}\\ \text{from vector}\\ \text{to human}\end{pmatrix}\begin{pmatrix}\text{Expected}\\ \text{infectious}\\ \text{life span}\\ \text{of vectors}\end{pmatrix} \quad (5)$$
$$= (m\alpha)(p_{vh})\left(\frac{1}{\varphi_h}\right)\left(e^{-\mu_v T_v}\right)(\alpha)(p_{hv})\left(\frac{1}{\mu_v}\right)$$
$$= \frac{m\alpha^2 p_{hv} p_{vh} e^{-\mu_v T_v}}{\varphi_h \mu_v}$$

This relatively simple Eqn (5) demonstrates that it is more effective to reduce the vector biting rate, α, and the adult vector death rate, μ_v, than to reduce the vector density, m. As such, this provided justification to the Global Malaria Elimination Program for its main control strategy of targeting adult vectors through indoor residual spraying.

The basic reproduction number has also been defined for vector-borne diseases as the number of new infections from one generation to the next (the geometric mean of the number of new infections from humans to vectors and from vectors to humans), following the next generation operator approach of Diekmann et al. (Diekmann et al., 1990). This definition of the basic reproductive ratio is the square root of that defined from humans through vectors and back to humans. Although the two definitions provide different numerical values for the basic reproduction number, they provide the same threshold condition at $R_0 = 1$.

Rogers (Rogers, 1988) expanded this DDE model to produce some of the first models of African trypanosomiases: for *T. vivax*, *T. congolense* and *T. brucei*. His main extensions to the malaria model were: to add a second species of vertebrate hosts; to add 'exposed' and 'immune' states to the vertebrate hosts; and to only allow a certain proportion of susceptible vectors to become infected when they bite infectious hosts (P_{hv} which was included in subsequent versions of the Ross—Macdonald model). We show a schematic of the corresponding model for the number of hosts and vectors in each class in Figure 3.

Rogers only presented the three differential equations for the proportions of infectious hosts (of both types) and vectors. He described the intrinsic incubation period, both as a fixed duration, and as exponentially distributed with a constant rate of becoming infectious (labelled as $1/\imath$), but only showed the equations for the fixed duration. We show the full system of equations (for the proportion of hosts and vectors), assuming a fixed duration for the intrinsic incubation period in Eqn (6),

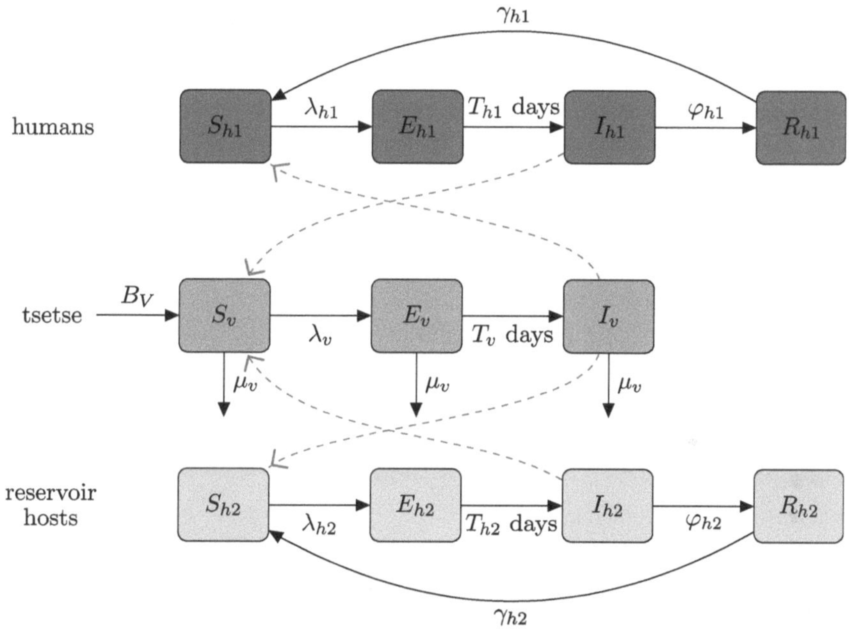

Figure 3 Compartmental diagram of the delay-differential equation model of HAT dynamics in Rogers (1988) showing the number of humans, reservoir hosts and vectors. Susceptible humans, S_{h1}, become latent with infection from the bites of infectious vectors. After a fixed period of time, latent humans, E_{h1} become infectious to tsetse flies. Infectious humans, I_{h1} recover at a constant rate to become immune. Recovered humans, R_{h1}, lose their immunity to infection at a constant rate (γ_{hi}) and return to the susceptible class. Birth and mortality in humans are ignored. Reservoir hosts are assumed to have similar dynamics to those for humans. Susceptible vectors, S_v emerge at a constant rate; become exposed to HAT when they bite either infectious humans or reservoir hosts; become infectious after a fixed time, T_v, and remain infectious for life. All vectors face a constant per capita mortality rate. The total population sizes of humans, reservoir hosts and vectors are assumed to be constant.

$$
\begin{aligned}
&\text{Hosts}(\{\mathrm{i}=1,2\}) \quad
\begin{cases}
\dfrac{\mathrm{d}s_{hi}}{\mathrm{d}t} = \gamma_{hi} r_{hi}(t) - \lambda_{hi}(t) s_{hi}(t) \\[2ex]
\dfrac{\mathrm{d}i_{hi}}{\mathrm{d}t} = \lambda_{hi}(t - T_{hi}) s_{hi}(t - T_{hi}) - \varphi_{hi} i_{hi}(t) \\[2ex]
\dfrac{\mathrm{d}r_{hi}}{\mathrm{d}t} = \varphi_{hi} i_{hi}(t) - \gamma_{hi} r_{hi}(t)
\end{cases} \\
&\text{Vectors} \quad
\begin{cases}
\dfrac{\mathrm{d}e_v}{\mathrm{d}t} = \lambda_v(t)(1 - e_v(t) - i_v(t)) - \mu_v e_v(t) \\
\quad - \lambda_v(t - T_v)(1 - e_v(t - T_v) - i_v(t - T_v)) e^{-\mu_v T_v} \\[2ex]
\dfrac{\mathrm{d}i_v}{\mathrm{d}t} = \lambda_v(t - T_v)(1 - e_v(t - T_v) - i_v(t - T_v)) e^{-\mu_v T_v} - \mu_v i_v(t)
\end{cases}
\end{aligned}
\tag{6}
$$

Here, the force of infection on hosts of humans ($i = 1$) and reservoir hosts ($i = 2$) is:

$$\lambda_{hi}(t) = m_i \alpha_i p_{hiv} i_v(t)$$

and the force of infection on tsetse flies is:

$$\lambda_v(t) = \sum_{i=1}^{2} \alpha_i p_{vhi} i_{hi}(t)$$

The equations for the proportion of exposed hosts are redundant because exposed hosts encounter no mortality and simply introduce a delay from the susceptible to the infectious class; and the proportion of exposed hosts is $e_{hi}(t) = 1 - s_{hi}(t) - i_{hi}(t) - r_{hi}(t)$ for $i = 1,2$.

The proportion of susceptible vectors is $s_v(t) = 1 - e_v(t) - i_v(t)$.

Rogers described the endemic equilibrium point of this system of equations as the implicit solution to a system of three simultaneous equations, and showed a phase portrait for a prescribed set of parameter values of the solutions approaching a globally asymptotically stable equilibrium point.

Due to the complications in deriving the basic reproduction number for HAT as derived for malaria in Eqn (5) with two different types of hosts, Rogers defined the basic reproductive number as the expected number of secondary infections in tsetse flies from one infected tsetse fly, through a generation of infection in humans and reservoir hosts, assuming fully susceptible host and vector populations.

Although this definition may provide a different expression than that defined from the next generation operator approach, both expressions provide the same threshold condition for the loss of stability of the disease-free equilibrium point and the existence of a positive asymptotically stable endemic equilibrium point:

$$R_0 = \frac{e^{-\mu_v T_v}}{\mu_v} \left(\frac{m_1 \alpha_1^2 p_{h1v} p_{vh1}}{\varphi_{h1}} + \frac{m_2 \alpha_2^2 p_{h2v} p_{vh2}}{\varphi_{h2}} \right) \tag{7}$$

again pointing to the importance of the vector adult mortality rate and biting rate. Rogers (Rogers, 1988) derived parameter values to model, what he called a typical village in West Africa, except for the transmission probability parameters, p_{vhi} and p_{hiv}, which he estimated from East African field data and laboratory experiments. Details of the parameter values and their subsequent revisions are provided in Section 3.2.

From this parameterization, Rogers derived equilibrium prevalence values for hosts and vectors and modelled the change in equilibrium infection rates as a function of tsetse mortality. Similar to previous analysis for malaria, Rogers' sensitivity analysis showed that the equilibrium prevalence of trypanosomiasis was most sensitive to the duration of the feeding cycle of tsetse flies (which relates to the biting rate on humans, α_1, and reservoir hosts, α_2), followed by the tsetse mortality rate, μ_v. However, his calculated values for equilibrium prevalence in humans (and reservoir hosts) were much higher than values measured in the field. This has remained an issue for most models of HAT.

Rogers used a separate model to include the teneral effect in flies for *T. brucei* (where tsetse flies are only susceptible for the first few days of their life). The details of this model and subsequent extensions and analysis by other authors are described in Section 3.4. Rogers also simulated seasonality in transmission by allowing the mortality rate of tsetse flies, μ_v to vary periodically, as described with subsequent extensions in Section 3.7.

Milligan and Baker (Milligan and Baker, 1988) published their first model of animal trypanosomiases in the same year as Rogers' paper. They included enhanced vector susceptibility in teneral flies, multiple host types (cattle and wild animals) and disease-induced mortality in both hosts and vectors. They derived the basic reproduction number and modelled the effects of chemoprophylaxis on cattle. They showed that the effectiveness of chemoprophylaxis depends on the frequency of application and duration of protective efficacy and that the effectiveness of vector control is reduced by the immigration of flies.

3.2 Parameter values for trypanosomiasis models

Parameter values and their ranges are typically based on a combination of estimates obtained from field data and literature, expert opinion and assumptions adopted from previous modelling exercises. Many of the model parameter values are reasonably well defined, such as those related to tsetse longevity, and baseline values and ranges are shown in Table 7. Here, we highlight two key components that have either used a wide range of values based on differing underlying assumptions, or have been based on assumptions that have been proven erroneous by more recent estimates.

3.2.1 Tsetse susceptibility to trypanosomes

The probability of a fly becoming infected when feeding on a parasitaemic individual will depend on a variety of factors and a fairly wide range of parameter values have been used in HAT models.

Table 7 Values for *T. b. gambiense* used in this paper and possible ranges of parameters (rate parameters have dimensions of days^{-1}). Many parameters depend on the temperature tsetse and humans are exposed to, and most values for these parameters assume a stable temperature of around 25 °C

Parameter	Value	Range	Source
μ_{h1}	4.4×10^{-5}		Assumption
μ_{hi}	Varies		
μ_v	0.034	0.014 – 0.047	Hargrove et al. (2004) and Davis et al. (2011)
α	0.33	0.25 – 0.5	Hargrove et al. (2004) and Davis et al. (2011)
δ_{h1}	0.002	0.0013 – 0.0029	Checchi et al. (2008b)
δ_{h2}	0		
σ_{hi}	0.083		Artzrouni and Gouteux (1996a)
σ_v	0.034	0.025 – 0.034	Artzrouni and Gouteux (1996a) and Davis et al. (2011)
p_{hiv}	0.62		Rogers (1988)
p_{vhi}	0.065	0.001 – 0.2	Rogers (1988) and Davis et al. (2011)
φ_{hi}	0.0019	0.0012 – 0.0028	Checchi et al. (2008b)
γ_{hi}	0.02		Rogers 1988)
m_i	Varies		

Most models have used the same assumption that only teneral flies can become infected with trypanosomes, but have differed in their assumptions about teneral duration and the infection risk associated with feeding during this period. Artzrouni and Gouteux (Artzrouni and Gouteux, 1996a) used a value of 0.1 and suggested a range of 0.05–0.14, and assumed that the teneral state lasted until the initial blood feeding. Rogers (Rogers, 1988) used a value of 0.065 but assumed that the window of infectivity lasted only a day for *T. b. gambiense*. Milligan and Baker (Milligan and Baker, 1988) included a teneral phase but used 0.05 as the probability for infection for both teneral flies and non-teneral flies, despite stating that this probability should be enhanced for the initial blood meal.

More recently a number of HAT models have foregone the modelling of an explicit teneral phase and simply used mean values for $\tilde{p}_{vhi}$. This could be interpreted as a constant:

$$\tilde{p}_{vhi} = p_{vhi}\frac{\mu_v}{\mu_v + \alpha}$$

which is mathematically equivalent at equilibrium if the tsetse suffer a constant mortality rate throughout their lifetime. However, it becomes a simplifying assumption when more realistic tsetse dynamics, seasonality or time-varying effects of vector control are included, since the proportion of teneral flies may change. Funk et al. (Funk et al., 2013) estimated a value of 0.18, applicable over the entire life span of tsetse, while Moore et al. (Moore et al., 2012) used a value of 0.0355 based on Baylis (Baylis, 1997). Walshe et al. (Walshe et al., 2009) provided experimental data on the teneral phenomenon for *Glossina morsitans morsitans*: for the first blood meal about 55% of flies established trypanosomes in the midgut, but these higher levels of establishment may be an artefact of laboratory settings where susceptible tsetse lines are used to increase rates of infection to aid study of tsetse–trypanosome interactions.

3.2.2 Duration of the infectious period in hosts

Rogers (Rogers, 1988) used values of between 50 (*T. brucei* infections) and 100 days (*T. congolense* and *T. vivax*) for the duration of the infectious period in hosts but admitted that these figures were 'crude average, or intermediate, values from the wide ranges in Hoare's book' (Hoare, 1972). Subsequent reviews (at least of HAT epidemiology in Uganda although there may be less or more virulent strains elsewhere) have shown the duration to be on the order of 17 months, so a reasonable value for Gambian HAT is closer to

1000 days (Checchi et al., 2008a). For *T. b. rhodesiense* most models have assumed an infectious duration between 1 and 3 months (Davis et al., 2011; Auty et al., 2012), which is closer to the observed data.

3.3 Modelling multiple host species

The terminology regarding reservoirs has evolved over the past years, but has not reached a widely accepted convention. Haydon et al. (Haydon et al., 2002) suggested that a 'maintenance' host should be defined as a host type that can sustain transmission in the absence of transmission events from other host types. A 'reservoir' would represent a species or community from which transmission events to a target population of interest (e.g. humans) occur, regardless of whether the reservoir species is negatively affected by the pathogen.

Whether the label of reservoir species should extend to those species that do contribute to transmission but are not essential for maintenance of the disease (incidental or liaison hosts) is a matter where opinions diverge. Ashford (Ashford, 1997) excluded these species, leading to a concept of a set of species required to maintain transmission indefinitely. This is comparable to the concept as used by Nishiura et al. (Nishiura et al., 2009) and Funk et al. (Funk et al., 2013), who argue that the reservoir should include all maintenance hosts and a minimal set of non-maintenance hosts.

Reservoir and maintenance hosts can be described using three threshold quantities derived from a next-generation matrix (NGM) approach described by Diekmann, Heesterbeek and Roberts (Diekmann et al., 2010): the basic reproduction number, R_0; the host-specific reproduction number, U; and the host-excluded reproduction number, Q (Roberts and Heesterbeek, 2003). A host species i can be considered a maintenance host if the host-specific contribution to R_0 is greater than unity, that is $U_i > 1$. A host or community of hosts can be considered a reservoir if in addition to $U_i > 1$, R_0 in the absence of transmission from this host species is less than 1, that is $Q_i < 1$.

The extent to which humans and nonhumans form maintenance or reservoir hosts for Gambian HAT remains a contentious issue. In several locations, *T. b. gambiense* has been isolated from nonhuman animals (Mehlitz et al., 1982; Njiokou et al., 2006; Simo et al., 2006). One study in a focus in Côte d'Ivoire reported *T. brucei* in both humans and pigs, but suggested these consisted of different zymodemes (that is, different populations or strains) indicating that there may have been little genetic exchange between the two transmission cycles (Jamonneau et al., 2004). In the laboratory, infection of various nonhuman animals and subsequent infection of tsetse has been accomplished. For instance, infected pigs harbouring low levels of *T. b. gambiense* were able to infect *Glossina mortisans submorsitans*, but interestingly not laboratory colony specimens of *G. p. gambiensis* (Wombou Toukam et al., 2011).

In spite of these observations, the commonly held belief is that *gambiense* HAT is mostly an anthroponosis, with only incidental transmission to and from nonhumans. This is partly based on clinical field experiences where treatment of the human population succeeded in eliminating HAT from certain foci (World Health Organization, 2013) and is one reason human screening and treatment remains the main tool employed against Gambian HAT.

A number of HAT modelling studies have suggested an opposite view: that the contribution of humans alone to R_0 would be below 1 and humans should therefore not be considered a maintenance host species (Rogers, 1988; Davis et al., 2011; Funk et al., 2013). In the model of Rogers (Rogers, 1988), where one alternative host type is modelled and could be interpreted as a population of domestic pigs, this species would constitute a maintenance host species (although Rogers assumed a very short duration of the infectious period in humans and R_0 for humans could be greater than 1 with a longer duration). Multiple species are considered by Funk et al. (Funk et al., 2013), none of which by themselves are capable of maintaining transmission, rather a mix of multiple domestic or wild animals, potentially in combination with humans, are required. Resolving these issues will require further field studies, particularly longitudinal measurements of prevalence in nonhumans potentially linked with genetic information on infecting trypanosomes.

Further modelling studies may also wish to keep the simplifying assumptions inherent to these models in mind. For instance, both Rogers and Funk assume a fixed proportion of bites for host species, rather than a proportion based on preference and relative population sizes (e.g. Milligan and Baker (Milligan and Baker, 1988); Moore et al. (Moore et al., 2012)). Further, Rogers assumes mortality of the hosts to be negligible and omits this factor, while Funk et al. present data based on a susceptible-infected model, omitting an incubating class for vectors (although they mention that inclusion of this state does not qualitatively change their outcomes).

The interplay between two host species (humans and animals) and the tsetse is crucial in understanding how HAT infection might be controlled and eliminated. It has been hypothesized that control methods such as culling animal populations in order to prevent cross-host infections (animal to tsetse to human), may result in the tsetse shifting their biting patterns to focus more on the human population, and potentially increase disease prevalence. Basic models of HAT, incorporating one host (human) population (e.g. (Artzrouni and Gouteux, 1996a,b)) are attractive due to their simplicity but they run the risk of failing to capture realistic dynamics potentially caused by secondary host populations. In order to ensure that control methods will both produce the desired results and be effective, the interdependence of HAT and the effects of two or more host species must be carefully modelled.

A first (perhaps naïve) postulate is that any nonhuman blood meals (from species such as lizards (Artzrouni et al., 2001a,b)) are not able to transmit HAT. In this case, the Ross–Macdonald model has a slightly altered force of infection for hosts:

$$\lambda_h = \frac{\alpha p_{hv} I_v}{N_h + N_o} \tag{8}$$

where N_h is the population of humans and N_o is the number of other hosts scaled by the proportion of blood meals that are taken from them by tsetse. Whilst these nonhuman hosts do not explicitly play a role in the transmission cycle, their presence could (in large enough numbers) avert tsetse biting away from the human population and hence reduce disease prevalence (see Figure 4). Work by Hargrove et al. suggests that if 80% or more of bites do not occur on humans then this is enough to reduce R_0 to less than 1 for Rhodesian HAT (Hargrove et al., 2012).

Another standard approach to multispecies modelling for HAT as taken by Rogers (Rogers, 1988), Milligan and Baker (Milligan and Baker, 1988)

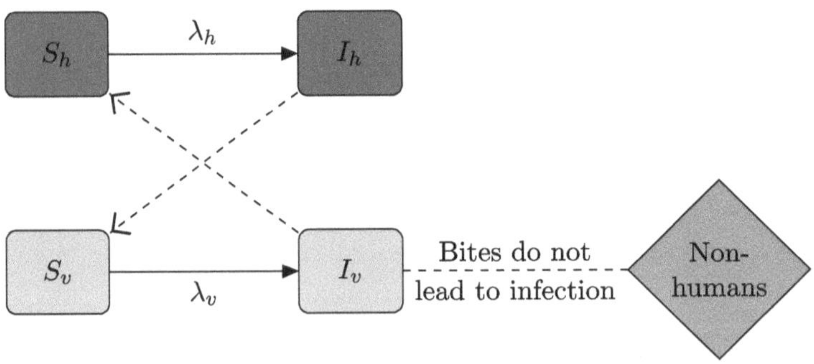

Figure 4 The simplest multi-host model with nonhuman hosts that do not contribute to onwards infection. These dead-end hosts do, however, reduce the tsetse biting rate upon humans and so effectively reduce forces of infection upon both hosts and vectors.

and many others is to partition the total host population into two categories, humans and (other) animals. Tsetse flies must select a single host from either category upon feeding and so the force of infection, both on the tsetse and to either host species, is dependent on this selection (see Figure 5).

$$\begin{aligned} &\text{Hosts, } i = \{1, \ldots, n\} && \frac{\mathrm{d}S_{hi}}{\mathrm{d}t} = B_{hi} - \lambda_{hi} S_{hi} - \mu_{hi} S_{hi} \\ & && \frac{\mathrm{d}I_{hi}}{\mathrm{d}t} = \lambda_{hi} S_{hi} - (\mu_{hi} + \delta_{hi}) I_{hi} \\ &\text{Vectors} && \frac{\mathrm{d}S_{v}}{\mathrm{d}t} = B_{v} - \lambda_{v} S_{v} - \mu_{v} S_{v} \\ & && \frac{\mathrm{d}I_{v}}{\mathrm{d}t} = \lambda_{v} S_{v} - \mu_{v} I_{v} \end{aligned} \tag{9}$$

Note that each species has its own demographic and disease parameters, based upon average life expectancies, disease susceptibility and progression. The force of infection for hosts is given by:

$$\lambda_{hi} = \frac{f_i \alpha p_{hi}}{N_{hi}} I_v \quad \text{where } i = \{1, \ldots, n\} \tag{10}$$

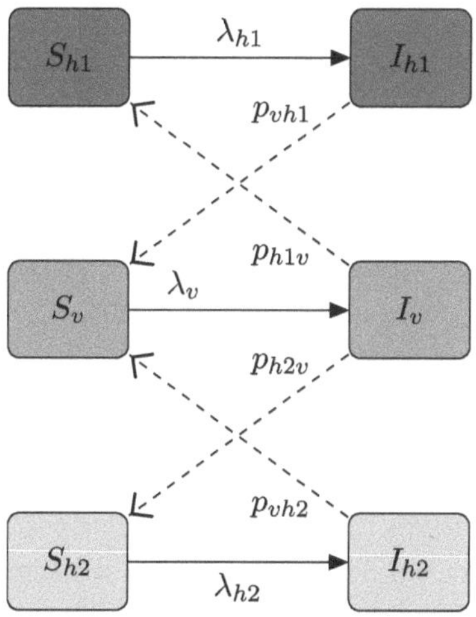

Figure 5 Interplay between the vectors and multiple host species. Here the vector selects host species i to feed on with probability f_i. Two host species are shown here but this could be easily extended for n host populations.

where f_i is the proportion of blood meals that a tsetse will take from host species i. The force of infection is calculated by considering the bite rate multiplied by the probability of selecting a host of type i from all hosts (f_i), multiplied by the probability of transmission from vector to a host i, times the probability of selecting a susceptible host from all hosts of type i and, finally, by the number of infectious vectors.

The force of infection for vectors is given by:

$$\lambda_v = \sum_i \frac{f_i \alpha p_{vhi}}{N_{hi}} I_{hi} \quad \text{where } i = \{1, \ldots, n\} \tag{11}$$

or (if $p_{vhi} = p_{vh}$ for all i):

$$\lambda_v = p_{vh}\alpha\left(\frac{f_1 I_{h1}}{N_{h1}} + \ldots + \frac{f_n I_{hn}}{N_{hn}}\right) \tag{12}$$

Here the literature diverges into two alternative (but relatable) ways of dealing with host selection by tsetse.

The first uses a fixed host preference, f_i, where the proportion of bites taken upon species i remains constant, regardless of host population sizes. The force of infection for this fixed preference is given in Eqn (10). This formulation may prove most useful in terms of parameter ascertainment as ingested blood meals can be used to find the proportion of feeds from different host species. Sometimes the term αf_i is written as α_i; this can be considered the biting rate of tsetse upon host species i.

The second formulation takes the tsetse's innate host preference into account along with relative host species availability so that f_i and α_i are dependent upon state variables, specifically the numbers of each host species, and so are expected to change dynamically. In the case where there are just two host species, the weighting of bites is given by the vector $s = (s_1, s_2)$ where:

$$\frac{s_i}{s_1} = \text{ratio of host preference from host } i \text{ to host } 1$$

when population sizes are the same

For example, when $s_1 = 1$ and $s_2 = 2$ there are twice as many bites on animals ($i = 2$) than humans ($i = 1$) (for equal population sizes). The biting preference weighting makes host species 2, s_2/s_1 times more preferable than species 1, and it is equivalent to increasing the secondary host population by a factor of s_2 then selecting one host out of the total.

This second formulation gives a variable biting preference for each host species, which is more suitable if host numbers fluctuate. For two host species we find:

$$f_i(N_1, N_2) = \frac{s_i N_{hi}}{s_1 N_{h1} + s_2 N_{h2}} \tag{13}$$

and, as before, this can be extended for n species:

$$f_i(N_1, \ldots, N_n) = \frac{s_i N_{hi}}{\sum_{j=1}^{n} s_j N_{hj}} \tag{14}$$

Conversely weighting s may be found if current population sizes and proportions of blood meals are known. For instance, for two species, s_1 is generally fixed at 1 and so:

$$s_2 = \frac{N_{h1} f_2}{N_{h2}(1 - f_2)} \tag{15}$$

This type of host preference is used by Milligan and Baker in their cattle/wild animal model (Milligan and Baker, 1988).

For fixed population sizes the two methods are equivalent. However, the advantage of the sliding preference over the fixed one becomes apparent when one or more of the host population's size changes (through disease, migration, culling, etc.) – it is assumed that a tsetse's innate preference will not change, but host availability will alter feeding patterns. Likewise, if f_i is unknown in some regions but the host population sizes are known, it may be reasonable to use the weighting of bites s (calculated from Eqn (15)) from a neighbouring area, utilizing the known host ratios, and extrapolate to find f_i for the new area.

To explore the potential impact of some of these assumptions of host biting preference, we write a NGM model for the case of two host species and one vector, using two different forces of infection. These formulations of the model ignore the teneral phenomenon and use an averaged value for p_{vhi} (following Funk et al. (Funk et al., 2013)). With a fixed biting preference, the force of infection on host species then is:

$$\begin{aligned} \lambda_{h1} &= \alpha f_1 p_{h1v} \frac{I_v}{N_{h1}} \\ \lambda_{h2} &= \alpha f_2 p_{h2v} \frac{I_v}{N_{h2}} \end{aligned} \tag{16}$$

where f_1 is the proportion of bites on host species 1 and we assume f_2 is equal to $1-f_1$. With a biting preference that scales with host numbers, the force of infection becomes:

$$\begin{aligned}\lambda_{h1} &= \alpha\frac{s_1 N_{h1}}{s_1 N_{h1} + s_2 N_{h2}} p_{h1v}\frac{I_v}{N_{h1}}\\ \lambda_{h2} &= \alpha\frac{s_2 N_{h2}}{s_1 N_{h1} + s_2 N_{h2}} p_{h2v}\frac{I_v}{N_{h2}}\end{aligned} \tag{17}$$

Here, as before, s_i describes the relative propensity or aversion to feed on host species i when the host species are equally abundant.

Writing the Jacobian of the system of equations, isolating the infected compartments, and separating into matrices T for transmission events and Σ for rates at which individuals exist the infected stages, the NGM is then given by $K = -T\Sigma^{-1}$. For a system with one vector species and two hosts:

$$K = \begin{pmatrix} 0 & 0 & k_{13} \\ 0 & 0 & k_{23} \\ k_{31} & k_{32} & 0 \end{pmatrix} \tag{18}$$

The basic reproduction number, R_0, is the dominant eigenvalue of K, squared to reflect transmission from humans to vectors back to humans, $R_0 = k_{13}k_{31} + k_{23}k_{32}$, while the host-specific contributions to R_0, are then $U_i = k_{i3}k_{3i}$.

For the first type of (fixed) host preference these are given by:

$$U_1 = \frac{\alpha^2 p_{h1v} p_{vh1} S_v \sigma_{h1} \sigma_v f_1^2}{N_{h1}(\sigma_{h1} + \mu_{h1})(\gamma_{h1} + \mu_{h1})\mu_v(\sigma_v + \mu_v)} \tag{19}$$

$$U_2 = \frac{\alpha^2 p_{h2v} p_{vh2} S_v \sigma_{h2} \sigma_v f_2^2}{N_{h2}(\sigma_{h2} + \mu_{h2})(\gamma_{h2} + \mu_{h2})\mu_v(\sigma_v + \mu_v)} \tag{20}$$

and for the second type of (variable) host preference these are given by:

$$U_1 = \frac{\alpha^2 p_{h1v} p_{vh1} S_v \sigma_{h1} \sigma_v N_{h1} s_1^2}{(s_1 N_{h1} + s_2 N_{h2})^2(\sigma_{h1} + \mu_{h1})(\gamma_{h1} + \mu_{h1})\mu_v(\sigma_v + \mu_v)} \tag{21}$$

$$U_2 = \frac{\alpha^2 p_{h2v} p_{vh2} S_v \sigma_{h2} \sigma_v N_{h2} s_2^2}{(s_1 N_{h1} + s_2 N_{h2})^2(\sigma_{h2} + \mu_{h2})(\gamma_{h2} + \mu_{h2})\mu_v(\sigma_v + \mu_v)} \tag{22}$$

Fixing the population sizes of humans and tsetse, Figure 6 considers the impact of changes in biting preference and population size of host species 2

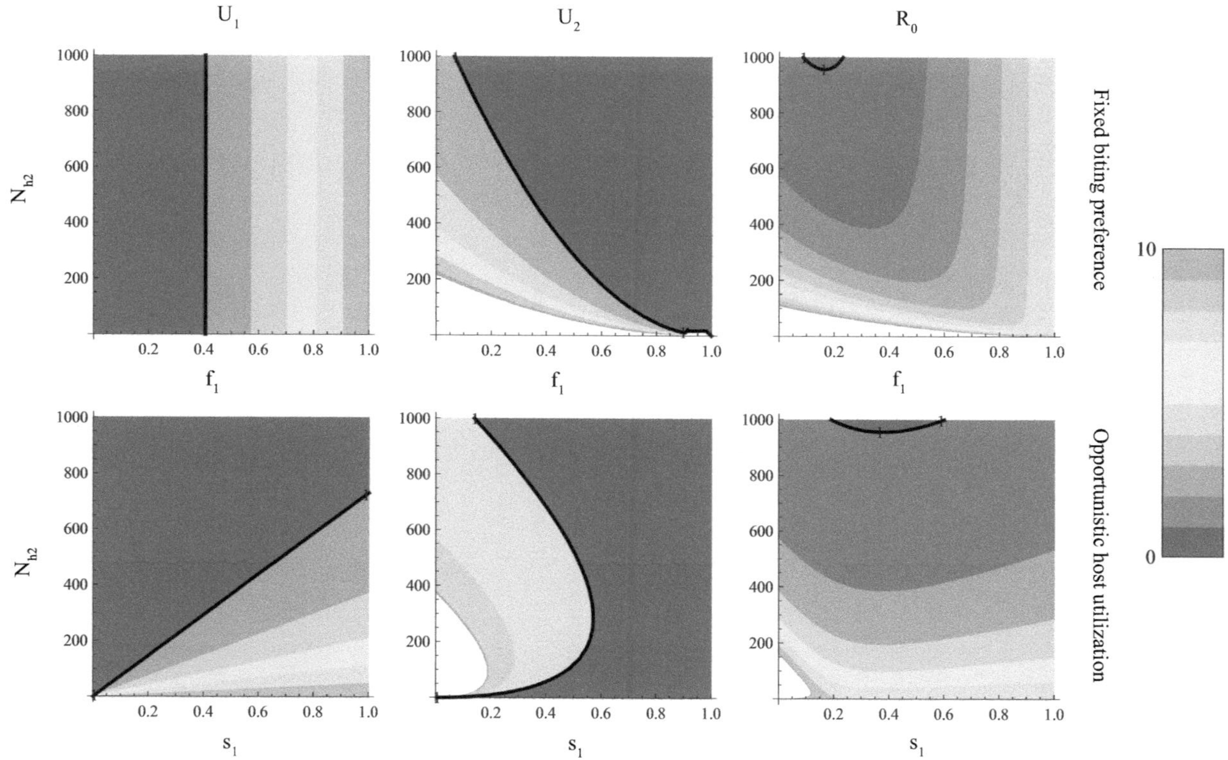

Figure 6 A comparison of implications on R_0 and the contribution of host types based on two approaches to modelling vector biting preferences (fixed or variable) for $N_{h1} = 500$ and $N_v = 2000$. The value of R_0 is indicated by contours, with the red colour corresponding to a value less than 1. The contribution of host types 1 and 2 to R_0 are indicated by U_1 and U_2, respectively. With the second assumption of variable biting, the presence of animals (e.g. pigs) that do not contribute to transmission (U_1) dilutes or acts as a protective screen: a greater number of an alternative host reduces R_0. If the second host type does contribute to transmission, a relatively large population size will be required for R_0 to remain below 1.

for the two formulations of biting preference (f — top row, s — bottom row). The columns show the effects on the total R_0 and the contributions of both hosts (U_1 and U_2). Figure 7 shows the impact of the background mortality rate of the second host type on R_0 for models based on s; for models based on fixed f, host mortality has limited impact.

The upper row of plots of Figure 6 was created assuming a fixed biting preference (force of infection as in (16)), while the lower row used the opportunistic or sliding biting preference (17). The plots graphically represent the numerical evaluation of R_0 (rightmost plots) and the host-specific contributions to R_0, U_1 and U_2 for host types 1 (e.g. humans) and 2 (e.g. pigs), respectively, when varying the preference to bite host type 1 (f_1 and s_1) and the size of the second host type population, but keeping all other parameter values constant. The thick black line indicates where (the host-specific contribution to) R_0 is equal to 1. If one assumes that only humans contribute to transmission, and nonhuman animals do not, then R_0 becomes equal to the host-specific contribution to R_0 of humans, U_1. The plots of U_1 highlight that it is theoretically feasible for transmission to occur without an animal reservoir, if the biting preference for humans is large enough (a point to the right of the black line). However, if we let the biting preference depend on relative population sizes (sliding or opportunistic host utilization), U_1 depends on an interaction between the population size of species 2 and the innate preference to bite humans, s_1, so that at higher densities of species 2, a level of anthropophily unknown to tsetse would be required for U_1 to be >1. If an animal reservoir does exist, there is only a small part of the parameter space where R_0 is <1 (the red area above the black line in the plots of R_0). Figure 7, which depicts the values of R_0 (assuming two host species contribute to transmission) when again varying the population size of the second host species and the innate preference for biting humans, highlights that the life span of the nonhuman host species also has an important influence on transmission: if this is short enough (e.g. 1 year, which may occur if pigs are being kept but slaughtered frequently), R_0 may be below 1 even with an animal reservoir, if sufficient animals are kept.

3.3.1 *Correlated bites on mammalian host species*

Some studies (Bouyer et al., 2007) suggest tsetse may select their hosts with a preference for the species which provided the previous blood meal. To incorporate this notion of correlated bites Funk et al. (Funk et al., 2013) assumed that vectors will feed upon one species, say host species 1 for an

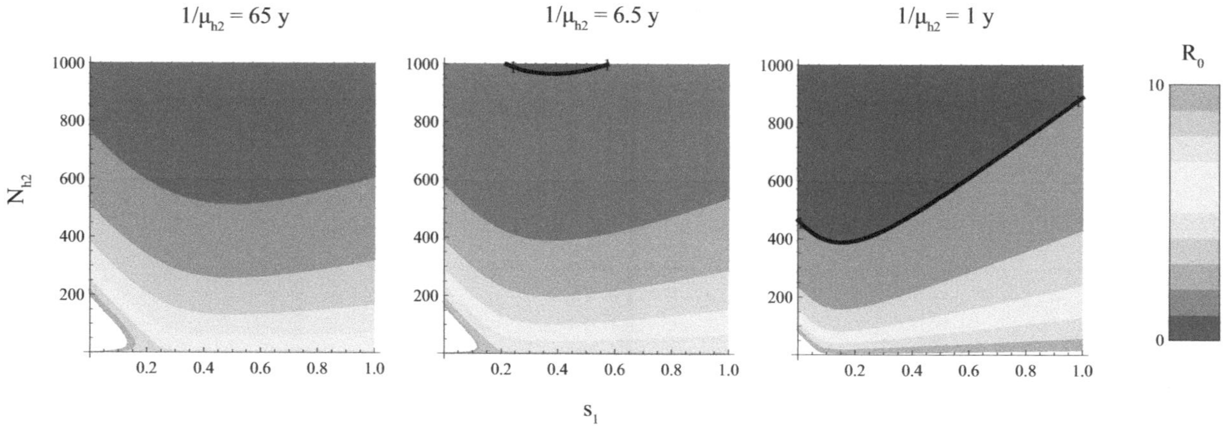

Figure 7 Contours of R_0 (red (dark grey in print versions) colour corresponding to $R_0 < 1$) illustrating the impact of the background mortality rate of the second host type in conjunction with population size (N_{h2}) and biting preference (where $s_1 = s_2$ represents a nonselective feeder whose choice depends merely on relative population sizes and values closer towards 0 representing an increasing aversion to biting humans). The shorter the life span of the second host, the lower the value of R_0. These results suggest that the keeping of peridomestic livestock such as pigs influence HAT transmission in several ways and depend on the proportion of humans to livestock and the rate at which livestock is slaughtered.

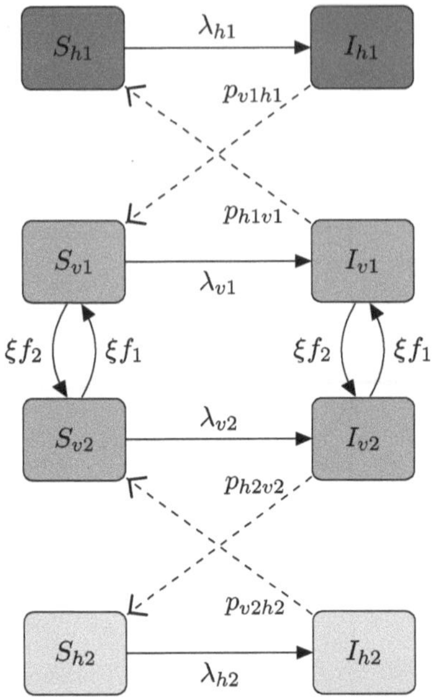

Figure 8 Interplay between the vectors and multiple host species with correlated bites (in this case $n = 2$). Here there are n classes of vectors, one for each host species. Vectors of class vi feed exclusively upon host species i until switching class which happens at a rate ξ and independently of infection status.

average period of time ξ^{-1} and then switch to another host of species 2, chosen by vectors host preference given by f_2.

Vectors are partitioned by the species from which their last blood meal came and teneral flies are assigned a class in proportion to the f_is. This gives rise to n susceptible and n infected vector classes. Note that whilst hosts and vectors both have n susceptible and n infected classes, hosts are born into one species and do not switch (other than from susceptible to infected). Vectors on the other hand switch at a rate ξ between classes independently of infection status (see Figure 8).

Equations for the vectors now become:

$$\frac{\mathrm{d}I_{vi}}{\mathrm{d}t} = \lambda_{vi} S_{vi} - \mu_v I_{vi} - \xi I_{vi} + \sum_j \xi I_{vj} f_j \tag{23}$$

The demographic parameter (μ_v) is independent of the last species bitten since mortality is independent of the source of the last blood meal and here, as previously, the force of infection is given by:

$$\lambda_{vi} = \alpha p_{vhi} \frac{I_{hi}}{N_{hi}} \tag{24}$$

Host equations will now become:

$$\frac{dI_{hi}}{dt} = \lambda_{hi} S_{hi} - \mu_{hi} I_{hi} - \varphi_{hi} I_{hi} \tag{25}$$

where the force of infection is simply:

$$\lambda_{hi} = \alpha p_{hiv} \frac{I_{vi}}{N_{hi}} \tag{26}$$

with the force of infection only associated with the class of vector that will bite that particular host species.

It is contentious whether or not correlated bites happen in general and there is only minimal corroboration for this assertion. Consequently, Funk et al. (Funk et al., 2013) have the only known model to incorporate this feature.

3.4 Modelling vector competence

Vector competence is a broad term that encompasses how amenable a vector species is to parasite uptake, establishment and development, typically denoted as a single parameter (p_{vhi} in our notation) in compartmental models. Vector competence represents a spectrum ranging from refractoriness to susceptibility and in reality is influenced by a wide range of immune and physical mechanisms within the vector. The outcomes differ not only between tsetse species and trypanosomiasis strains, but also between and within individual tsetse.

Typical assumptions in models have been that tsetse flies have increased susceptibility during their first blood meal and lower susceptibility during subsequent feeding; however, some models have also included variations in susceptibility and mortality in flies due to the nutritional status of the fly and the presence of symbionts.

3.4.1 Teneral and age-dependent susceptibility

To capture teneral susceptibility, the Ross–Macdonald model (1) was adapted by adding another class of flies, G_v, that consists of non-teneral flies whose first blood meal did not lead to infection and who consequently cannot become infected (or may only become infected at a lower probability). This allows for an alternative formulation in which a different compartmental model is used for the tsetse population (see Figure 9). This

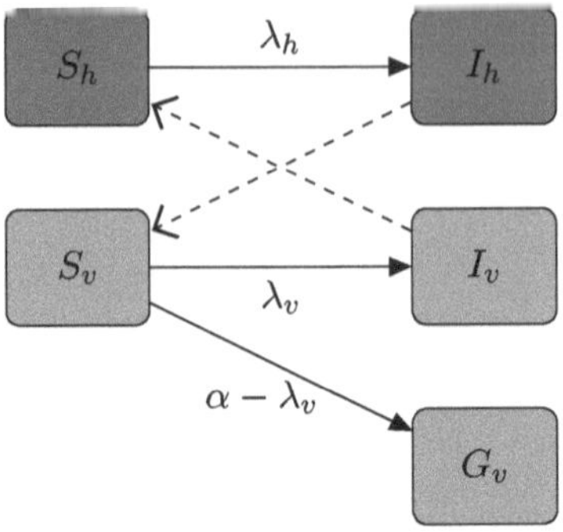

Figure 9 Schematic of model for teneral susceptibility. Here the G_v compartment represents flies that have fed at least once and were not infected on their first feed (and are subsequently outside of the disease dynamics).

formulation and slight variations follow naturally from the biology and implementation discussed by several authors (including Milligan and Baker (Milligan and Baker, 1988), Rogers (Rogers, 1988), Artzrouni and Gouteux (Artzrouni et al., 2001a), Medlock et al. (Medlock et al., 2013) and Funk et al. (Funk et al., 2013)).

The new ODEs are:

$$\text{Hosts} \quad \begin{aligned} \frac{\mathrm{d}S_h}{\mathrm{d}t} &= B_h + \gamma_h I_h - \lambda_h S_h - \mu_h S_h \\ \frac{\mathrm{d}I_h}{\mathrm{d}t} &= -\gamma_h I_h + \lambda_h S_h - (\mu_h + \delta_h) I_h \end{aligned}$$

$$\text{Vectors} \quad \begin{aligned} \frac{\mathrm{d}S_v}{\mathrm{d}t} &= B_v - \lambda_v S_v - \mu_v S_v \\ \frac{\mathrm{d}I_v}{\mathrm{d}t} &= \lambda_v S_v - \mu_v I_v \\ \frac{\mathrm{d}G_v}{\mathrm{d}t} &= (\alpha - \lambda_v) S_v - \mu_v G_v \end{aligned} \tag{27}$$

where the force of infection, λ_i, remains the same as previously (i.e. Eqn (8) where there are no reservoir hosts). New refractory vectors (class G_v) arise from susceptible vectors biting (at rate α) whose first bite did not lead to infection (due to either biting a noninfected host or from failure of transmission). The refractoriness is assumed to be complete and life long and since biting is assumed to be frequency rather than density dependent, the bite rate is independent of the host(s) population(s). Many authors do not explicitly model the refractory, non-teneral class. Artzrouni and Gouteux (Artzrouni et al., 2001a) use an approximation method to compute the teneral population (S_v) by assuming the population is approximately at equilibrium and so:

$$\frac{dS_v}{dt} = B_v - \alpha S_v - \mu_v S_v \approx 0 \tag{28}$$

which gives:

$$S_v^* = \frac{B_v}{(\mu_v + \alpha)} \tag{29}$$

Artzrouni and Gouteux note that $\alpha > \mu_v$ (significantly, $\alpha \approx 1/3$ days^{-1} whereas $\mu_v \approx 1/33$ days^{-1}) and so $S_v^* = \frac{B_v}{\alpha}$. It is this value S_v^* which is used throughout their work as an approximation of S_v.

Rogers (Rogers, 1988) also uses this type of equilibrium argument to compute the teneral population and, again, eliminate the need for explicit modelling of refractory flies. However, an additional requirement of susceptibility in the model is that unfed flies must be less than t days old to be able to transmit infection at the next blood meal. The number of flies less than t days old (given that bites and deaths are considered Poisson processes and hence are exponentially distributed) is given by:

$$\mathbb{P}(\text{un- fed fly} \leq t \text{ days old}) = \int_0^t (\mu_v + \alpha) e^{-(\mu_v+\alpha)x}\, dx = 1 - e^{-(\mu_v+\alpha)t} \tag{30}$$

and so:

$$S_v^* = \frac{B_v}{\alpha + \mu_v}\left(1 - e^{-(\alpha+\mu_v)t}\right) \tag{31}$$

Using parameters given by Rogers (Rogers, 1988), this extra requirement on age reduces the susceptible population by around 33% compared to the unfed population total yielding about 8% of the tsetse population susceptible (rather than 11%). Arguably the stipulation that teneral flies must be less than t days old may represent not only the loss of susceptibility of these flies, but the expected mortality associated with newly emergent tsetse failing to take a blood meal. Blood feeding in tsetse is a necessity for survival, not only for reproduction as with the mosquito (although female mosquitoes also use blood for energy and survival), and newly emerged tsetse are more prone to starvation than tsetse which have fed at least once (Kubi et al., 2006). Models of this increased mortality are described in more detail below in Section 3.4.2.

Both Rogers (Rogers, 1988) and Artzrouni and Gouteux (Artzrouni et al., 2001a) assume fixed vector population sizes so this removes the need to model the refractory class explicitly. Milligan and Baker (Milligan and Baker, 1988) take another approach based upon a similar principle. In their model, tsetse either become infected and are considered exposed (E_v), upon the first blood meal, or their susceptibility to HAT reduces by a factor, denoted here by ε, for subsequent blood meals and so they move to a 'non-teneral' class (G_v) (see Figure 10) so the immunity to infection is not complete.

While under Rogers' formulation, the number of susceptible flies is less than the number of unfed flies, the modelling of teneral flies described in Figure 10 has the opposite effect: the number of flies which may acquire infection upon biting is greater than the number of unfed flies.

Refractoriness was also prominently featured in a study based on an NGM approach to obtain expression for the basic reproduction number (Davis et al., 2011), where in addition to the teneral phenomenon, they assumed that only a small fraction of teneral flies are susceptible to infection. This is different from a more intuitive interpretation of having a single low value of p_{vhi}, where all (teneral) flies have a small probability of becoming infected, although the distinction only matters when it is tied to genetics or to the harbouring of intracellular symbionts (see below). The NGMs

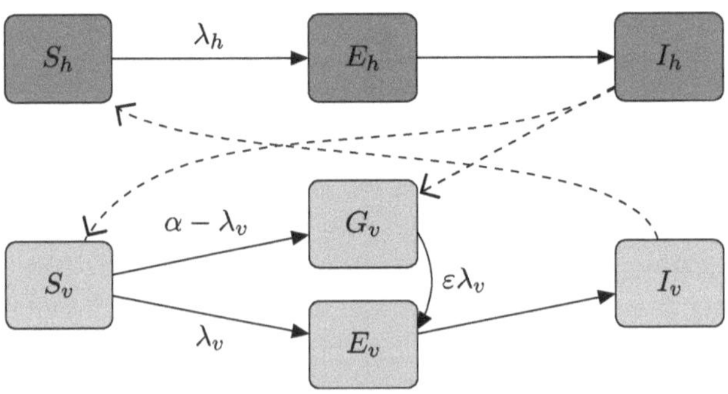

Figure 10 Transmission patterns including partial non-teneral immunity. Here then fed flies may still play a role in the transmission cycle, however, they have a reduced force of infection (by ε) compared their unfed counterparts.

they derived for *T. b. gambiense* and *T. b. rhodesiense* differed in their inclusion of livestock and wild animal host types, though both allowed for transmission by two vector species, *G. f. fuscipes* and *G. pallidipes* or *G. morsitans*. A global sensitivity analysis revealed that the proportion of blood meals taken on humans was the parameter that most strongly influenced R_0 for *T. b. gambiense*, while for *T. b. rhodesiense* refractoriness to trypanosomes ranked highest. Surprisingly, tsetse survival was only of moderate importance, which deviates from the typical interpretation of vectorial capacity (Garrett-Jones, 1964), where vector survival, together with the biting rate, are thought to affect transmission potential most strongly (as described above in Section 3.1). An interesting outcome, which potentially lends further support to the notion of an animal reservoir, was that over the range of their defined parameter space, R_0 for *T. b. gambiense* was below 1, but above 1 for *T. b. rhodesiense*.

In their agent-based model, Muller et al. (Muller et al., 2004) explicitly included an age-dependent susceptibility that increased over the first 3 days of a fly's life but then decreased rapidly as a function of age. As is typically the case with agent-based models, the lack of an explicit algebraic formulation makes it difficult to interpret the effect of this assumption on their simulation results.

3.4.2 Effect of nutritional status

Blood is the only food source of tsetse flies, so without it, tsetse flies starve. Formulating starvation in an ODE model framework can be achieved by introducing an additional death rate, μ_{vT}, which affects all members of the tsetse population equally unless the population is further subdivided. Mathematically this does not change the equations, because this term may be considered to be included in μ_v, the 'natural' mortality of tsetse, μ_v (through a reparameterization).

It is well documented that teneral flies must feed soon after emergence (Hargrove et al., 2004, 2011a; Kubi et al., 2006). The size and nutritional reserves of teneral flies vary (Hargrove et al., 2004) and this seems likely to contribute to variation in not only mortality but also in the feeding behaviour of these flies. We can include faster teneral starvation by amending only the susceptible fly population:

$$\frac{\mathrm{d}S_v}{\mathrm{d}t} = B_v - \alpha S_v - \mu_v S_v - \boldsymbol{\mu_{vT} S_v} \tag{32}$$

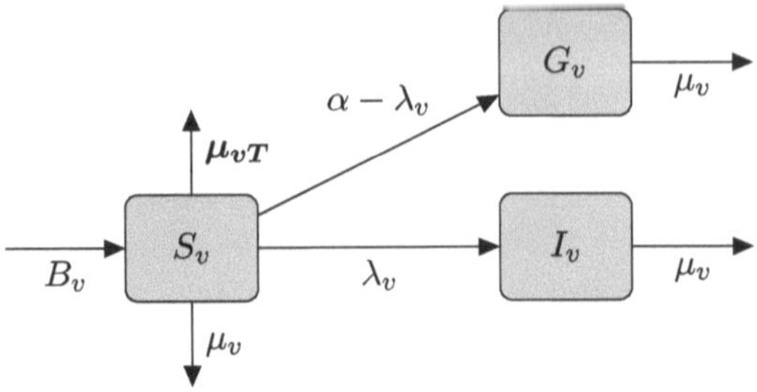

Figure 11 Compartmental diagram showing a possible implementation of teneral tsetse starvation. Here previously unfed tsetse experience an additional starvation pressure of μ_{vT} compared to their non-teneral counterparts.

whilst leaving the other equations as before (see Figure 11). Here μ_{vT} is the additional starvation rate experienced by teneral flies only. 'Normal' starvation of all tsetse is included in the μ_v term.

If the births into the population, B_v, compensate for this extra teneral starvation (as would be expected for stable populations), there is no change to the number of either susceptible or 'immune' tsetse, consequently there is no change from teneral starvation to infection in either hosts or vectors.

Rogers' model (Rogers, 1988) does not include faster starvation of teneral flies, but has a similar assumption by removing teneral flies older than t days old from the system. However, whilst not explicitly stated, these flies remain in the population, rather than dying. The formulation of this model calculates the number of surviving teneral tsetse, younger than t days using Eqn (30).

Using the given parameters of Rogers (Rogers, 1988), the number of teneral flies is 66% less than that predicted when assuming that flies only lose their susceptibility through biting alone. This is since $t < 1/\mu_v$ and so:

$$1 - \exp(-(\alpha + \mu_v)t) < 1 - \exp\left(-\left(\frac{\alpha + \mu_v}{\alpha}\right)t\right) \approx 0.67 \tag{33}$$

A visualization of the type of reduction in teneral flies given by Rogers' model is given in Figure 12.

Realistically the feeding frequency α would also be expected to be higher for newly emergent tsetse than non-teneral flies leading to a higher exit rate from the susceptible class into the 'immune' class. This would have a similar type of impact to Rogers' loss of susceptibility of teneral flies.

Modelling the increase in susceptibility due to starvation is difficult to capture in ODE models because of the assumption of a homogeneous fly population and exponentially distributed biting times. This would be

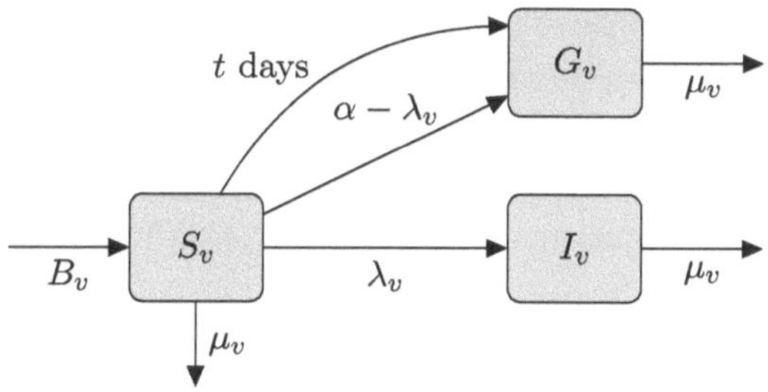

Figure 12 Compartmental diagram showing a possible implementation of teneral tsetse loss of susceptibility through ageing. Here susceptible tsetse are lost through biting at a rate α or through becoming older than t days.

relatively easy to include in an agent-based model that tracks individual flies, such as the one by Muller et al. (Muller et al., 2004) but they do not alter susceptibility by time since last feed.

3.4.3 Effects of symbionts

Tsetse flies are host to a number of intracellular symbionts: *Wigglesworthia glossinidius*, *Wolbachia pipientis* and *Sodalis glossinidius*. Both *Wigglesworthia* and *Sodalis* are thought to be symbiotic by contributing B-complex vitamins that are lacking in the blood-only diet of the tsetse and longevity is reduced in flies lacking these symbionts. *Sodalis*, initially described as a *Rickettsia*-like organism, has been incriminated as having a role in refractoriness. In certain, but not all tsetse populations, the presence of *Sodalis* in flies was found to be correlated to their susceptibility to infection. One hypothesis for this is that *Sodalis*, particularly in teneral flies, can inhibit certain lectins within the midgut environment that normally disrupt trypanosomes, although this notion has fallen out of favour in recent years (Walshe et al., 2009; Welburn et al., 1993). A model incorporating the role of *Rickettsia*-like organisms (*S. glossinidius*) was developed by Baker et al. (Baker et al., 1990) to explain periodic oscillations in human prevalence. They make the critical assumptions that *Sodalis*-infected tsetse flies enjoy a higher pupal survivorship; that only teneral *Sodalis*-harbouring flies are susceptible to infection with trypanosomes; and that trypanosome infection negatively affects tsetse fitness (reduced fecundity rather than pathogen-induced mortality). The change in the proportion of infected hosts (only one population is considered) is:

$$\frac{\mathrm{d}i_h}{\mathrm{d}t} = \left(\frac{\alpha_h^2 m_h p_{hv} \exp(-\mu_v T_v)}{\mu_v + \alpha_h}\right)\theta i_h(1 - i_h) - \varphi_h i_h. \tag{34}$$

This is a modification of the Ross–Macdonald model, where α_h is the biting rate on humans, m_h the number of tsetse per human, p_{hv} the probability of an infectious bite leading to infection, $\exp(-\mu_v T_v)$ the probability of a tsetse surviving the extrinsic incubation period, μ_v the rate of tsetse mortality, θ the proportion of flies harbouring *Sodalis*, φ_h the rate at which humans lose infection and i_h the fraction of humans that are infected with HAT. The novel component of their model is a differential equation for the fraction of *Sodalis*-infected flies,

$$\frac{d\theta}{dt} = \frac{1}{\tau_g}\theta(1-\theta)\frac{(k - gi_h)}{(1 - k + \theta(k - gi_h))} \tag{35}$$

where k is the fractional gain in pupal survival due to the symbiont and g the fractional loss in fecundity due to trypanosome infection and τ_g the time between generations. The equilibrium solution of Eqn (35) leads to the interpretation that when *Sodalis* is not at fixation ($\theta \neq 1$) nor entirely absent ($\theta \neq 0$) in the population, the fraction of infected humans depends only on the proportional increase of larval survival due to *Sodalis* and the cost due to trypanosome infection,

$$i_h^* = \frac{k}{g} \tag{36}$$

and not directly through vector density, biting rate or adult mortality. Lack of *Sodalis* in this model thus functions as a form of tsetse immunity, and gives rise to oscillations and periodic (c.20 year cycles) epidemics if the vectorial capacity is high enough. An increase in trypanosomiasis prevalence will serve to reduce the fraction of flies infected with *Sodalis* until prevalence reaches a low level and *Sodalis* will again increase, but very slowly due to the low-reproductive rate of tsetse and the small fitness advantage associated with the symbiont. A strange conclusion that follows from this work is that vector control (barring elimination of the tsetse population) would not in the long run affect disease prevalence, because a lowered vectorial capacity will merely result in a higher proportion of *Sodalis*-infected flies. These results all hinge on the assumptions of the model, which may not hold for all field populations.

3.5 Modelling stages of disease progression in mammalian hosts

Under Rogers' formulation decreasing mammalian host immunity to infection is included, but host demography (births and deaths) is not. The various models in the literature are not consistent in how disease progression in mammalian hosts is modelled. For instance, whilst Rogers' 'removed' class accounts for recovered individuals with immunity to disease, Artzrouni and Gouteux (Artzrouni and Gouteux, 1996b) use four host stages where the latter two correspond to first and second stages of the disease (asymptomatic and symptomatic respectively) and there is no immune effect (recovery is immediately into the susceptible class).

Interestingly, although these disease progression models are biologically very different, they are mathematically similar in formulation and the schematic shown for humans in Figure 3 is unchanged as Artzrouni and Gouteux assume that humans with secondary stage disease are not available for blood feeding of tsetse and are therefore 'removed' and analogous to an immune class. However, Artzrouni and Gouteux also exclude these individuals as a source of blood so the removed class, R_h, does not contribute to infection and similarly, the total number of humans to feed from is decreased.

Consequently the force of infection onto the tsetse population is given by:

$$\frac{\alpha_1 p_{vh1} I_{h1}}{(S_{h1} + E_{h1} + I_{h1} + R_{h1})}$$

for Rogers and

$$\frac{\alpha_1 p_{vh1} I_{h1}}{(S_{h1} + E_{h1} + I_{h1})}$$

for Artzrouni and Gouteux. Similarly, the measured human prevalence is given by:

$$\frac{I_{h1}}{S_{h1} + E_{h1} + I_{h1} + R_{h1}}$$

for Rogers' model of progression and

$$\frac{I_{h1} + R_{h1}}{S_{h1} + E_{h1} + I_{h1} + R_{h1}}$$

for Artzrouni and Gouteux.

3.6 Modelling spatial heterogeneity

Tsetse and human populations are highly spatially structured. This spatial heterogeneity is not explicitly captured in any of the models developed using a differential equation approach; and these limitations were discussed in more detail by Peck and Bouyer (Peck and Bouyer, 2012). The impact of spatial heterogeneity and tsetse/human movement has long been recognized by the tsetse community (e.g. (Williams et al., 1992)) and by researchers investigating HAT (e.g. (Chalvet-Monfray et al., 1998)). In many ways spatial structure makes disease more easy to control (if more difficult to model): 'hot spots' of locally intense transmission may be identified and control efforts focussed on these populations and the riverine distribution of some tsetse species means that reinvasion after control efforts (the scourge of most control programmes) occur along predictable corridors.

One notable model that does attempt to incorporate this heterogeneity is the 'HAT-trick' model (Torr et al.). This is based on a 50 × 50 grid of 'habitat cells' each of which is characterized for its habitat and consequent effect on tsetse ecology, and the presence of animals and livestock. The user can overlay a habitat map on this grid, including riverine habitats, define tsetse ecology in each of the habitat types, allow migration between cells, and then impose control measures to gauge their impact on tsetse number, age-structure and HAT transmission. It is a highly sophisticated model, particularly on tsetse ecology, and its mapping functions make it appealing to policy makers who need to consider particular scenarios in detail. However, the main cost of this sophistication is that this model cannot generate the generic insights possible from a differential equation approach.

An analogous approach to incorporate spatial heterogeneity is to use agent-based modelling. This has become a fairly widespread technique in computer sciences with dedicated software suites for its implementation. In the current case, the 'agents' are humans and tsetse who move across a spatially-structured environment while the simulation tracks their individual properties such as age, infection status, and so on, as implemented by Muller et al. (Muller et al., 2004). As in HAT trick, the cost of this more realistic spatially-centred approach is the lack of insights that may be generated through a simpler tractable approach.

A final way of capturing spatial heterogeneity is through the use of metapopulation models (Hanski and Gaggiotti, 2004) where the populations of humans and vectors are split into semi-isolated subpopulations with movement between them; this can realistically reflect a fragmented habitat. The chief drawback of this approach is that tsetse populations may be structured but are rarely fragmented: the savannah form usually occurs in large contiguous populations while the riverine tsetse species tend to be contiguous linear populations with extensive movement along rivers. Despite these restrictions, it remains a valuable attempt to understand spatial heterogeneity and may be useful in future situations as HAT nears extinction and the trypanosomes form semi-isolated populations.

3.7 Modelling tsetse fly population dynamics and seasonality

A useful measure of exposure for vector-borne diseases is the entomological inoculation rate, or the number of infectious bites received per person per day. This is a compelling argument for linking disease transmission models to realistic vector population dynamics. To date, models focussing on HAT transmission have frequently ignored tsetse population dynamics completely (assuming fixed population sizes) (Artzrouni and Gouteux, 1996a; Funk et al., 2013)), or have used simple models with some seasonal variation (e.g. (Milligan and Baker, 1988; Rogers, 1988)); while more sophisticated models of tsetse fly population dynamics have been derived but have not been integrated with HAT models.

The components of a realistic population dynamics model and the dependency of the parameters to temperature and climate are fairly well defined (Hargrove et al., 2004). One study linked the temperature-dependent life history traits of tsetse to an expression for R_0 in order to predict possible range expansions or contractions of Rhodesian HAT due to global climate change (Moore et al., 2012), although this was based on predicted mean annual temperatures rather than seasonal fluctuations.

It is not immediately obvious how important it is to include seasonal variation of tsetse abundance in HAT models, because although population sizes do vary considerably (Rogers, 1990), this variation is relatively minor compared to that found in other vectors such as mosquitoes, where seasonality is often ignored despite its obvious importance. Additionally, the duration of infection in both humans and vectors is relatively long, which may temper the impact of seasonal variation.

It has been noted that there is significant variation in the seasonality of tsetse abundance between tsetse species, with certain species showing only moderate or barely any fluctuations and others, such as *G. m. submorsitans* vary considerably with resulting seasonal parasitaemia in mammalian hosts (Milligan and Baker, 1988). It is likely that seasonality influences HAT transmission in manners beyond fluctuations in tsetse abundance, for example, through changes in agricultural activities that may affect human tsetse contact rates. Seasonality has subsequently been modelled through different means. For instance, Milligan and Baker incorporated seasonality by including a sinusoidal form for the emergence rate (Milligan and Baker, 1988), while Rogers simulated the effects of seasonality by varying adult mortality and biting rates with temperature (Rogers and Randolph, 1984; Rogers, 1988) and Artzrouni and Gouteux used a linear relationship between temperature and emergence rate (Artzrouni and Gouteux, 2006).

Mortality in juvenile stages also appears to have a seasonal component, but it may be that this represents a density-dependent effect instead (Hargrove, 1999; Phelps and Burrows, 1969a; Phelps and Clarke, 1974). Various other density-dependent effects have been included in tsetse models, for instance, Rogers (Rogers, 1990) needed to include a form of density-dependent mortality for both the adult and pupal stages in his population model to match field data. Density-dependent migration has also been included in models of tsetse dynamics (Artzrouni and Gouteux, 2003). A minimal population dynamics model including seasonality would therefore account for both temperature and density-dependent losses during the immature stages as well as the teneral phase.

Models that focused on the population dynamics of tsetse have been used to assess impact of natural factors (e.g. population density (Rogers, 1990), environmental temperature (Hargrove and Williams, 1998)) and vector control (Hargrove, 1988) on the growth of tsetse populations (see review by Hargrove (Hargrove et al., 2004)). Other models have been developed to analyse the growth of open populations (Williams et al., 1992) by assuming logistic growth and diffusive movement of populations. This approach formed the basis of a series of studies concerned with analysing the impact of various tsetse control operations (Hargrove, 2000; Hargrove et al., 2003).

More recently, simulation models of tsetse populations have been developed to allow not only distribution, spread and abundance of tsetse to be modelled but also age-structure of the tsetse population (Vale and Torr, 2005). These simulation models have been used to assess the likely impact and cost of insecticidal and sterile insect techniques (SIT) of vector control (Vale and Torr, 2005; Barclay and Vreysen, 2011a; Hargrove et al., 2011b; Barclay and Vreysen, 2011b, 2013), impact of aerial spraying (Kgori et al., 2006), insecticide-treated livestock (Torr and Vale, 2011) and the relative costs of tsetse control (Shaw et al., 2013). The 'Tsetse Muse' model's (Vale and Torr, 2005) outputs includes age-structure and hence can be used to provide an indication of the density of tsetse that are old enough to be potential vectors. There is a need to combine these models with models of HAT transmission.

4. MODEL COMPARISONS

We now explore an extended model formulation incorporating various elements of the models previously discussed (see Figure 13 for model outline and Figure 14 for simulation dynamics) to elucidate the effects of key parameters and model assumptions. In order to retain some sense of continuity, the model structure (with I_{hi} and R_{hi} corresponding in this case to first- and second-stage infection respectively) is used both here and in the previous Section 3.3 and the type of disease progression in the human population is that used by Artzrouni and Gouteux (Artzrouni and Gouteux, 1996b).

The total prevalence in the human population comprises individuals in both stage I and II (even though only stage I cases transmit infection to tsetse).

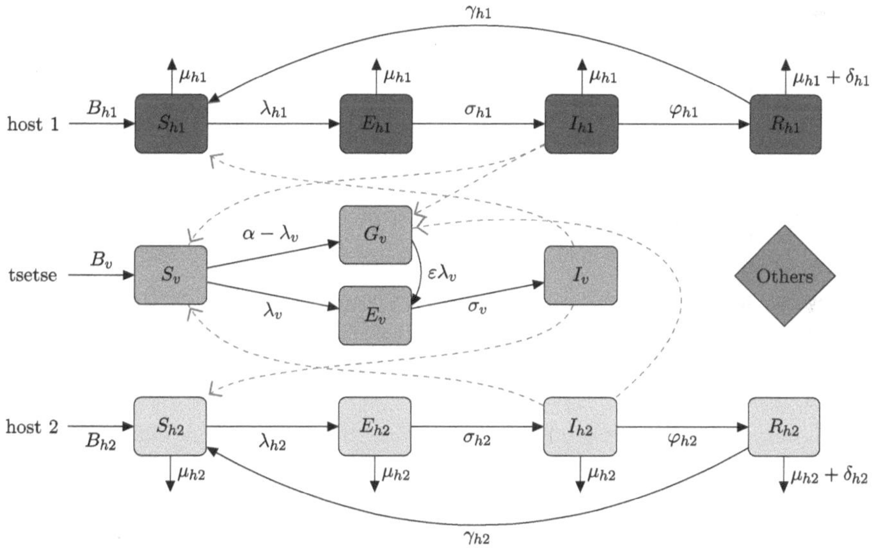

Figure 13 Interplay between the vectors and multiple host species. This model includes both reservoir (host 2) and non-reservoir (other) hosts in addition to humans (host 1) and tsetse. It is assumed that there are four possible infection statuses for hosts: susceptible, exposed, stage I and stage II. Upon entering stage II, hosts are not infectious to tsetse through hospitalization. Tsetse also have 4 stages corresponding to teneral flies, exposed flies, infectious flies and non-teneral noninfected flies. Non-teneral, noninfected tsetse have a reduced probability by a factor ε of becoming infected. Vector mortality which is at a constant per capita rate, μ_v out of all classes, regardless of infection or teneral status, is not shown. Dashed grey lines indicate transmission paths.

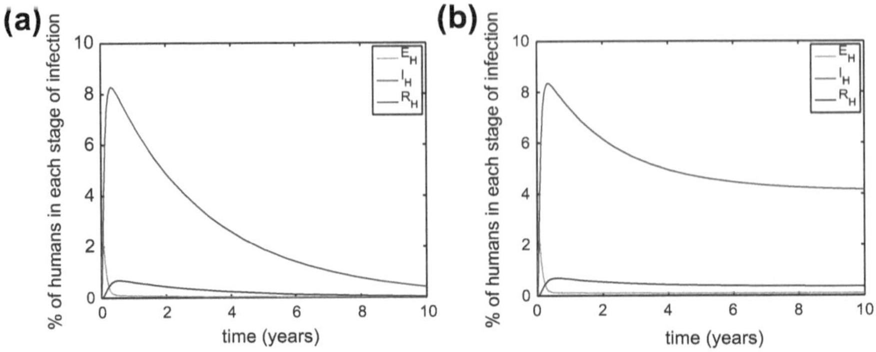

Epidemic, $f_2 = 0$ Endemic, $f_2 = 0.051$

Figure 14 Examples of the possible dynamics generated by this model. Both simulations were run using $f_1 = 0.1$, $N_{h2} = 300$ and $\varepsilon = 0.05$. Initially all hosts were susceptible with 1% of tsetse infected. Figure (a), which shows an epidemic followed by eventual disease extinction (after around 50 years), was generated using $f_2 = 0$ as the secondary host feeding preference (i.e. all blood meals not from humans were from non-reservoir species). Figure (b), which shows an initial peak in infection followed by an endemic level of infection, was generated using $f_2 = 0.051$. Throughout, the frequency of bites on the non-reservoir species is $f_3 = 1 - f_1 - f_2$.

The model can be used to study the effect of certain parameters upon the endemic equilibrium. The two main variations in modelling have typically concerned:

- Host choice/preference and multi-host models including either secondary (reservoir) host populations, non-reservoir populations or a mixture of the two
- The teneral effect: models range from no teneral effect (Funk et al., 2013) through partial immunity (Milligan and Baker, 1988) to full 'immunity' (Rogers, 1988).

All other parameters are kept constant with values taken from Table 7.

4.1 Host preference

We assume that there are three species of mammalian hosts: humans, secondary reservoir hosts (which can transmit infection) and non-reservoir hosts. Additionally we assume that while the tsetses' host preference (s) is not explicitly known, it is possible to find the proportion of blood meals taken from a certain host species (f_i) by analysis of blood meals. Here, f_1 corresponds to proportion of bloods meals on humans; f_2 to proportion of blood means on reservoir hosts; and f_3 to proportion of blood meals on all other hosts that are not part of the HAT transmission cycle; with $f_1 + f_2 + f_3 = 1$.

It is important to note that both the relative size of the secondary population and host preference will affect the prevalence of disease in the human population. Smaller reservoir host population sizes lead to more infection for the same proportion of blood feeding as there is more chance of the same reservoir host being bitten at least twice to complete its role in the transmission cycle. Conversely the size of the non-reservoir population is of no consequence if the proportion of blood meals taken is known. The relationship between secondary host population size, N_{h2}, the proportion of blood meals taken from them f_2 and the prevalence in the human population at equilibrium is shown in Figure 15.

In these simulations we assume that a proportion, f_1, of blood meals are from the human population and the remaining $f_2 + f_3 = 1 - f_1$ are from either the reservoir (f_2) or the non-reservoir (f_3) hosts. The results show that increasing f_2 increased the human prevalence. Funk et al. (Funk et al., 2013) and Rogers (Rogers, 1988) found that without secondary reservoir hosts (equivalent to $f_2 = 0$), HAT infection cannot be sustained within the human population. The results here indicate that if the proportion of blood feeding on humans is low (around 10% or less) then disease will become extinct in the absence of a reservoir animal population, however, if the proportion feeding on humans increases, an endemic state occurs even when $f_2 = 0$.

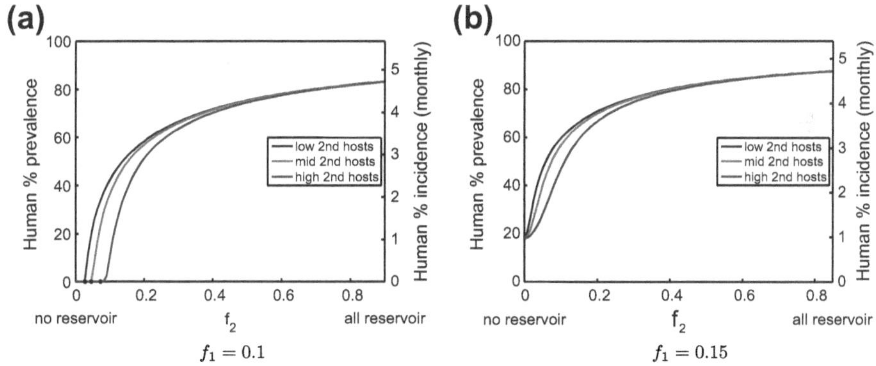

Figure 15 The relationship between host selection and human prevalence (or incidence) is shown for two choices of f_1: Figure a) $f_1 = 0.1$; and Figure b) $f_1 = 0.15$. For each choice, the human population is fixed ($N_{h1} = 300$) and the secondary (reservoir) host population is varied in size, where 'low' represents $N_{h2} = 100$, 'mid' represents $N_{h2} = 300$ and 'high' represents $N_{h2} = 1000$. Bites not occurring on human or reservoir hosts arise from non-reservoir animals. In all cases shown, $\varepsilon = 0.05$ so non-teneral susceptibility is low and most new tsetse infections occur in teneral flies. The black dots show where the human prevalence/incidence drops to 0%.

We note that a host prevalence of 20% or more is unnaturally high so it seems likely that $f_1 < 0.15$. Likewise, many other parameter choices including large f_2 or low N_{h2} are unlikely due to the high levels of host infection they create.

Endemic equilibria are highly sensitive to ratios of human to vectors and also humans to other hosts as well as the proportion of blood meals taken. As with all vector-borne disease models, the greater the ratio of vectors to hosts, the higher the disease prevalence/incidence. Likewise, Figure 15 demonstrates how the endemic prevalence may be increased dramatically (from 0% to 40%) by just a small increase in secondary host feeding preference. This causes problems in reconciling model outputs with field observations that measured prevalence levels are extremely low (around 1% as compared to the values shown here).

4.2 Teneral effect

In order to study the effect of no, partial or full susceptibility of non-teneral tsetse, we vary the parameter ε. At $\varepsilon = 0$ full 'immunity' is acquired upon the first blood meal, whereas at $\varepsilon = 1$, all tsetse have the same susceptibility (similar to a standard Ross—Macdonald model.

Assuming that Gambian HAT is not zoonotic (no nonhuman bites lead to onwards transmission) simulations (Figures 16(a) and (b)) show the relationship between ε and the equilibrium prevalence in humans. As intuition would dictate, prevalence in the human population increases as 'immunity' of non-teneral flies decreases. For a low proportion of bites upon humans, there are values of ε for which there is no positive endemic equilibrium, and values where one exists. As the feeding upon humans increases, even $\varepsilon = 0$ (no chance of infection after the first blood meal) gives endemic prevalence.

If reservoir hosts are included (Figures 16(c) and (d)), the results are similar, but indicate how large an increase in a secondary reservoir population could have on human prevalence.

Figure 16 shows that higher values of ε are unfeasible as host prevalence of 80% are never seen with any form of trypanosomiasis. From these results it would be expected that ε lies somewhere in the range 0—0.2. Likewise, since prevalence is usually very low, the results here indicate that $f_1 \geq 0.2$ is unrealistic, even if tsetse can only acquire infection upon their first blood meal.

If $f_1 \leq 0.1$ and there are no reservoir hosts, then there must be some non-teneral flies which become infected since the disease cannot persist otherwise. The higher f_2 is, the lower ε needs to be to generate the same level

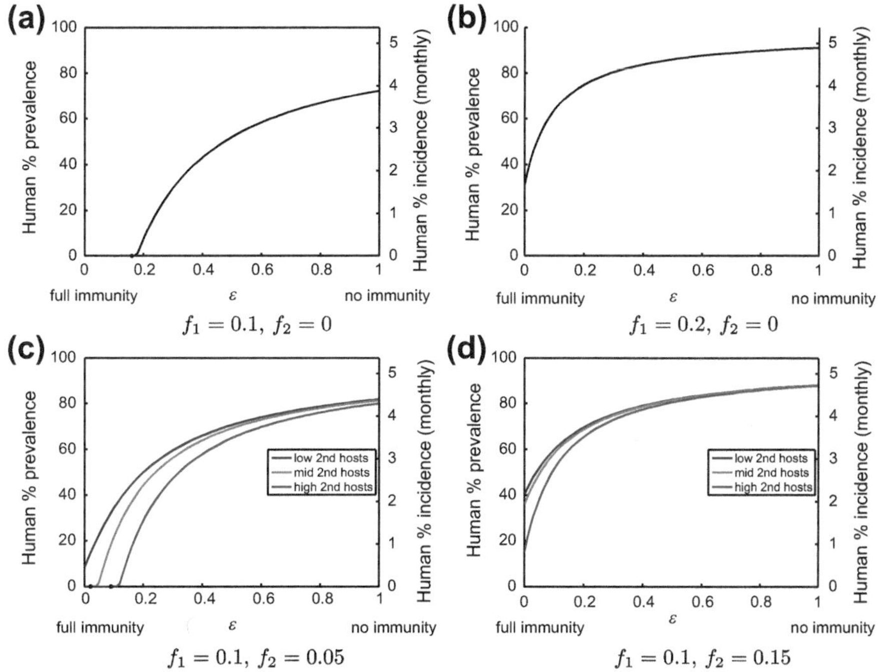

Figure 16 The relationship between the teneral effect parameter, ε, and human prevalence (or incidence) is shown for various choices of host preference. As in Figure 15, the remaining bites not taken on humans or secondary reservoir hosts occur on non-reservoir animals. In cases (a) and (b) there is no animal reservoir population, however only 10% of bites occur on humans in (a) compared to 20% in (b). In cases (c) and (d), a reservoir population is present and accounts for 5% and 15% of the blood-meals taken respectively. Finally it is noted that cases (a), (c) and (d) all have 10% of bites occurring on humans, however the figures show how the increase in bites taken upon reservoir hosts leads to an increase in human prevalence.

of infection in the human population. In order to generate realistic low levels of prevalence in humans, it seems most likely that there is some low probability of non-teneral flies becoming infected and that at least a small proportion of bites occur upon a reservoir host population.

5. MODELS OF CONTROL INTERVENTIONS AND THEIR COST EFFECTIVENESS

There are two main ways to control HAT: eliminate the infections in humans, or reduce tsetse fly densities and/or biting rates on humans. Tsetse densities and biting rates on humans can be reduced intentionally through vector control or a consequence of changes in the densities and availability of nonhuman hosts. An important operational question is therefore to decide which is the more effective strategy and whether and how this choice depend on local HAT epidemiology.

Artzrouni and Gouteux (Artzrouni and Gouteux, 1996b) directly addressed this key operational issue of whether elimination can best be achieved by case detection in humans or by reducing tsetse numbers by control programmes. They developed a Ross–Macdonald type model (as described above), derived the basic reproductive number, R_0, for the model and examined which type of control, human or tsetse, would most easily push R_0 below 1. Their methodology is discussed in more detail in Sections 5.1 and 5.2. They created an example showing that control measures depend on local epidemiology. In their model setting, Village A has high tsetse density but a small decrease in human duration of infectiousness would push HAT R_0 below 1. Conversely Village B has low tsetse numbers and a small decrease in their numbers would push HAT R_0 below 1. Their approach was to conclude that human screening and treatment would be favoured in Village A, while vector control would be more appropriate in Village B. There are likely to be other scenarios where both controls are necessary to control HAT and the optimal strategy is dependent upon parameterization.

We consider this an important paper because it explicitly compares the different strategies and, importantly, relates them to local epidemiology. However, it is important to note that it has several important limitations in both its calibration and its model implementation. Its parameterization took no account of how feasible it actually is to reduce the duration of human infectiousness compared to reducing tsetse density. Field data suggest tsetse number can be quickly reduced to very low levels so it is possible that this option is better even in villages where large reductions in tsetse numbers would be required to bring R_0 below 1. The problem inherent in its implementation is that it investigated equilibrium values with a simple target of $R_0 < 1$. In other words, it had no time component and could not distinguish strategies that eliminate the epidemic in 4 weeks, from a policy that eliminates it in 4 years. Similarly it did not distinguish between a policy that reduces R_0 to 0.9 and one that reduces it to 0.1; if 100 people are currently infected, the first scenario would result in 90 new infections, while the second would lead to only 10 new cases.

The important message is that time is essential when controlling an epidemic and a rapid reduction in incidence (i.e. the number of new infection per unit time) is the main objective of an intervention rather that its long-term equilibrium conditions. We would argue that tsetse control and human screen and treat operate on very different time scales. Field data suggest tsetse numbers can be reduced to low levels over a very short period while humans screen and treat programs take significant periods of time to implement and that changes in the proportion of humans infectious takes a much longer period to decline. In many practical ways, it is not an 'either/or' decision in an epidemic situation as it would be morally unacceptable to ignore humans infected in a HAT epidemic so human mass screen and treat will occur; we would argue that rapid deployment of focussed tsetse control be implemented as a priority because models suggest that changes in tsetse density and longevity of the magnitude achieved in previous interventions have the ability to rapidly reduce R_0 and hence incidence in the early stages on an epidemic. Vector control also has fewer logistical implications, fewer ethical implications (we cannot force people to be screened or to take the drugs), and appears to be much more cost-effective. These considerations are invariably absent from simple models and we would argue that there is a clear research gap between the relatively large number of basic models already constructed based on equilibrium conditions and the operational research questions posed by the control community which have a clear temporal element.

5.1 Detection and treatment of humans

The detection and treatment of cases in humans has been a primary control strategy for HAT. Cases can be detected through either large-scale screening programmes which occur periodically or via continuous (but smaller scale) screening at health care centres.

The simplest way to model continuous detection and treatment is to assume that this increases the exit rate, φ_{h1} from the infectious stage (I_{h1}), although the manner in which the intervention will affect this parameter must be considered. Artzrouni and Gouteux (Artzrouni and Gouteux, 1996b) formulate a model with a new parameter, C_h that is, the monthly percent detection of stage I cases. If the parameter φ_{h1}, the rate of movement of infectious individuals (stage I) to recovered individuals (stage II), is considered to be a composite of the intrinsic underlying disease progression and the removal rate by treatment (extrinsic) such that:

$$\varphi_{h1} = \varphi_{\text{int}} + \varphi_{\text{ext}} \tag{37}$$

then the monthly percent detection is given by:

$$C_h = 100[1 - \exp(-30\varphi_{\text{ext}})] \tag{38}$$

(Note that in this paper under their parameterization, rates are given with the unit of time taken as 3 days rather than 1). Consequentially, the exit rate from the class I_{h1} is given by:

$$\varphi_{h1} = \varphi_{\text{int}} - \frac{1}{30}\ln\left(1 - \frac{C_h}{100}\right) \tag{39}$$

Now that φ_{h1} is written as a function of the monthly percent detection, it can easily be seen that linear changes in detection do not produce linear changes in φ_{h1}.

This type of approach yields a meaningful way in which the parameter φ_{h1} may be controlled and the impact of such a control method upon both the disease dynamics or upon the basic reproductive ratio (R_0) can now be explored. Artzrouni and Gouteux (Artzrouni and Gouteux, 1996b) examine the effect of altering the monthly percent detection (in addition to another control) upon R_0 to find the threshold at which disease no longer occurs ($R_0 = 1$) in terms of this controllable parameter C_h.

We note, however, that this assumption of incorporating the treatment rate into φ_{hi} is only valid for analysis of R_0 and not for model simulations or equilibrium values because in Artzrouni and Gouteux's formulation, all humans in R_h, including the newly detected ones, experience an additional disease mortality rate (corresponding to stage II disease). A more consistent formulation would include an additional class of detected and treated humans.

5.2 Control of tsetse

Whilst there are several methods to control tsetse, such as aerial spraying and the deployment of natural or artificial baits, there are essentially three governing parameters for tsetse populations: the number of tsetse N_v, the birth rate B_v and the death rate μ_v. It is not hard to see that these are interlinked – if the birth rate were reduced, it would be expected that the population size would also fall. Likewise, controls which reduce the population size (such as trapping) essentially increase a tsetse's daily mortality.

Modelling work, in some tsetse control scenarios, has made the assumption that altering tsetse mortality only changes the flux through the system rather than the population size, in this case the birth rate would have to increase in order for the system to equilibrate at the same level as before or, more likely, there is some natural relocation of tsetse from other areas to retain the original population size.

Artzrouni and Gouteux (Artzrouni and Gouteux, 1996b) model tsetse control in this way; hypothesizing that tsetse controls will affect mortality but not population size. In a similar fashion to modelling treatment of humans, (described in Section 5.1) the vector death rate μ_v is considered to comprise of two parts: the underlying 'natural' mortality experienced by tsetse in their environment $\mu_{v,int}$ and an additional death rate, $\mu_{v,ext}$, imposed by some control strategy such that:

$$\mu_v = \mu_{v,int} + \mu_{v,ext} \tag{40}$$

This death rate is related to the daily percentage of flies killed, denoted here by C_v, by:

$$C_v = 100\left[1 - \exp\left(-\mu_{v,ext}\right)\right] \tag{41}$$

(in the paper, Artzrouni and Gouteux use rates with 3 days as the unit of time, so their equations account for this) and so:

$$\mu_v = \mu_{v,\mathrm{int}} - \ln\left(1 - \frac{C_v}{100}\right) \tag{42}$$

This, along with the corresponding formulation for φ_{h1}, were used by Artzrouni and Gouteux (Artzrouni and Gouteux, 1996b) to examine the expected change in R_0 as the two controllable parameters C_h (treatment of asymptomatic humans) and C_v (killing of tsetse) are varied.

Here similar results are reproduced with the same methodology, but with parameter values as defined in Table 7. Simulation results show that with this parameterization, when the proportion of bites on humans is low, $f_1 \leq 0.1$, R_0 is always less than 1. Figure 17 demonstrates how, as the percentage of bites taken upon humans is increased, increased control efforts are needed to eliminate HAT transmission. As f_1 becomes larger, the more unlikely it is that a single strategy alone will not be sufficient to control and eliminate HAT.

Other modelling work on control of tsetse flies has focused on the tsetse population alone. Hargrove (Hargrove, 2005) used stochastic branching process theory to predict the impact of control measures on tsetse populations. As tsetse population numbers become low, stochastic effects will dominate both population extinction and the timescale over which this may occur. Hargrove uses two key controllable parameters: adult (female) tsetse daily mortality (referred to as Δ but is similar to C_v) and the probability of a successfully mating with a fertile male, η. The latter

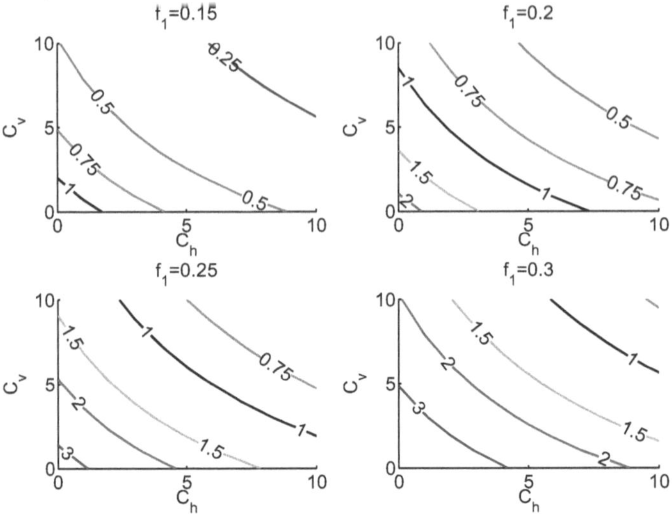

Figure 17 The plots show level sets of R_0 for various values above and below the $R_0 = 1$ threshold (shown in black). To the right of and above the $R_0 = 1$ contour, disease will die out, to the left of and below, there will be either an epidemic or endemic scenario. C_h is the monthly percentage of detection and treatment of humans. C_v is the daily percentage of tsetse killed.

may be controlled through the introduction of sterile males into the population, known as the SIT.

Results produced by Hargrove have indicated that if the probability of insemination is less than 10%, the population will become extinct. However, SIT is more costly than other methods such as trapping, so it likely to be more cost-effective to use trapping alone or in combination with SIT. Additionally, the increase in adult tsetse mortality due to trapping would reduce the time to extinction, even if it would eventually occur with SIT alone.

5.3 Paratransgenesis

Theoretical work has been done to investigate the potential use of tsetse symbionts for trypanosomiasis control based on paratransgenesis (Alam et al., 2011; Medlock et al., 2013). Paratransgenesis relies on the ability to genetically modify one of the tsetse symbionts to release an effector molecule that would inhibit trypanosomes within the vector. Genetic modification of *Sodalis* has been accomplished, and its distribution within the tsetse midgut would allow for the effector to be expressed in a relevant location where it can interact with the pathogen. One suggestion has been to use *W. pipientis* as a drive mechanism to allow for a population replacement strategy, which would hinge on the effector and drive mechanism not becoming disassociated over time. The potential for using *W. pipientis* as a drive mechanism was brought into focus when Alam et al. (Alam et al., 2011) revealed that *Wolbachia*, as in mosquitoes, induces cytoplasmic incompatibility in tsetse: eggs fertilized by a male harbouring a different strain of *Wolbachia* result in greatly reduced larval deposition. Further, *Wolbachia* was thought to result in a fitness benefit to tsetse, in addition to the benefits associated with *Wigglesworthia*.

The authors constructed a continuous time model of the proportion of *Wolbachia*-infected tsetse which led to the two major conclusions that *Wolbachia* is likely to reach fixation from any initial release ratio, due to the fitness benefits associated with the bacteria, and that the time to fixation is relatively short, a median of 529 days in their simulations. The latter result may depend on the assumption that females remain receptive to mates throughout life. Although remating is thought to be common in tsetse, mating with more than two to three males is unlikely, and remating may be most common in the first few days of life (Hargrove, 2012).

These issues were addressed in a follow-up study (Medlock et al., 2013), where age-structure was incorporated into the model in 10 day increments and mating assumed to occur within the first, with a proportion of females mating a second time. The spread of *Wolbachia* was hampered only if the females that had mated with an incompatible male remated. This population dynamics model was linked to a HAT transmission model, corresponding to that of Rogers (Rogers, 1988). The teneral period during which tsetse are susceptible was taken as 24 h. The results indicated that HAT prevalence would decrease at a rate comparable to the spread of the transgene and depended on the initial release ratio. Multiple tsetse species being present, while only one is targeted by the transgenic release was not found to be a major impediment to the interruption of transmission, unless the secondary species was present at a proportion greater than c.15%.

Although enticing as an additional tool in the arsenal against African trypanosomiasis, a key consideration will have to be whether such population replacement strategies can be more cost-effective than other control methods. Although our understanding of the microsymbionts of tsetse, their occurrence in field populations, impacts on tsetse susceptibility to trypanosomes and associated fitness costs or benefits, has developed greatly (though many questions remain) since the modelling study by Baker et al. (Baker et al., 1990), our understanding of the role of these symbionts on HAT epidemiology remains limited.

6. OUTLOOK

The many models summarized here have demonstrated the variety of ways that modelling HAT can be approached. For *T. b. gambiense* some of the key modelling themes have been the potential existence and effect of animal reservoir populations upon human disease prevalence and the impact of teneral susceptibility upon disease transmission. However, there are distinct gaps as would be expected for any neglected tropical disease. Some of the issues that need addressing in future work are highlighted here.

6.1 Transient disease dynamics

Most models of HAT published in the literature are derived and given as systems of ODEs (and sometimes DDEs) which may be solved to yield temporal disease dynamics, but temporal behaviour is rarely discussed. Instead, the primary focus of the literature has been the final endemic equilibrium state rather than changes in transmission over time or capturing epidemic dynamics.

Over 95% of human cases of HAT occur for Gambian HAT in foci where there are often small epidemics rather than a stable endemic situation, so it is important that models are able to encapsulate both sets of behaviour. A clear gap in the understanding of Gambian HAT is the generation of these foci and so studying the temporal dynamics rather than solely equilibria is crucial.

Even in endemic situations, if control strategies are executed, it is crucial to understand the timescale over which measures will take effect and for how long such controls will need to be implemented. Until this is known, it is impossible to determine the potential cost of an intervention. Fortunately, the type of models already developed provides a suitable starting point from which to hypothesize the impact and timescale of control upon disease dynamics. These temporal, rather than equilibrium, considerations are vital in choosing optimal control strategies as discussed above in Section 5.

6.2 Tsetse fly biting

At present, the impact of the frequency and timing of vector bites upon disease prevalence is not well understood. Most HAT models rely on the Ross–Macdonald type force of infection generated by using a mean time between bites and assuming these times are exponentially distributed.

Some work by Hargrove and Williams, however, addressed this problem by creating suitable models for patterns of tsetse feeding (Hargrove and Williams, 1995). Hargrove and Williams performed cost-benefit analysis by assuming only that tsetse must replicate themselves in order for the population to survive using the knowledge that around three blood meals are needed during pregnancy to produce a viable larva. Other work, based upon field data of tsetse caught in odour-baited traps, used the amount of blood in the tsetse's gut as a proxy for the time since last blood meal, by fitting to an ODE metabolism model (Hargrove and Packer, 1993).

These types of feeding models must now be integrated into disease models in order to discern the interaction of bite distribution and human disease prevalence and control.

6.3 Age-structure in tsetse flies

It has long been known that, like other vectors such as the mosquito, tsetse do not have constant mortality. Hargrove (Hargrove et al., 2011a) proposed that tsetse exhibit a 'U-shaped' mortality function of the form:

$$\mu_v(a) = k_1[k_2 \exp(-k_2 a) + k_3 \exp(k_3 a)] \tag{43}$$

where k_1, k_2, k_3 are constants and a is age. This function was formulated using mark-recapture data and was shown to be a good fit to wild flies (Hargrove et al., 2011a).

The impact of vector senescence upon disease transmission has been explored for the mosquito (Bellan, 2010), however, there are currently no such studies for the tsetse. Mortality patterns in mosquitoes have been found to increase with age and be described well by Gompertz functions (Clements and Paterson, 1981), or logistic functions (Styer et al., 2007). It is possible that under certain field situations, a higher mortality for newly emerged mosquitoes also occurs as it does for tsetse: for instance, Lyimo and Takken (Lyimo and Takken, 1993) found that the average size of newly emerged female *Anopheles gambiae* s.l. was significantly smaller than that of indoor-resting females, suggesting a higher mortality for smaller females in the days following emergence, while in nectar-deprived situations, early life mortality is increased (Stone et al., 2012), in which case Weibull functions described mortality patterns best. Bellan (Bellan, 2010) found that the basic 'constant mortality' assumption overestimated the impact of control strategies which reduces mosquito survival since constant mortality consistently overestimates vectorial capacity (Dietz, 1988; Styer et al., 2007). It is difficult to say if the same qualitative effects would be true for the tsetse as it has a different age-dependent mortality and, additionally for HAT, the teneral phenomenon needs to be incorporated. This highlights a need for more work in this area.

It is important to note the interchangeable use of the terms 'teneral', 'unfed' and 'susceptible' throughout the literature. These are not necessarily synonymous; it has been shown that tsetse susceptibility to trypanosome infection or the 'teneral effect' is correlated with age (Walshe et al., 2011). Susceptibility to trypanosomes has been shown to decrease with age upon emergence, however, if pushed to starvation, older tsetse can experience an increase in susceptibility. In order for models to truly capture the 'teneral phenomenon', the effects of age, feeding status and starvation upon susceptibility must first be disentangled. Future modelling work with age-structured tsetse populations would be capable of exploring waning susceptibility as a function of age rather than (or in addition to) whether they have fed or not.

Ultimately the way in which teneral susceptibility is modelled must be biologically motivated to generate a true picture of its effects upon disease transmission and prevalence, and here agent-based models which track the state of individual tsetse may be more appropriate.

6.4 Existence of foci and heterogeneity

Models are only as good as the assumption upon which they are based, which in turn reflect our knowledge of the underlying biology and epidemiology of the disease. There are two features of HAT epidemiology which are not fully explained and remain an area for future research: (1) low prevalence rates in hosts and tsetse flies and (2) spatially stable disease foci of the *gambiense* form.

HAT infection of *gambiense* is overwhelmingly restricted to about 300 disease foci that have been historically stable over the last 150 years. The underlying reasons for their existence and stability remain a matter of speculation. One putative explanation is the presence of an animal reservoir in these foci that sustain the disease in humans as supported by Funk et al. (Funk et al., 2013). Alternatively a more effective, cryptic vector species or subspecies may be present in these foci. The problem with both these suggestions is that the foci have remained stable despite the huge environmental changes that have occurred over the last 150 years and it seems a reasonable assumption that such widespread changes would have led to the elimination of some foci and the creation of some new ones. In general, foci are spatially stable with the notable exception of SE Uganda where movement of livestock has led to movement of foci (Fèvre et al., 2001). Additionally, many experienced field biologists have studied HAT in these foci and have not noted any environmental similarities between the foci that would support the contention of a local animal reservoir or tsetse subspecies.

The second epidemiological factor is the low prevalence of HAT in both the human and tsetse. It is generally necessary to dissect around 4000 tsetse before finding an infection. Molecular detection techniques are more sensitive, but even under these conditions, the prevalence of gambiense remains remarkably low. Infection rates in humans may occasionally go as high as 10% as noted in a very large epidemic in Uganda but generally the prevalence rates peak at around 1 or 2% in a human epidemic. There are several ways of reconciling the observations. There may be a large amount of heterogeneity in transmission such that most people are extremely unlikely to become infected while the highly exposed (and susceptible) subpopulation are all infected. Most infectious diseases are heterogeneous but the scale of the heterogeneity required to explain a prevalence of 1–2%, suggests that other factors may also be responsible. However, although most trypanosomiases models have included multiple host species, with a few exceptions (such as Muller et al. (Muller et al., 2004)), they have assumed homogeneous host populations within each species. This implies that each human is equally likely to get bitten and subsequently get infected and contract the disease and ignores the substantial heterogeneity in HAT transmission between villages and within villages. Also, no mathematical models have accounted for trypanotolerance, which was recently shown in humans (Jamonneau et al., 2012) where some proportion of the population can be infectious for multiple years before self-curing and returning to a susceptible stage.

Further model development, either through individual-based models, spatially explicit models or meta-population models are needed to understand the role of population heterogeneity and spatial structure in HAT transmission dynamics. This should lead to an improved understanding of the reasons for the existence and persistence of HAT transmission foci, may explain why mathematical models frequently overestimate prevalence of infection in humans and tsetse flies, and could point to more efficient strategies for HAT control and elimination.

6.5 Stochastic models

The majority of HAT models have been deterministic in nature. Exceptions are Muller et al. (Muller et al., 2004) who showed that prevalence in humans is sensitive to human densities and the initial number of infected flies, implying sensitivity to initial conditions; and Hargrove (Hargrove, 2005) who examined tsetse population extinctions using a stochastic framework but did not explicitly include disease.

As control programmes further reduce prevalence and elimination becomes the goal, the effects of individual level and low-probability events will become even more important. Stochastic models will become even more necessary to help map out a path towards elimination, and additionally to find explanations for localized disease take offs and extinctions in individual foci, which are characteristic of the Gambian form of HAT.

6.6 Towards predictive models of HAT

Mathematical models of infectious diseases come in a wide variety ranging from more general, tractable models that by necessity make vast simplifying assumptions, to more complex simulation-based models that are more closely tied to the biology of the system and possess greater predictive power, at the cost of analytical tractability. Most HAT models to date lean towards the more general models, often including simplifications such as assuming perfect mixing of vectors and hosts; using a prevalence-based rather than an infection burden-based approach to modelling infection; ignoring density dependence in parasite establishment or uptake; and assuming exponential exit rates for most, if not all, compartments. One area where these models have struggled is in capturing the incredibly low-prevalence rates in humans that are typically recorded in the field, where 1% prevalence is considered as very high. For *T. b. gambiense*, Rogers predicted an equilibrium prevalence of 7% in humans, 28.7% in nonhuman animals and 0.61% in tsetse (Rogers, 1988). Fitting this model to age—prevalence curves resulted in seemingly unrealistic estimates, for instance, the duration of infection was estimated as high as 58 years, and each human expected to receive a bite only once every 4 years (Rogers, 1989). Artzrouni and Gouteux (1996a) mention this as a justification for the development of their compartmental model. Although they, unlike Rogers, find that R_0 can be greater than 1 in the absence of animal reservoirs, the equilibrium prevalence predicted by their model is far higher than that of Rogers. The null-clines of their collapsed model when assuming there are 3000 tsetse biting 300 humans in a village are depicted in Figure 18 (redrawn from their paper), show equilibrium values at 0% for humans and tsetse, and 67% infection in humans and 1% in tsetse. Like Rogers, their estimate for the duration of infection was short at 4 months, which they based on the average duration during an epidemic in Niari, Congo. Adjusting this to the current

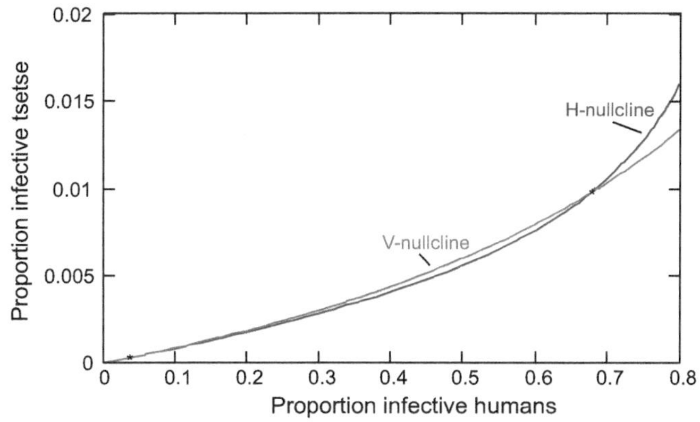

Figure 18 Nullclines for the simplified two-variable (infective humans and tsetse) model of Artzrouni and Gouteux (1996a), at a vector:human ratio of 10:1 and their specified parameter values. *(Redrawn from Artzrouni and Gouteux (1996a).)* The red (dark grey in print versions) line indicates values where $di_h/dt = 0$ and the green (light grey in print versions) line where $di_v/dt = 0$. There are three equilibrium points: trivial (0,0), unstable (0.039, 0.00032) and stable (0.67, 0.0097). This accentuates an issue with the predictive ability of HAT models lacking an animal reservoir, heterogeneity or other modifications, as the unstable equilibrium point at 3.9% prevalence in humans would be considered an epidemic situation with extremely high prevalence in reality.

(longer) estimates for the duration of gambiense HAT would shift the nontrivial equilibrium to even higher prevalence levels.

A later paper used a similar model but with different fly death rates, human population sizes, and allowed immigration of tsetse to predict ranges of prevalence of infection in humans between 13 and 48% (Artzrouni et al., 2001a). An agent-based model that incorporated a greater amount of biological detail nevertheless predicted prevalence ranging from 10 to 16% in simulations modelled after the Bipindi focus (Muller, Grebaut & Gouteux (Muller et al., 2004)), only dropping to 4–5% prevalence when assuming a short stage I duration of 4–12 months. Funk et al. (2013) (Funk et al., 2013), based on a susceptible-infected-susceptible model, found that under an assumption of random mixing they could only recreate realistic prevalence patterns (1.2% prevalence in humans) when allowing for animal reservoirs, in line with Rogers, 1988. However, they also showed that a human-vector only cycle without reservoir hosts was possible in their model under fairly strong heterogeneous exposure to bites.

Further investigations on the impact of the simplifying assumptions present in HAT models and model fitting and validation exercises appear the way forward, though a bottleneck may be the availability of high-quality field data on infection rates in humans, animals and vectors.

ACKNOWLEDGMENTS

The authors would like to thank Harriet Auty, Reto Brun, John Hargrove, Michael Lehane, Liam Morrison and Glyn Vale for helpful discussions and comments on the manuscript. KR was funded by an EPSRC PhD grant. CS and NC were supported by the Bill and Melinda Gates Foundation through grant number, OPP1037660. NC also acknowledges financial support from the Research and Policy for Infectious Disease Dynamics (RAPIDD) program of the Science & Technology Directorate, Department of Homeland Security and Fogarty International Center, National Institutes of Health. ST is supported by the Biotechnology and Biological Sciences Research Council, the Department for international Development, The Economic & Social Science Research Council, The Natural Environment Research Council and the Defence Science and Technology Laboratory, under the Zoonosis and Emerging and Livestock Systems (ZELS) programme (Grant no. BB/L019035/1).

REFERENCES

Akoda, K., Bossche, P.V.D., Lyaruu, E.A., Deken, R.D., 2009. Maturation of a *Trypanosoma brucei* infection to the infectious metacyclic stage is enhanced in nutritionally stressed tsetse flies. J. Med. Entomol. 6, 1446–1449.

Aksoy, S., 2011. Sleeping sickness elimination in sight: time to celebrate and reflect, but not relax. PLoS Negl. Trop. Dis 5 (2), e1008. http://dx.doi.org/10.1371/journal.pntd.0001008.

Alam, U., Medlock, J., Brelsfoard, C., Pais, R., Lohs, C., Balmand, S., Carnogursky, J., Heddi, A., Takac, P., Galvani, A.P., Aksoy, S., 2011. *Wolbachia* symbiont infections induce strong cytoplasmic incompatibility in the tsetse fly *Glossina morsitans*. PLoS Pathog. 7, e1002415.

Allsopp, R., Hursey, B.H., 2004. Insecticidal control of tsetse. In: Maudlin, I., Holmes, P.H., Miles, M.A. (Eds.), The Trypanosomiases. CABI, Wallingford.

Anderson, R.M., May, R.M., 1991. Infectious Diseases of Humans: Dynamics and Control. Oxford University Press, Oxford.

Apted, F.I.C., 1970. Clinical manifestations and diagnosis of sleeping sickness. In: Mulligan, H.W., Potts, W.H. (Eds.), The African Trypanosomiases. Wiley Interscience, London, pp. 645–660.

Aron, J.L., May, R.M., 1982. The population dynamics of malaria. In: Anderson, R.M. (Ed.), The Population Dynamics of Infectious Disease: Theory and Applications. Chapman and Hall, London, pp. 139–179.

Artzrouni, M., Gouteux, J.-P., 1996a. A compartmental model of sleeping sickness in Central Africa. J. Biol. Syst. 4 (4), 459–477.

Artzrouni, M., Gouteux, J.-P., 1996b. Control strategies for sleeping sickness in Central Africa: a model-based approach. Trop. Med. Int. Health 1 (6), 753–764.

Artzrouni, M., Gouteux, J.-P., 2001a. A model of Gambian sleeping sickness with open vector populations. Math. Med. Biol. 18 (2), 99–117.

Artzrouni, M., Gouteux, J.-P., 2001b. Population dynamics of sleeping sickness: a microsimulation. Simul. Gaming 32 (2), 215–227.

Artzrouni, M., Gouteux, J.-P., 2003. Estimating tsetse population parameters: application of a mathematical model with density-dependence. Med. Vet. Entomol. 17 (3), 272–279.

Artzrouni, M., Gouteux, J.-P., 2006. A parity-structured matrix model for tsetse populations. Math. Biosci. 204 (2), 215–231.

Ashcroft, M.T., Burtt, E., Fairbairn, H., 1959. The experimental infection of some African wild animals with *Trypanosoma rhodesiense, T. brucei* and *T. congolense*. Ann. Trop. Med. Parasitol. 53, 147–161.

Ashford, R.W., 1997. What it takes to be a reservoir host. Belg. J. Zool 127 (S1), 85–90.

Auty, H.K., Malele, I., Cleaveland, S., Raj, P. and Lembo, T. Host feeding preferences of tsetse (*Glossina swynnertoni* and *Glossina pallidipes*) in Serengeti National Park, Tanzania. (Unpublished data).

Auty, H.K., Picozzi, K., Malele, I., Torr, S.J., Cleaveland, S., Welburn, S.C., 2012. Using molecular data for epidemiological inference: assessing the prevalence of *Trypanosoma brucei rhodesiense* in tsetse in Serengeti, Tanzania. PLoS Negl. Trop. Dis. 6 (1), e1501.

Baker, R.D., Maudlin, I., Milligan, P.J.M., Molyneux, D.H., Welburn, S.C., 1990. The possible role of *Rickettsia*-like organisms in trypanosomiasis epidemiology. Parasitology 100 (2), 209–217.

Barclay, H.J., Hargrove, J.W., 2005. Probability models to facilitate a declaration of pest-free status, with special reference to tsetse (Diptera: Glossinidae). Bull. Entomol. Res. 95 (1), 1–11.

Barclay, H.J., Vreysen, M.J.B., 2011a. A dynamic population model for tsetse (Diptera: Glossinidae) area-wide integrated pest management. Popul. Ecol 53 (1), 89–110.

Barclay, H.J., Vreysen, M.J.B., 2011b. Conclusions from a dynamic population model for tsetse: response to comments. Popul. Ecol 53 (2), 417–420.

Barclay, H.J., Vreysen, M.J.B., 2013. The interaction of dispersal and control methods for the riverine tsetse fly *Glossina palpalis gambiensis* (Diptera: Glossinidae): a modelling study. Popul. Ecol 55 (1), 53–68.

Barrett, M.P., Vincent, I.M., Burchmore, R.J.S., Kazibwe, A.J.N., Matovu, E., September 2011. Drug resistance in human African trypanosomiasis. Future Microbiol. 6 (9), 1037–1047.

Barry, J.D., McCulloch, R., 2001. Antigenic variation in trypanosomes: enhanced phenotypic variation in a eukaryotic parasite. Adv. Parasitol. 49, 1–70.

Basáñez, M.G., Ricárdez-Esquinca, J., 2001. Models for the population biology and control of human onchocerciasis. Trends Parasitol 17, 430–438.

Baylis, M., Mbwabi, A.L., Stevenson, P., 1994. The feeding success of tsetse flies, *Glossina pallidipes* (Diptera: Glossinidae), on oxen treated with pyrethroid pour-ons at Galana Ranch, Kenya. Bull. Entomol. Res. 84 (4), 447–452.

Baylis, M., Mbwabi, A.L., 1995. Feeding behaviour of tsetse flies (*Glossina pallidipes austen*) on *Trypanosoma*-infected oxen in Kenya. Parasitology 110 (3), 297–305.

Baylis, M., Nambiro, C.O., May 1993. The effect of cattle infection by *Trypanosoma congolense* on the attraction, and feeding success, of the tsetse fly *Glossina pallidipes*. Parasitology 106 (4), 357–361.

Baylis, M.M., May 1997. The daily feeding rate of tsetse (Diptera: Glossinidae) on cattle at Galana Ranch, Kenya and comparison with trypanosomiasis incidence. Acta Trop. 65 (2), 81–96.

Bellan, S.E., 2010. The importance of age dependent mortality and the extrinsic incubation period in models of mosquito-borne disease transmission and control. PLoS One 5 (4), e10165.

Bouyer, J., Pruvot, M., Bengaly, Z., Guerin, P.M., Lancelot, R., 2007. Learning influences host choice in tsetse. Biol. Lett. 3 (2), 113–117.

Brun, R., Blum, J., Chappuis, R., Burri, C., 2010. Human african trypanosomiasis. Lancet 375 (9709), 148–159.

Brun, R., Don, R., Jacobs, R.T., Wang, M.Z., Barrett, M.P., 2011. Development of novel drugs for human African trypanosomiasis. Future Microbiol. 6 (6), 677–691.

Burri, C., Stich, A., Brun, R., 2004. Current chemotherapy of human African trypanosomiasis. In: Maudlin, I., Holmes, P.H., Miles, M.A. (Eds.), The Trypanosomiases. CABI, Wallingford, pp. 403–419.

Chalvet-Monfray, K., Artzrouni, M., Gouteux, J.-P., Auger, P., Sabatier, P., 1998. A two-patch model of Gambian sleeping sickness: application to vector control strategies in a village and plantations. Acta Biotheor. 46 (3), 207–222.

Checchi, F., Filipe, J.A.N., Barrett, M.P., Chandramohan, D., 2008. The natural progression of gambiense sleeping sickness: what is the evidence? PLoS Negl. Trop. Dis. 2 (12), e303.

Checchi, F., Filipe, J.A.N., Haydon, D.T., Chandramohan, D., Chappuis, F., 2008. Estimates of the duration of the early and late stage of gambiense sleeping sickness. BMC Infect. Dis. 8 (1), 16.

Clausen, P.-H., Adeyemi, I., Bauer, B., Breloeer, M., Salchow, F., Staak, C., 1998. Host preferences of tsetse (Diptera: Glossinidae) based on bloodmeal identifications. Med. Vet. Entomol. 12 (2), 169–180.

Clements, A.N., Paterson, G.D., 1981. The analysis of mortality and survival rates in wild populations of mosquitoes. J. Appl. Ecol. 18, 373–399.

Corson, J.F., June 1939. The infections produced in sheep and antelopes by a strain of *Trypanosoma rhodesiense*. Trans. R. Soc. Trop. Med. Hyg. 33 (1), 37–46.

Curtis, C.F., Jordan, A.M., 1970. Calculations of the productivity of *Glossina austeni* Newst. maintained on goats and on lop-eared rabbits. Bull. Entomol. Res. 59 (04), 651–658.

Dale, C., Welburn, S.C., Maudlin, I., Milligan, P.J.M., 1995. The kinetics of maturation of trypanosome infections in tsetse. Parasitology 111 (2), 187–191.

Davies, H., 1971. Further eradication of tsetse in the Chad and Gongola river systems in north-eastern Nigeria. J. Appl. Ecol. 8 (2), 563–578.

Davis, S., Aksoy, S., Galvani, A., April 2011. A global sensitivity analysis for African sleeping sickness. Parasitology 138 (4), 516–526.

Deborggraeve, S., Büscher, P., September 2012. Recent progress in molecular diagnosis of sleeping sickness. Expert Rev. Mol. Diagn. 12 (7), 719–730.

Diekmann, O., Heesterbeek, J.A.P., Metz, J.A.J., 1990. On the definition and the computation of the basic reproduction ratio R_0 in models for infectious diseases in heterogeneous populations. J. Math. Biol. 28 (4), 365–382.

Diekmann, O., Heesterbeek, J.A.P., Roberts, M.G., April 2010. The construction of next-generation matrices for compartmental epidemic models. J. R. Soc. Interface 7 (47), 873–885.

Dietz, K., 1988. Mathematical models for transmission and control of malaria. In: Wernsdorfer, W.H., McGregor, I. (Eds.), Malaria: Principles and Practice of Malariology, vol. 2. Churchill Livingstone, Edinburgh, pp. 1091–1133.

Dransfield, R.D., Brightwell, R., Kyorku, C., Williams, B., July 2009. Control of tsetse fly (Diptera: Glossinidae) populations using traps at Nguruman, south-west Kenya. Bull. Entomol. Res. 80 (03), 265.

Dwinger, R.H., Grootenhuis, J.G., Murray, M., Moloo, S.K., Gettinby, G., 1986. Susceptibility of buffaloes, cattle and goats to infection with different stocks of *Trypanosoma vivax* transmitted by *Glossina morsitans centralis*. Res. Vet. Sci. 41 (3), 307–315.

Dyer, N.A., Rose, C., Ejeh, N.O., Acosta-Serrano, A., April 2013. Flying tryps: survival and maturation of trypanosomes in tsetse flies. Trends Parasitol. 29 (4), 188–196.

Eisler, M.C., Torr, S.J., Coleman, P.G., Machila, N., Morton, J.F., July 2003. Integrated control of vector-borne diseases of livestock—pyrethroids: panacea or poison? Trends Parasitol. 19 (8), 341–345.

Fèvre, E.M., Coleman, P.G., Odiit, M., Magona, J.W., Welburn, S.C., Woolhouse, M.E., 2001. The origins of a new *Trypanosoma brucei rhodesiense* sleeping sickness outbreak in eastern Uganda. Lancet 358 (9282), 625–628.

Fèvre, E.M., Picozzi, K., Jannin, J.G., Welburn, S.C., Maudlin, I., January 2006. Human African trypanosomiasis: epidemiology and control. Adv. Parasitol. 61, 167–221.

Funk, S., Nishiura, H., Heesterbeek, H., Edmunds, W.J., Checchi, F., 2013. Identifying transmission cycles at the human-animal interface: the role of animal reservoirs in maintaining gambiense human African trypanosomiasis. PLoS Comput. Biol. 9 (1), e1002855.

Garrett-Jones, C., 1964. Prognosis for interruption of malaria transmission through assessment of the mosquito's vectorial capacity. Nature 204 (4964), 1173–1175.
Gibson, G., Torr, S.J., 1999. Visual and olfactory responses of haematophagous Diptera to host stimuli. Med. Vet. Entomol. 13 (1), 2–23.
Glasgow, J.P., 1961. The feeding habits of *Glossina swynnertoni Austen*. J. Anim. Ecol. 30, 77–85.
Glasgow, J.P., 1963. The Distribution and Abundance of Tsetse. Pergamon, Oxford, p. 241.
Grootenhuis, J.G., Dwinger, R.H., Dolan, R.B., Moloo, S.K., Murray, M., March 1990. Susceptibility of African buffalo and Boran cattle to *Trypanosoma congolense* transmitted by *Glossina morsitans centralis*. Vet. Parasitol. 35 (3), 219–231.
Habbema, J.D.F., van Oortmarssen, G.J., Plaisier, A.P., 1996. The ONCHOSIM model and its use in decision support for river blindness control. In: Isham, V., Medley, G. (Eds.), Models for Infectious Human Diseases: Their Structure and Relation to Data. Cambridge University Press, pp. 360–380.
Hackett, F., Berrang Ford, L., Fèvre, E., Simarro, P., 2014. Incorporating scale dependence in disease burden estimates: the case of human African trypanosomiasis in Uganda. PLoS Negl. Trop. Dis. 8 (2), e2704.
Haines, L.R., 2013. Examining the tsetse teneral phenomenon and permissiveness to trypanosome infection. Front. Cell. Infect. Microbiol. 3, 84.
Hanski, I.A., Gaggiotti, O.E. (Eds.), 2004. Ecology, Genetics and Evolution of Metapopulations: Standard Methods for Inventory and Monitoring. Elsevier Academic Press.
Hargrove, J.W., July 1975. Some changes in the flight apparatus of tsetse flies, *Glossina morsitans* and *G. pallidipes*, during maturation. J. Insect Physiol. 21 (8), 1485–1489.
Hargrove, J.W., 1976. Effect if human presence on the behaviour of tsetse (*Glossina* spp) (Diptera, Glossinidae) near a stationary ox. Bull. Entomol. Res. 66, 173–178.
Hargrove, J.W., 1981. Tsetse dispersal reconsidered. J. Anim. Ecol. 50, 351–373.
Hargrove, J.W., June 1988. Tsetse: the limits to population growth. Med. Vet. Entomol. 2 (3), 203–217.
Hargrove, J.W., 1995. Towards a general rule for estimating the stage of pregnancy in field-caught tsetse flies. Physiol. Entomol. 20, 213–223.
Hargrove, J.W., 1999. Reproductive abnormalities in tsetse flies in Zimbabwe. Entomol. Exp. Appl. 92 (1), 89–99.
Hargrove, J.W., 2000. A theoretical study of the invasion of cleared areas by tsetse flies (Diptera: Glossinidae). Bull. Entomol. Res. 90, 201–209.
Hargrove, J.W., 2003. Tsetse Eradication: Sufficiency, Necessity and Desirability. DFID Animal Health Programme, Centre for Tropical Veterinary Medicine.
Hargrove, J.W., 2004. Tsetse population dynamics. In: Maudlin, I., Holmes, P.H., Miles, M.A. (Eds.), The Trypanosomiases. CABI Publishing, Oxfordshire, UK, pp. 113–138.
Hargrove, J.W., 2005. Extinction probabilities and times to extinction for populations of tsetse flies Glossina spp. (Diptera: Glossinidae) subjected to various control measures. Bull. Entomol. Res. 95 (1), 13–21.
Hargrove, J.W., July 2012. Age-specific changes in sperm levels among female tsetse (*Glossina* spp.) with a model for the time course of insemination. Physiol. Entomol. 37 (3), 278–290.
Hargrove, J.W., Holloway, M.T.P., Vale, G.A., Gough, A.J.E., Hall, D.R., 1995. Catches of tsetse (*Glossina* spp.) (Diptera: Glossinidae) from traps and targets baited with large doses of natural and synthetic host odour. Bull. Entomol. Res. 85, 215–227.
Hargrove, J.W., Ouifki, R., Ameh, J.E., 2011a. A general model for mortality in adult tsetse (*Glossina* spp.). Med. Vet. Entomol. 25 (4), 385–394.
Hargrove, J.W., Ouifki, R., Kajunguri, D., Vale, G.A., Torr, S.J., May 2012. Modeling the control of trypanosomiasis using trypanocides or insecticide-treated livestock. PLoS Negl. Trop. Dis. 6 (5), e1615.

Hargrove, J.W., Packer, M.J., 1993. Nutritional states of male tsetse flies (*Glossina* spp.) (Diptera: Glossinidae) caught in odour-baited traps and artificial refuges: models for feeding and digestion. Bull. Entomol. Res. 83, 29–46.
Hargrove, J.W., Torr, S.J., Kindness, H.M., May 2003. Insecticide-treated cattle against tsetse (Diptera: Glossinidae): what governs success? Bull. Entomol. Res. 93 (3), 203–217.
Hargrove, J.W., Torr, S.J., Vale, G.A., 2011b. Comment on Barclay and Vreysen: published dynamic population model for tsetse cannot fit field data. Popul. Ecol 53 (2), 413–415.
Hargrove, J.W., Williams, B.G., 1995. A cost-benefit analysis of feeding in female tsetse. Med. Vet. Entomol. 9, 109–119.
Hargrove, J.W., Williams, B.G., 1998. Optimized simulation as an aid to modelling, with an application to the study of a population of tsetse flies, *Glossina morsitans morsitans* (Diptera: Glossinidae). Bull. Entomol. Res. 88 (04), 425–435.
Haydon, D.T., Cleaveland, S., Taylor, L.H., 2002. Identifying reservoirs of infection: a conceptual and practical challenge. Emerg. Infect. Dis 8 (12), 1468–1473.
Heaversedge, R.C., 1969. Levels of insect parasitism in puparia of glossina morsitans orientalis vanderplank (diptera) in rhodesia. Entomol. Soc. South. Afr. J. 32, 231–235.
Hoare, C.A., 1972. The Trypanosomes of Mammals. A Zoological Monograph. Blackwell Scientific Publications.
Hoppenheit, A., Bauer, B., Steuber, S., Terhalle, W., Diall, O., Zessin, K.-H., Clausen, P.-H., May 2013. Multiple host feeding in *Glossina palpalis gambiensis* and *Glossina tachinoides* in southeast Mali. Med. Vet. Entomol. 27 (2), 222–225.
Jackson, C.H.N., 1933. The causes and implications of hunger in tsetse flies. Bull. Entomol. Res. 24, 443–482.
Jackson, C.H.N., 1954. The hunger-cycles of *Glossina morsitans* Westwood and *G. swynnertoni* Austen. J. Anim. Ecol. 23, 368–372.
Jamonneau, V., Ilboudo, H., Kaboré, J., Kaba, D., Koffi, M., Solano, P., Garcia, A., Courtin, D., Laveissière, C., Lingue, K., Büscher, P., Bucheton, B., 2012. Untreated human infections by *Trypanosoma brucei gambiense* are not 100% fatal. PLoS Negl. Trop. Dis. 6 (6), e1691.
Jamonneau, V., Ravel, S., Koffi, M., Kaba, D., December 2004. Mixed infections of trypanosomes in tsetse and pigs and their epidemiological significance in a sleeping sickness focus of Cote d'Ivoire. Parasitology 129 (06), 693–702.
Jordan, A.M., January 1985. Tsetse eradication plans for southern Africa. Parasitol. Today 1 (5), 121–123.
Jordan, A.M., 1986. Trypanosomiasis Control and African Rural Development. Longman Group Ltd.
Jordan, A.M., Curtis, C.F., 1968. Productivity of *Glossina austeni* newst. maintained on lop-eared rabbits. Bull. Entomol. Res. 58 (02), 399–410.
Jordan, A.M., Curtis, C.F., 1972. Productivity of *Glossina morsitans morsitans* westwood maintained in the laboratory, with particular reference to the sterile-insect release method. Bull. World Health Organ. 46 (1), 33.
Kappmeier, K., Nevill, E.M., Bagnall, R.J., 1998. Review of tsetse flies and trypanosomosis in South African. Onderstepoort J. Vet. Res. 65, 195–203.
Kgori, P.M., Modo, S., Torr, S.J., 2006. The use of aerial spraying to eliminate tsetse from the Okavango Delta of Botswana. Acta Trop 99 (2–3), 184–199.
Kubi, C., van den Abbeele, J., de Deken, R., Marcotty, T., Dorny, P., van den Bossche, P., December 2006. The effect of starvation on the susceptibility of teneral and non-teneral tsetse flies to trypanosome infection. Med. Vet. Entomol. 20 (4), 388–392.
Langley, P.A., Wall, R., 1990. The implications of hunger in the tsetse-fly, *Glossina-pallidipes*, in relation to its availability to trapping techniques. J. Insect Physiol. 36, 903–908.
Leak, S.G.A., 1999. Tsetse Biology and Ecology: Their Role in the Epidemiology and Control of Trypanosomosis. CABI Publishing, p. 592.

Lyimo, E.O., Takken, W., 1993. Effects of adult body size on fecundity and the pre-gravid rate of *Anopheles gambiae* females in Tanzania. Med. Vet. Entomol. 7 (4), 328–332.

Macdonald, G., 1950. The analysis of infection rates in diseases in which superinfection occurs. Trop. Dis. Bull. 47, 907–915.

Macdonald, G., 1952. The analysis of the sporozoite rate. Trop. Dis. Bull. 49, 569–585.

Maudlin, I., December 2006. African trypanosomiasis. Ann. Trop. Med. Parasitol. 100 (8), 679–701.

Maudlin, I., Green, C.H., 1981. The potential for insecticide resistance in *Glossina* (Diptera: Glossinidae) – an investigation by computer simulation and chemical analysis. Bull. Entomol. Res. 71, 691–702.

Maudlin, I., Holmes, P.H., Miles, M.A. (Eds.), 2004. The Trypanosomiases. CABI, Wallingford.

Medlock, J., Atkins, K.E., Thomas, D.N., Aksoy, S., Galvani, A.P., 2013. Evaluating paratransgenesis as a potential control strategy for African trypanosomiasis. PLoS Negl. Trop. Dis. 7 (8), e2374.

Mehlitz, D., Zillmann, U., Scott, C.M., 1982. Epidemiological studies on the animal reservoir of gambiense sleeping sickness. Part III. Characterization of trypanozoon stocks by isoenzymes and sensitivity to human serum. Tropenmedizin Parasitol. 33 (2), 113–118.

Milligan, P.J.M., Baker, R.D., 1988. A model of tsetse-transmitted animal trypanosomiasis. Parasitology 96 (1), 211–239.

Mitashi, P., Hasker, E., Lejon, V., Kande, V., Muyembe, J.-J., Lutumba, P., Boelaert, M., November 2012. Human African trypanosomiasis diagnosis in first-line health services of endemic countries, a systematic review. PLoS Negl. Trop. Dis. 6 (11), e1919.

Moloo, S.K., Olubayo, R.O., Kabata, J.M., Okumu, I.O., 1992. A comparison of African buffalo, N'Dama and Boran cattle as reservoirs of Trypanosoma congolense for different Glossina species. Med. Vet. Entomol. 6 (3), 225–230.

Moloo, S.K., Orinda, G.O., Sabwa, C.L., Minja, S.H., Masake, R.A., 1999. Study on the sequential tsetse-transmitted *Trypanosoma congolense, T. brucei brucei* and *T. vivax* infections to African buffalo, eland, waterbuck, N'Dama and Boran cattle. Vet. Parasitol. 80 (3), 197–213.

Moloo, S.K., Sabwa, C.L., Baylis, M., 2000. Feeding behaviour of *Glossina pallidipes* and *G. morsitans centralis* on Boran cattle infected with *Trypanosoma congolense* or *T. vivax* under laboratory conditions. Med. Vet. Entomol. 14 (3), 290–299.

Moloo, S.K., Sabwa, C.L., Kabata, J.M., 1992. Vector competence of *Glossina pallidipes* and *G. morsitans centralis* for *Trypanosoma vivax, T. congolense* and *T. b. brucei*. Acta Trop. 51 (3–4), 271–280.

Moore, S., Shrestha, S., Tomlinson, K.W., Vuong, H., 2012. Predicting the effect of climate change on African trypanosomiasis: integrating epidemiology with parasite and vector biology. J. R. Soc. Interface 9 (70), 817–830.

Morrison, L.J., Majiwa, P., Read, A.F., Barry, J.D., 2005. Probabilistic order in antigenic variation of *Trypanosoma brucei*. Int. J. Parasitol. 35 (9), 961–972.

Mulla, A.F., Rickman, L.R., December 1988. How do African game animals control trypanosome infections? Parasitol. Today 4 (12), 352–354.

Muller, G., Grébaut, P., Gouteux, J.-P., 2004. An agent-based model of sleeping sickness: simulation trials of a forest focus in southern Cameroon. Comptes Rendus Biol. 327 (1), 1–11.

Murray, M., January 1989. Factors affecting duration and intensity of trypanosome infection of domestic animals. Ann. Soc. Belg Med. Trop. 69, 189–196.

Nishiura, H., Hoye, B., Klaassen, M., Bauer, S., 2009. How to find natural reservoir hosts from endemic prevalence in a multi-host population: a case study of influenza in waterfowl. Epidemics 1 (2), 118–128.

Njagu, Z., Mihok, S., Kokwaro, E., Verloo, D., 1999. Isolation of it *Trypanosoma brucei* from the monitor lizard (*Varanus niloticus*) in an endemic focus of Rhodesian sleeping sickness in Kenya. Acta Trop. 72 (2), 137–148.

Njiokou, F., Laveissière, C., Simo, G., Nkinin, S.W., 2006. Wild fauna as a probable animal reservoir for *Trypanosoma brucei gambiense* in Cameroon. Infection 6 (2), 147–153.
Nsoesie, E.O., Brownstein, J.S., Ramakrishnan, N., Marathe, M.V., 2014. A systematic review of studies on forecasting the dynamics of influenza outbreaks. Influenza Other Respir. Viruses 8 (3), 309–316.
Omolo, M.O., Hassanali, A., Mpiana, S., Esterhuizen, J., Lindh, J., Lehane, M.J., Solano, P., Rayaisse, J.B., Vale, G.A., Torr, S.J., Tirados, I., January 2009. Prospects for developing odour baits to control *Glossina fuscipes* spp., the major vector of human African trypanosomiasis. PLoS Negl. Trop. Dis. 3 (5), e435.
Peacock, L., Cook, S., Ferris, V., Bailey, M., Gibson, W., January 2012. The life cycle of *Trypanosoma (Nannomonas) congolense* in the tsetse fly. Parasit. Vectors 5 (1), 109.
Peck, S.L., Bouyer, J., 2012. Mathematical modeling, spatial complexity, and critical decisions in tsetse control. J. Econ. Entomol. 105 (5), 1477–1486.
Phelps, R.J., Burrows, P.M., 1969a. Lethal temperatures for puparia of *glossina morsitans orientalis*. Entomol. Exp. Appl. 12 (1), 23–32.
Phelps, R.J., Burrows, P.M., 1969b. Prediction of the pupal duration of *Glossina morsitans orientalis Vanderplank* under field conditions. J. Appl. Ecol 57, 227–257.
Phelps, R.J., Burrows, P.M., April 1969c. Puparial duration in *Glossina morsitans orientalis* under conditions of constant temperature. Entomol. Exp. Appl. 12 (1), 33–43.
Phelps, R.J., Clarke, G.P.Y., 1974. Seasonal elimination of some size classes in males of *Glossina morsitans morsitans* Westw. (Diptera, Glossinidae). Bull. Entomol. Res. 64 (02), 313–324.
Phelps, R.J., Vale, G.A., 1978. Studies on populations of *Glossina morsitans morsitans* and *G. pallidipes* (Diptera: Glossinidae) in Rhodesia. J. Appl. Ecol 15, 743–760.
Randolph, S.E., Rogers, D.J., 1978. Feeding cycles and flight activity in field populations of tsetse (Diptera: Glossinidae). Bull. Entomol. Res. 68 (04), 655.
Randolph, S.E., Rogers, D.J., Dransfield, R.D., Brightwell, R., 1991. Trap-catches, nutritional condition and the timing of activity of the tsetse fly *Glossina longipennis* (Diptera: Glossinidae). Bull. Entomol. Res 81, 455–464.
Randolph, S.E., Williams, B.G., Rogers, D.J., Connor, H., 1992. Modelling the effect of feeding-related mortality on the feeding strategy of tsetse (Diptera: Glossinidae). Med. Vet. Entomol. 6 (3), 231–240.
Ravel, S., Patrel, D., Koffi, M., Jamonneau, V., Cuny, G., 2006. Cyclical transmission of *Trypanosoma brucei gambiense* in *Glossina palpalis gambiensis* displays great differences among field isolates. Acta Trop 100 (1–2), 151–155.
Rayaisse, J.B., Tirados, I., Kaba, D., Dewhirst, S.Y., Logan, J.G., Diarrassouba, A., Salou, E., Omolo, M.O., Solano, P., Lehane, M.J., Pickett, J.A., Vale, G.A., Torr, S.J., Esterhuizen, J., 2010. Prospects for the development of odour baits to control the tsetse flies *Glossina tachinoides* and it G. palpalis s.l. PLoS Negl. Trop. Dis. 4 (3), e632.
Reiner, R.C., Perkins, T.A., Barker, C.M., Niu, T., Chaves, L.F., Ellis, A.M., George, D.B., Le Menach, A., Pulliam, J.R.C., Bisanzio, D., Buckee, C., Chiyaka, C., Cummings, D.A.T., Garcia, A.J., Gatton, M.L., Gething, P.W., Hartley, D.M., Johnston, G., Klein, E.Y., Michael, E., Lindsay, S.W., Lloyd, A.L., Pigott, D.M., Reisen, W.K., Ruktanonchai, N., Singh, B.K., Tatem, A.J., Kitron, U., Hay, S.I., Scott, T.W., Smith, D.L., 2013. A systematic review of mathematical models of mosquito-borne pathogen transmission: 1970–2010. J. R. Soc. Interface 10 (81), 20120921.
Roberts, M.G., Heesterbeek, J.A.P., 2003. A new method for estimating the effort required to control an infectious disease. Proc. R. Soc. B: Biol. Sci. 270 (1522), 1359–1364.
Rogers, D., 1977. Study of a natural population of *Glossina fuscipes fuscipes* Newstead and a model of fly movement. J. Anim. Ecol 46, 309–330.
Rogers, D.J., August 1988. A general model for the African trypanosomiases. Parasitology 97 (1), 193–212.

Rogers, D.J., 1989. The development of analytical models for human trypanosomiasis. Ann. Soc. Belg. Méd. Trop. 69 (Suppl. 1), 73–88.
Rogers, D.J., 1990. A general model for tsetse populations. Insect Sci. Its Appl. 11 (3), 331–346.
Rogers, D.J., Randolph, S.E., 1984. Local variation in the population dynamics of *Glossina palpalis palpalis* (Robineau-Desvoidy) (Diptera: Glossinidae). I. Natural population regulation. Bull. Entomol. Res 74 (3), 403–423.
Rogers, D.J., Randolph, S.E., 1986. Detection of activity cycles of tsetse from capture–recapture data. Ecol. Entomol 11, 95–109.
Rogers, D.J., Randolph, S.E., April 1990. Estimation of rates of predation on tsetse. Med. Vet. Entomol. 4 (2), 195–204.
Ross, R., 1911. The Prevention of Malaria, second ed. John Murray, London.
Ross, R., February 1916. An application of the theory of probabilities to the study of a priori pathometry. Part I. Proc. R. Soc. A: Math. Phys. Eng. Sci. 92 (638), 204–230.
Rurangirwa, F.R., Musoke, A.J., Nantulya, V.M., Nkonge, C.C., Njuguna, L.L., Mushi, E.Z., Karstad, L.L., Grootenhuis, J.J., 1986. Immune effector mechanisms involved in the control of parasitaemia in *Trypanosoma brucei*-infected wildebeest (*Connochaetes taurinus*). Immunology 58 (2), 231–237.
Shaw, A.P.M., Torr, S.J., Waiswa, C., Cecchi, G., Wint, G.R.W., Mattioli, R.C., Robinson, T.P., June 2013. Estimating the costs of tsetse control options: an example for Uganda. Prev. Vet. Med. 110 (3–4), 290–303.
Simarro, P.P., Cecchi, G., R Franco, J., Paone, M., Diarra, A., Ruiz-Postigo, J.A., èvre, E.M.F., Mattioli, R.C., Jannin, J.G., January 2012. Estimating and mapping the population at risk of sleeping sickness. PLoS Negl. Trop. Dis. 6 (10), e1859.
Simarro, P.P., Jannin, J.G., Cattand, P., 2008. Eliminating human African trypanosomiasis: where do we stand and what comes next? PLoS Med. 5 (2), e55.
Simo, G., Asonganyi, T., Nkinin, S.W., Njiokou, F., Herder, S., June 2006. High prevalence of *Trypanosoma brucei gambiense* group 1 in pigs from the Fontem sleeping sickness focus in Cameroon. Vet. Parasitol. 139 (1–3), 57–66.
Smith, D.L., Battle, K.E., Hay, S.I., Barker, C.M., Scott, T.W., McKenzie, F.E., 2012. Ross, Macdonald, and a theory for the dynamics and control of mosquito-transmitted pathogens. PLoS Pathog. 8 (4), e1002588.
Solano, P., Torr, S.J., Lehane, M.J., 2013. Is vector control needed to eliminate gambiense human African trypanosomiasis? Front. Cell. Infect. Microbiol. 3.
Steverding, D., February 2008. The history of African trypanosomiasis. Parasit. Vectors 1 (1).
Stone, C.M., Jackson, B.T., Foster, W.A., October 2012. Effects of plant-community composition on the vectorial capacity and fitness of the malaria mosquito *Anopheles gambiae*. Am. J. Trop. Med. Hyg. 87 (4), 727–736.
Styer, L.M., Carey, J.R., Wang, J.L., 2007. Mosquitoes do senesce: departure from the paradigm of constant mortality. Am. Soc. Trop. Med. Hyg. 76 (1), 111–117.
Du Toit, R., 1954. Trypanosomiasis in Zululand and the control of tsetse flies by chemical means. Onderstepoort J. Vet. Res. 26, 317–387.
Torr, S., Vale, G., Mangwiro, C. Laveissiere, C., Hall, D., Morton, J., van Munster, B., Hargrove, J. HAT Trick. website http://www.tsetse.org/trick/ (accessed 15.05.14.).
The malERA Consultative Group on Modeling, 2011. A research agenda for malaria eradication: modeling. PLoS Med. 8 (1), e1000403.
Torr, S.J., July 2009. Responses of tsetse flies (Diptera: Glossinidae) to warthog (*Phacochoerus aethiopicus* Pallas). Bull. Entomol. Res. 84 (03), 411–419.
Torr, S.J., Chamisa, A., Clemen Mangwiro, T.N., Vale, G.A., January 2012. Where, when and why do tsetse contact humans? Answers from studies in a national park of Zimbabwe. PLoS Negl. Trop. Dis. 6 (8), e1791.
Torr, S.J., Hargrove, J.W., March 1998. Factors affecting the landing and feeding responses of the tsetse fly *Glossina pallidipes* to a stationary ox. Med. Vet. Entomol. 12 (2), 196–207.

Torr, S.J., Hargrove, J.W., Vale, G.A., November 2005. Towards a rational policy for dealing with tsetse. Trends Parasitol. 21 (11), 537–541.
Torr, S.J., Maudlin, I., Vale, G.A., March 2007. Less is more: restricted application of insecticide to cattle to improve the cost and efficacy of tsetse control. Med. Vet. Entomol. 21 (1), 53–64.
Torr, S.J., Prior, A., Wilson, P.J., Schofield, S., December 2007. Is there safety in numbers? the effect of cattle herding on biting risk from tsetse flies. Med. Vet. Entomol. 21 (4), 301–311.
Torr, S.J., Solano, P., 2010. Olfaction in Glossina–host interactions: a tale of two tsetse. In: Olfaction in Vector-host Interactions.
Torr, S.J., Vale, G.A., October 2011. Is the even distribution of insecticide-treated cattle essential for tsetse control? Modelling the impact of baits in heterogeneous environments. PLoS Negl. Trop. Dis. 5 (10), e1360.
Turner, D.A., Golder, T.K., 1986. Susceptibility to topical applications of endosulfan and dieldin in sprayed and unsprayed populations of *Glossina pallidipes* Austen (Diptera: Glossinidae) in Kenya. Bull. Entomol. Res. 76, 519–527.
Vale, G.A., 1974. The responses of tsetse flies (Diptera: Glossinidae) to mobile and stationary baits. Bull. Entomol. Res 64, 545–588.
Vale, G.A., 1977. Feeding responses of tsetse flies (Diptera: Glossinidae) to stationary hosts. Bull. Entomol. Res 67, 635–649.
Vale, G.A., Hargrove, J.W., Solano, P., Courtin, F., Rayaisse, J.-B., et al., 2014. Explaining the host-finding behavior of blood-sucking insects: Computerized simulation of the effects of habitat geometry on tsetse fly movement. PLoS Negl. Trop. Dis. 8 (6), e2901. http://dx.doi.org/10.1371/journal.pntd.0002901.
Vale, G.A., Lovemore, D.F., Flint, S., Cockbill, G.F., March 1988. Odour-baited targets to control tsetse flies, *Glossina* spp. (Diptera: Glossinidae), in Zimbabwe. Bull. Entomol. Res. 78 (01), 31–49.
Vale, G.A., Torr, S.J., 2004. Development of bait technology to control tsetse. In: The Trypanosomiases. CABI, Wallingford, pp. 509–523.
Vale, G.A., Torr, S.J., August 2005. User-friendly models of the costs and efficacy of tsetse control: application to sterilizing and insecticidal techniques. Med. Vet. Entomol. 19 (3), 293–305.
Vreysen, M.J., Saleh, K.M., Ali, M.Y., Abdulla, A.M., Zhu, Z.R., Juma, K.G., Dyck, V.A., Msangi, A.R., Mkonyi, P.A., Feldmann, H.U., February 2000. Glossina austeni (Diptera: Glossinidae) eradicated on the island of Unguja, Zanzibar, using the sterile insect technique. J. Econ. Entomol. 93 (1), 123–135.
Walshe, D.P., Lehane, M.J., Haines, L.R., November 2011. Post eclosion age predicts the prevalence of midgut trypanosome infections in *Glossina*. PLoS One 6 (11), e26984.
Walshe, D.P., Ooi, C.P., Lehane, M.J., Haines, L.R., 2009. The enemy within: Interactions between tsetse, trypanosomes and symbionts. In: Advances in Insect Physiology. Academic Press, pp. 119–175.
Welburn, S.C., Arnold, K., Maudlin, I., Gooday, G.W., 1993. Rickettsia-like organisms and chitinase production in relation to transmission of trypanosomes by tsetse flies. Parasitology 107, 141–145. http://dx.doi.org/10.1017/S003118200006724X.
Welburn, S.C., Maudlin, I., October 1999. Tsetse–trypanosome interactions: rites of passage. Parasitol. Today 15 (10), 399–403.
Welburn, S.C., Maudlin, I., January 2012. Priorities for the elimination of sleeping sickness. Adv. Parasitol. 79, 299–337.
Welburn, S.C., Maudlin, I., Milligan, P.J.M., 1995. Trypanozoon: infectivity to humans is linked to reduced transmissibility in tsetse I. Comparison of human serum-resistant and human serum-sensitive field isolates. Exp. Parasitol 81 (3), 404–408.

Welburn, S.C., Picozzi, K., Coleman, P.G., Packer, C., January 2008. Patterns in age-seroprevalence consistent with acquired immunity against *Trypanosoma brucei* in Serengeti lions. PLoS Negl. Trop. Dis. 2 (12), e347.
Willemse, L.P.M., 1991. A trial of odour baited targets to control the tsetse fly, Glossina morsitans centralis (Diptera: Glossinidae) in west Zambia. Bull. Entomol. Res. 81 (3), 351–357.
Williams, B., Dransfield, R., Brightwell, R., 1992. The control of tsetse flies in relation to fly movement and trapping efficiency. J. Appl. Ecol. 29, 163–179.
Wombou Toukam, C.M., Solano, P., Bengaly, Z., Jamonneau, V., Bucheton, B., November 2011. Experimental evaluation of xenodiagnosis to detect trypanosomes at low parasitaemia levels in infected hosts. Parasite 18 (4), 295–302.
World Health Organization, 2012. Accelerating Work to Overcome the Global Impact of Neglected Tropical Diseases: A Roadmap for Implementation. WHO/HTM/NTD/2012.2.
WHO, November 2013. Control and Surveillance of Human African Trypanosomiasis. Technical Report 984.
World Health Organization, 2013. Control and surveillance of human African trypanosomiasis. In: World Health Organization Technical Report Series.

CHAPTER FOUR

Ecology, Evolution and Control of Chagas Disease: A Century of Neglected Modelling and a Promising Future

Pierre Nouvellet*, Zulma M. Cucunubá[†,§], Sébastien Gourbière[¶,1]
*Medical Research Council Centre for Outbreak Analysis and Modelling, Department of Infectious Disease Epidemiology, Imperial College London, London, UK
[†]Grupo de Parasitología, Instituto Nacional de Salud, Colombia
[§]Department of Infectious Disease Epidemiology, Imperial College London, London, UK
[¶]Institut de Modélisation et d'Analyse en Géo-Environnements et Santé (IMAGES), Université de Perpignan Via Domitia, Perpignan, France
[1]Corresponding author: E-mail: gourbiere@univ-perp.fr

Contents

Advances in Parasitology, Volume 87
ISSN 0065-308X
http://dx.doi.org/10.1016/bs.apar.2014.12.004

Abstract

More than 100 years after its formal description, Chagas disease remains a major public health concern in Latin America with a yearly burden of 430,000 Disability-Adjusted Life Years (DALYs). The aetiological agent, a protozoan named *Trypanosoma cruzi*, is mainly transmitted to mammalian hosts by triatomine vectors. Multiple species of mammals and triatomines can harbour and transmit the parasite, and the feeding range of triatomine species typically includes many noncompetent hosts. Furthermore, the transmission of the pathogen can occur via several routes including the typical vector's faeces, but also oral, congenital and blood transfusion routes. These ecological and epidemiological complexities of the disease have hindered many control initiatives. In such a context, mathematical models provide invaluable tools to explore and understand the dynamics of *T. cruzi* transmission, and to design, optimize and monitor the efficacy of control interventions. We intend here to provide the first review of the mathematical models of Chagas disease, focussing on how they have contributed to our understanding of (1) the population dynamics and control of triatomine vectors, and (2) the epidemiology of *T. cruzi* infections. We also aim at suggesting promising lines of modelling that could further improve our understanding of the ecology, evolution, and control of the disease.

1. INTRODUCTION

The American trypanosomiasis is a 'Neglected Tropical Disease' (NTD) endemic in many areas of Latin American countries (WHO, 2014a). The pathogen, *Trypanosoma cruzi*, is a generalist protozoan parasite able to infect a broad range of mammalian hosts that include wild and domesticated animals (Orozco et al., 2013; Rocha et al., 2013) as well as human beings. The main mode of transmission is by contact with the faeces of haematophageous bugs of the Triatominae subfamily (Hemiptera: Reduviidae) (Gourbière et al., 2012; Wood, 1951). However, congenital (Howard et al., 2014) and oral transmissions (Shikanai-Yasuda and Carvalho, 2012) also contribute to the spread of the pathogen to both human and nonhuman hosts (Rabinovich et al., 2001) and blood transfusions are responsible for additional infections in and out of the area of endemicity (Wendel and Gonzaga, 1993). Around 7–8 million people are estimated to be infected worldwide, most of them living in Central and South America. In these parts of the world, the disease remains as a serious public health concern with vast areas of active transmission and a high potential of emergence and re-emergence in unstable socioeconomic contexts (WHO, 2014b). Incidence have been reduced in places where campaigns of vector control and the screening of blood transfusions have been undertaken consistently (Dias et al., 2002; Schofield et al., 2006), demonstrating the potential efficacy of active and sustained public health policies. However, *T. cruzi* still causes 10,000 deaths and the loss of 430,000 Disability-Adjusted Life Years (DALYs) every year (Lee et al., 2013).

The ecology and epidemiology of the disease have been considered together from the very early stage of its discovery. A century ago, Carlos Chagas first identified and named, after his mentor Oswaldo Cruz, the pathogen *T. cruzi*. Remarkably, he simultaneously described the whole transmission cycle, including mammal hosts and triatomine vectors, and the clinical implications of the disease (Chagas, 1909; Goldbaum and Barreto, 2008), thereby setting the stage for today's 'Eco-health' approach of Chagas' disease (Boischio et al., 2009; Briceño-León, 2009; Lucero et al., 2013). This has led to a large amount of entomological field studies to assess the temporal and spatial distributions of the over 40 species of triatomines able to transmit *T. cruzi*, and to relate them to various ecological and socioeconomic factors to better understand the local dynamic of transmission (e.g. Dumonteil et al., 2013) and to identify geographic areas at high risk of transmission (Guhl et al., 2005). Those scientific achievements have contributed to the design of national and international control initiatives that have reduced the impact of the disease on populations of Latin America over the last 30 years (Dias et al., 2002). In many aspects, the ecoepidemiological knowledge of the transmission of Chagas disease has thus developed through field and experimental approaches similar to what was contemporaneously achieved for other human vector-borne diseases.

Field and experimental studies have always been the two most prominent approaches in biology as they measure the actual components of biological systems. Mathematical modelling has proved to be a very informative 'third road' in many fields of biology. Mathematical models are defined here as sets of equations or computing rules that represent a biological system through a specific quantitative description of the processes driving its dynamical changes. As such, those models are different from statistical models that are meant to analyse data sets using functions such as linear or quadratic relationships. Accordingly, mathematical models are key research tools when one is interested in investigating the link between biological processes and spatial or temporal patterns in the status of a biological system, e.g. the number of individuals in a population. They can contribute to (1) identifying the natural conditions that allow reproducing observed trends, (2) predicting future variations, and (3) anticipating the impact of human activities on biological systems. The high potential of this approach has been used to obtain a better understanding and control of various communicable diseases and these contributions have been reviewed for several of them such as malaria (Mandal et al., 2011), tuberculosis (Aparicio and Castillo-Chavez, 2009) or dengue (Andraud et al., 2012). We aim here at producing a first and comprehensive review of the mathematical models of Chagas disease, focussing on how they have contributed to our understanding of (1) the population dynamics and control of triatomine vectors, and (2) the epidemiology of *T. cruzi* infections. Meanwhile, we intend to provide a simple introduction to what types of modelling can be used in these fields, so that a broader range of scientists working on Chagas disease can incorporate modelling into their practice. We screened the modelling literature on Chagas disease in PubMed and the Web of Science removing (1) papers that we could not assess because they are published in journals with no online access, and (2) mathematical papers that required a too strong mathematical background and were not informed by data. The results of the search conducted here most likely constitute the core literature on the modelling of Chagas disease that is likely to effectively help improving our understanding of the disease transmission. Finally, this chapter aims at suggesting promising lines of modelling that could further improve current understanding of the ecology, evolution and control of the disease.

2. MODELS OF TRIATOMINES POPULATION DYNAMICS AND CONTROL

In endemic countries, vector transmission remains the main route of *T. cruzi* infections in humans. Substantial research is focussed on the spatial and/or temporal distribution of triatomines with the ultimate objective of reducing transmission through rational and (ideally) locally adapted vector-control programmes. Mathematical models can then provide highly valuable tools to describe, understand and predict the presence and/or abundance of vectors, as well as to estimate the potential effect of various vector control strategies. Two main modelling approaches have been used in this context. First, ecological niche modelling (Costa and Peterson, 2012) that allows the identification of environmental factors associated to triatomine species presence—absence and proceeds by such clues to map their distribution (Costa et al., 2002; Gurgel-Gonçalves et al., 2012). This literature falls outside of the scope of this review, as those models typically are statistical ones. The second approach consists of mathematical models that link the processes of vector survival, reproduction and dispersal to the spatial and/or temporal variation in their abundance or presence–absence in order to better describe and understand its key determinants.

2.1 Modelling survival and reproduction

Dynamical models of vector abundance account for at least two basic processes: survival and reproduction.

Under the simplest assumptions these two processes are described by vital rates (i.e. survival and fertility) that (1) are the same for all vector individuals, (2) do not vary with time or in space, and (3) do not depend on the size of the population. The resulting 'basic' model (Box 1**(A)**) is well-known to produce exponential decrease or increase of the size of the population (Figure 1(a)) because it does not account for any density-dependent regulatory processes or compensatory dispersal. This sets an obvious difference between populations that have the potential to sustain themselves and those that do not, i.e., 'source' and 'sink' populations, respectively. While the above simplifying assumptions are commonly used in epidemiological models of *T. cruzi* transmission (see 'Epidemiological model of *T. cruzi* transmission'), all published models focussing on the variation in vector abundance account for differences of vital rates between individuals according to their stage of development (Table 1).

When vital rates depend on the stage of development of individuals, the dynamics of the vector population is typically investigated using 'matrix' population models (Box 1**(B)**). These models also lead to an exponential decrease or increase in abundance because vital rates are still assumed not to vary with population size. As with the 'basic' model, they permit identification of 'sink' and 'source' populations, but with respect to all stage-specific rates of survival, reproduction and development (Box 1**(B)**). When applied to domestic populations of triatomines, matrix models thus offer simple quantitative tools to define and identify 'domiciliated' and 'nondomiciliated' populations (Gourbière et al., 2008). Additionally, they facilitate prediction of the expected age/stage structure in a focal population, which typically converges towards a stable distribution (Figure 1(b), Caswell, 2000). This mathematical formalism is directly related to key population statistics that can be calculated from standard life-tables, such as the finite population growth rate, generation time or life-time reproductive value (see Caswell, 2000; for an in-depth

Box 1 Triatomines Population Modelling

A

Basic model:

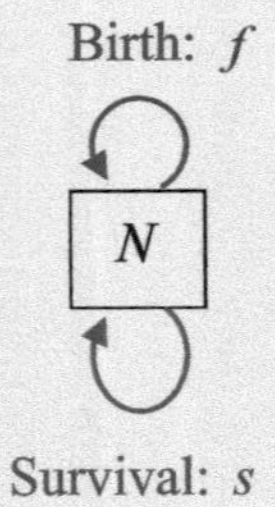

$$N_t = (s + f) N_{t-1}$$
$$N_t = R\,N_{t-1}$$

B

Matrix model:

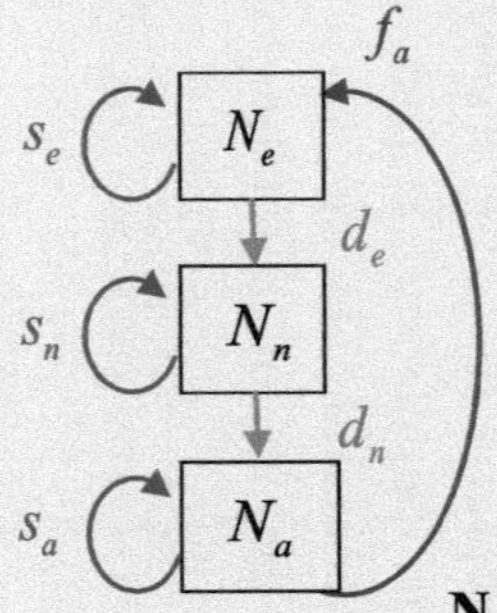

$$\begin{cases} N_{e,t} = s_e N_{e,t-1} + f_a N_{a,t-1} \\ N_{n,t} = d_e N_{e,t-1} + s_n N_{n,t-1} \\ N_{a,t} = d_n N_{n,t-1} + s_a N_{a,t-1} \end{cases} \text{, or}$$

$$\mathbf{N_t} = \mathbf{A} \cdot \mathbf{N_{t\text{-}1}} \text{ with } \mathbf{N_t} = \begin{bmatrix} N_{e,t} \\ N_{n,t} \\ N_{a,t} \end{bmatrix} \text{ and } \mathbf{A} = \begin{bmatrix} s_e & 0 & f_a \\ d_e & s_n & 0 \\ 0 & d_n & s_a \end{bmatrix}$$

C

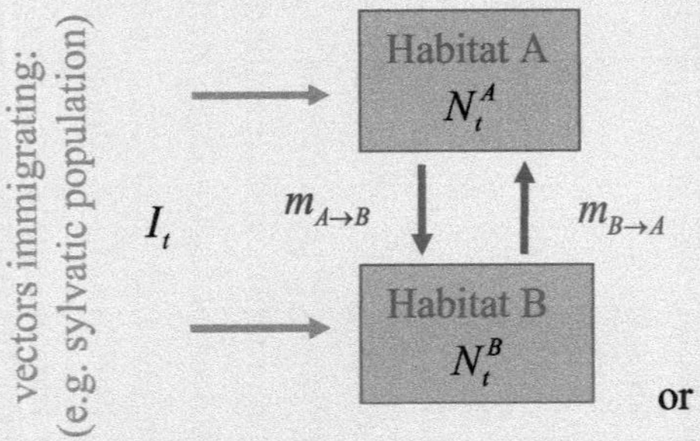

'Habitat' or 'patch' model:

$$\begin{cases} N_t^A = R\left(N_{t-1}^A\right) N_{t-1}^A + I_t - m_{A\to B} N_{t-1}^A + m_{B\to A} N_{t-1}^B \\ N_t^B = R\left(N_{t-1}^B\right) N_{t-1}^B + I_t + m_{A\to B} N_{t-1}^A - m_{B\to A} N_{t-1}^B \end{cases}$$

$$\text{or } \mathbf{N_t} = R(\mathbf{N_{t-1}})\mathbf{N_{t-1}} + \mathbf{I_t} + \mathbf{M} \cdot \mathbf{N_{t\text{-}1}} \text{ with } \mathbf{M} = \begin{bmatrix} -m_{A\to B} & m_{B\to A} \\ m_{A\to B} & -m_{B\to A} \end{bmatrix}$$

Box 1 Triatomines Population Modelling (*cont'd*)

Vector population dynamic models

(A) Basic model. Population size at time t is calculated with respect to survival s and fertility f that are assumed to be the same for all individuals alive at time $t-1$. These combine to define the population growth rate $R = s + f$. The basic discrete time ('geometric') model can be transformed into a basic continuous time ('exponential') model by considering very small time steps. The basic continuous-time model is written as a linear differential equation: $dN/dt = rN$, where $r = log(R)$ is the intrinsic population growth rate. **(B)** Matrix model. When individuals go through discrete developmental stages (e.g. eggs, nymphs and adults), several equations link the abundance in each stage $(N_{e,t}, N_{n,t}, N_{a,t})$ to the abundance in each stage at time $t-1$, and to the stage-specific survival (s_e, s_n, s_a), fertility (f_a) and development (d_e, d_n) rates. These equations can be lumped together in a vector N_t, which contains the abundance of each subpopulation at time t, and a transition matrix A, which contains all transition rates. **(C)** Model with dispersal and biotic regulation. The simplest way to model dispersal is to add a flow of immigration (I_t) into the focal population(s) (here patches A and B) sometimes referred to as an external forcing. Alternatively, dispersal can be modelled by assuming that a proportion of individual of each patch $(m_{B\rightarrow A}, m_{A\rightarrow B})$ disperse into another patch, which potentially account for mortality during dispersal. Population size limitation typically requires biotic regulation that often occurs within-patch. The survival, fertility and development of an individual then depend on the number of individuals it interacts with, so that the population growth rate becomes a function of the population density, $R(N)$. Finally, these models can easily be expanded to include seasonal forcing to mimic seasonality, e.g. in dispersal.

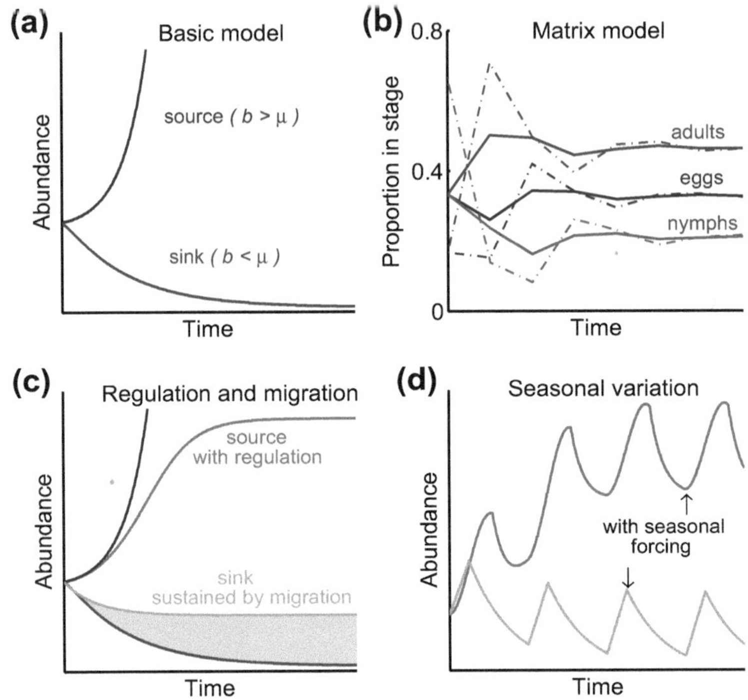

Figure 1 Visual representation of standard triatomine population models. (a) Basic model. Basic exponential (or geometric) models admit solutions: $N_t = N_0\ e^{rt}$ (or $N_t = N_0\ R^t$), see Box 1(**A**) and (**B**), where $r = b - \mu$ (or $R = s + f$ and $r = \log R$), and N_0 is initial abundance. Population size can then either increase ($r > 0$ or $R > 1$) or decrease ($r < 0$ or $R < 1$) exponentially, which corresponds to 'source' (blue) or 'sink' (red) populations. Hypothetically, $R = 1$ leads to a constant population size as births perfectly balance deaths. Such 'neutral' stability is unlikely to last and cannot explain the regulation of natural populations. (b) Matrix model. Most matrix models show a simple dynamical behaviour because transition matrices **A** typically are nonprimitive (Caswell, 2000). After a *transitory* stage, where the proportion of individuals in each stage fluctuate as it depends on initial abundances, the system reaches an *asymptotic* regime that only relies on transmission rates. The population then grows exponentially at a constant rate and the stage structure is stable, which are given by the dominant eigenvalue λ of **A** and its left eigenvector ω, respectively (Caswell, 2000). (c) Model with dispersal and biotic regulation. Density-dependent regulation can limit the exponential increase expected in source populations (blue curves) and lead them to reach a stable finite size where births exactly balance deaths. A sink population can also be maintained at an equilibrium level where the local exponential decrease is balanced by immigration (red curves). (d) Seasonal forcing. Populations can experience external forcing that leads demographic rates to vary with time. Here, fecundity/migration are forced to reproduce seasonal variation in abundance of domiciliated (blue, Castañera et al., 2003) and nondomiciliated (red, Gourbière et al., 2008) triatomines.

Table 1 Models of triatomine population dynamics and control

	Variation in vital rates								
Stage(1)	**Time(2)**	**Space(3)**	**Density(4)**	**Data(5)**	**Habitat(6)**	**Scale(7)**	**Control strategy(8)**	**Species(9)**	**Reference**
E,5N,A (Age)	–	–	Dp	Lab	–		–	*T. infestans*	Rabinovich (1972)
E,5N,A (Age)	–	–	–	Lab	–		–	*T. maculata*	Feliciangeli and Rabinovich (1985)
Age	–	–	–	Lab	–		–	*R. prolixus*	Chaves et al. (2004)
5N,A	–	SE	–	Fexp	D, P	V	Bn	*T. infestans*	Levy et al. (2008)
2N,A	Seas	SI	SwM	F	D, P	V	–	*T. dimidiata*	Gourbière et al. (2008)
2N,A	Seas	SI	SwM	F	D, P	V	Is, Sc, Bn	*T. dimidiata*	Barbu et al. (2009)
preA, A	Seas	SE	SwP, Dp*	–	D	V	Is	*T. dimidiata*	Slimi et al. (2009)
–	–	SE	–	F	D	V	Is	*T. infestans*	Zu Dohna et al. (2009)
Age	–	–	–	Lab	–		–	*R. neglectus*	Rabinovich et al. (2010)
A	Seas	SE	SwM	F	D, P	V	–	*T. dimidiata*	Barbu et al. (2010)
A	Seas	SE	SwM	F	D, P	V	Is, Sc, Cl, Tr, Hi	*T. dimidiata*	Barbu et al. (2011)
E,5N,A	–	–	Dp	La, Sn	–		–	*R. neglectus*	Rabinovich and Nieves (2011)
–	–	SE	–	F	D	V	–	*T. infestans*	Barbu et al. (2013)
E,5N,A	Seas	–	Dm	–	D	H	–	*T. infestans*	Castañera et al. (2003)
preA, A	–	–	–	Lab	–	–	–	*M. longipennis*	Martinez-Ibarra et al. (2013)

The table focuses on studies modelling the abundance of vectors. (1) Age or stage structure modelled: E = eggs, N = nymphs, A = adults, preA = pre-Adults. (2) Time variation in vital rates: seas = seasonal forcing. (3) Spatial variation and dispersal: SI = spatially implicit, SE = spatially explicit. (4) Regulation: Dp = density dependence – phenomenological model, DM = density dependence – mechanistic model, SwM = sink population with migration. (5) Lab = laboratory data, Sn = seminatural, F = field data, Fexp = field experiment. (6) D = domiciliated, P = Peri-domiciliated. (7) V = village, H = household. (8) Bn = Bednet, Is = insecticide spraying, Sc = insect screen, Cl = clearing peridomicile, Tr = insect trap, Hi = house improvement. (9) T = Triatoma, R = Rhodnius, M = Meccus. Three additional studies modelled vector dynamics (Lardeux, 2013; Levy et al., 2010; Stevens et al., 2013) with an alternative approach akin to COM models for infestation and are not included here (see main text).

derivation of those measures from matrix models). However, while many such life-tables have been built up in the 1970s for key species of triatomines (see Table 1 for a nonexhaustive list of studies), the estimates of the stage-specific vital rates that they provide have never been used to build matrix models. This is somewhat unfortunate as such an approach would provide predictions on the patterns of temporal variation in vector abundance according to the life-history specificities of triatomine species, which can be very useful in anticipating their relative speed of invasion/reinfestation when density-dependent processes remain of secondary importance. Alternatively, matrix models can be coupled with longitudinal follow-up or cross-sectional survey of field populations to estimate vital rates (Caswell, 2000). These fitting methods provide simultaneous estimates of all vital rates that allow the best reproduction of the variations in vector abundance and/or stage-structure observed in the field. Surprisingly, such alternative approaches have not been attempted yet for emblematic domiciliated species, such as *Triatoma infestans*, while large data sets with good quantification of the stage structure can easily be obtained. The only attempt to do so was made on nondomiciliated populations of *Triatoma dimidiata* using a seasonally forced matrix model to account for the seasonal immigration of those vectors into human dwellings (see Modelling Dispersal below).

Basic and matrix models are sometimes called 'linear' models as they do not account for the interaction between individuals that generate negative feedbacks and 'nonlinear' regulatory processes (see below). They are thus good quantitative tool to identify 'domiciliated' and 'nondomiciliated' triatomine populations, but they cannot contribute to explaining how sources are down-regulated and how sink populations persist. To do so, models have been expanded to include a description of dispersal and/or density-dependent regulation.

2.2 Modelling dispersal

The role of dispersal in the dynamics of natural populations is central in ecology (Colbert et al., 2001; Gilpin and Hasnky, 1991; Otto and Day, 2011) and various models have been developed to account for dispersal, which can be classified into three main categories (Table 1).

2.2.1 'Spatially Implicit' models

These models account for emigration and immigration while still focussing on a unique population (Gilpin and Hasnky, 1991). Emigration from the focal population is then usually modelled as an additional vital rate similar to mortality, and immigration appears as an incoming flow of vectors that is mathematically referred to as an external 'forcing' (see Box 1**(C)**). Such a modelling of dispersal has been adopted to study the population dynamics of *T. dimidiata* in the Yucatan peninsula, Mexico (Gourbière et al., 2008). A stage-structured matrix described survival and reproduction within a typical house and an external 'forcing' was added to the model to account for the seasonal immigration of triatomines into the domestic environment. A fit of this model to a 2-year follow-up of vector abundance inside houses of different villages of the Yucatan peninsula provided estimates of the demographic and dispersal parameters that best explained the observed transient infestation of domestic habitats. Such a simple model demonstrated that the studied populations were not sustainable in the absence of compensatory immigration, and that up to 90% of the vector individuals found in houses were immigrants. The model was subsequently adapted to optimize vector control strategies for such nondomiciliated 'sink' populations (see Modelling vector control).

2.2.2 'Habitat' or 'Patch' models

In such models, two or more populations inhabiting different habitats are represented separately and connected by dispersal (Gilpin and Hasnky, 1991). Population dynamics is then described within each patch and dispersal is modelled as a proportion of each population moving from one patch to another per unit-time (Box 1**(C)**, Figure 1(c)). There is actually no explicit account of distances between patches or dispersal distances, and the focus is rather on the importance of the fragmentation of the population (e.g. Gourbiere and Gourbiere, 2002) and/or the heterogeneity of the habitat (e.g. Hanski, 1991) on the overall dynamics. While this approach would be especially relevant to model the exchanges between the domestic, peridomestic and sylvatic habitats, it has never been attempted for the modelling of vector population dynamic on their own. This would be a valuable addition to the field that would help to better understand the quite complex epidemiological models of *T. cruzi* transmission that have accounted for such 'habitat' or 'patch' structure (see Rabinovich and Himschoot, 1990; 'Epidemiological model of *T. cruzi* transmission'). However, two original lines of modelling related to 'habitat' or 'patch' models have recently been developed by adapting standard models of transmission of macro- and microparasites (Anderson and May, 1992). In both cases, authors used the same analogies to relate the habitats of triatomines, parasites, and vertebrate hosts. Lardeux (2013) considered triatomines as macroparasites 'infecting' human dwellings, and Stevens et al. (2013) assumed that 'susceptible' sites can be colonized by triatomines to become 'infected'. From these modelling approaches, authors derived the basic reproduction number of triatomines (R_0) and used it to (1) identify the conditions for niche invasion and coexistence between wild and domestic populations of triatomines (Lardeux, 2013) and to (2) quantify the effect of peridomestic and sylvatic populations on the rate of household infestation (Stevens et al., 2013). Noteworthy, these contributions provide general 'analytical' expression of R_0 with respect to the parameters of the models. Therefore such formulations can be used for any population, as long as the assumptions of the modelling are satisfactory. Producing general results is an important feature of 'patch' or 'habitat' models whose use should thus be encouraged as there are few general theoretical predictions on triatomine population dynamics and infestation. This could provide a basic theory to be tested on various biological systems and give the opportunity to compare more efficiently the dynamics of different populations or species of triatomines.

2.2.3 'Spatially-Explicit' models

These models provide a complete description of space through a system of coordinates. Their key interest is in obtaining an explicit description of dispersal distances through the use of a dispersal kernel (see Figure 2(a)), i.e. a statistical distribution describing the probability of dispersal from one place to another according to distance, habitat 'matrix' permeability and biases in dispersal. Such models are typically more specific and focus on particular triatomine populations. They have been used to investigate (1) the dispersal of nondomiciliated *T. dimidiata* populations within a village of the Yucatan peninsula (Barbu et al., 2009; Slimi et al., 2009, Figure 2(b)); (2) the invasion of the city of Arequipa in Peru by *T. infestans* (Barbu et al., 2013; Bowman et al., 2008) and (3) the re-establishment of *T. infestans* local populations after insecticide spraying in villages of northwest Argentina (Zu Dohna et al., 2009). In none of these studies, the a priori knowledge about triatomine dispersal was sufficient to provide independent estimates of the parameters of the model — in a similar way as life-tables do for survival and fertility rates of matrix models — mostly because field studies about triatomine dispersal remain very scarce. The methodology adopted in all these contributions was then to infer the shape and parameters of the dispersal kernel from a model selection approach relying on field data describing the spatiotemporal dynamics of triatomine infestation or reinfestation. Interestingly, despite the similarity in the general approach, these studies show

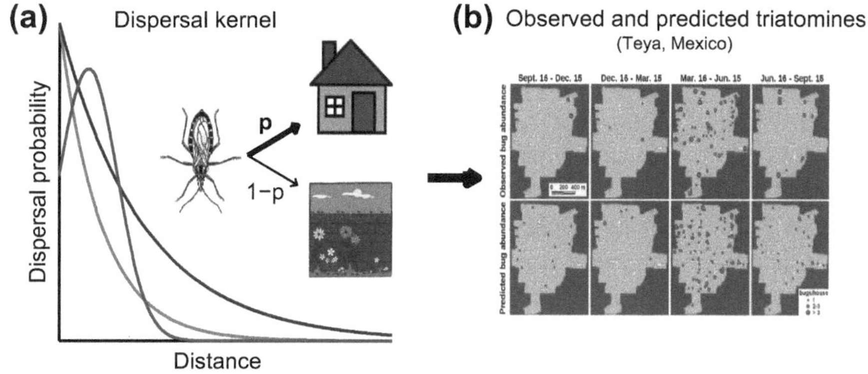

Figure 2 Spatially-explicit models of triatomine population dynamics. (a) The dispersal kernel describes the probability of dispersal over a given distance per time step. It usually is a standard distribution, e.g. exponential, log-normal, or Cauchy for 'fat tails' with high probabilities of long-range dispersal. More elaborate kernels account for seasonal variation in dispersal and/or matrix heterogeneity and biased displacements toward habitats, e.g. domestic vs. peridomestic (*p*: probability of dispersal into houses). (b) Dispersal kernels can be inferred by comparing the predictions of dynamical models to the observed dynamics of spatial triatomine distribution. The resulting parameterized models, once validated, can be used to predict the dynamics of infestation in subsequent years (Barbu et al., 2010, 2011). Similar model selection approaches can be used with projection models, and dispersal kernels can be estimated from two snapshots of the spatial distribution of triatomines.

significant differences as some used 'dynamical' and other 'projection' models. Since these alternatives illustrate well today's state of the art in modelling vector spatial population dynamics, we describe them in more depth.

2.2.3.1 Dynamical models

To study *T. dimidiata* infestation in Teya, Mexico, the modelling choice was to expand a population dynamics model into space to predict the within-year spatial variation in vector abundance inside this typical village of the Yucatan peninsula (Barbu et al., 2009). The vector population dynamics were mapped onto a grid representing the village and the cells of the grid were classified into domestic, peridomestic and sylvatic according to satellite images to produce a GIS-based realistic landscape. The modelling facilitated calculation of the number of bugs in each cell according to 'local' rules describing birth and death within cells, and rules of bugs' dispersal between cells. Different hypotheses about (1) the origin (sylvatic and/or peridomestic) of the dispersing vectors, (2) the distribution of dispersal distances and (3) the propensity of bugs to move toward houses were implemented producing different dispersal kernels. Each set of hypotheses (1 to 3) thus led to a distinct model that predicted different spatiotemporal dynamics of infestation inside the village. A model selection approach was then adopted to compare model approach was then adopted to compare how well different models fitted the data and, accordingly, how much support the different hypotheses (underlying the competing models) received from the data. This provided key quantitative insights into the dispersal of these nondomiciliated vectors that complemented those obtained from the simpler 'spatially implicit' model designed for a single house (Gourbière et al., 2008). Triatomines infesting houses came from colonies established in both the peridomestic (55%) and sylvatic (45%) habitats, and dispersing vectors spread over rather small distances, typically 40 to 60 m biweekly, being 5—15 times more attracted by houses than by peridomestic areas. These dynamical models were further adapted to evaluate the potential of spatially targeted vector control strategies (see 'Modelling vector control').

2.2.3.2 Projection models

To investigate the infestation of the city of Arequipa, Levy et al. (2008) set up a second type of spatially-explicit models. Such models can be referred to as 'projection' models as they relate the spatial distribution of vectors at a given time to a previous distribution, but are not meant to forecast the distribution at many other points in time — which is necessary to fully investigate the dynamics of a system. Trained entomology collectors surveyed all households in the community to inform on the initial distribution of *T. infestans*, and sentinel animal enclosures provided data on the subsequent change in this distribution. As the infestation of a given enclosure is likely to depend on the presence of surrounding 'source' populations of triatomines, all of them were georeferenced and various kernels of dispersion were considered to link the probability of infestation of a given enclosure with both the proximity and the initial density of surroundings sources. These data were combined with the spatially-explicit projection models in a model selection approach to identify the kernel of dispersal best reproducing the observed changes in the spatial distribution of triatomines. Noteworthy, the biweekly distance of triatomines' dispersal was found to be of the same order of magnitude as that obtained from the dynamical modelling described above. The model was further coupled with field data collected after placing insecticide impregnated nets on some of the animal enclosures to quantify their impact on infestation (see 'Modelling vector control').

Comparable analyses were performed by Zu Dohna et al. (2009) to predict the dynamics of re-establishment of triatomine colonies following an insecticide spraying campaign. Similar dispersal kernels were tested, and parameters were estimated from observations of the changes in the spatial distribution of infested houses. This contributed to demonstrate a 6-month time-lag between bug establishment on a site and dispersal from this site. Meanwhile, the fitted models produced quantitative evidence in favour of the importance of seasonal dispersal with both re-infestation risk and dispersal distances increasing during the May—November period. While the two studies above (Levy et al., 2008; Zu Dohna et al., 2009) relied on the observation of changes in the spatial patterns of infestation; kernel of dispersal and factors influencing dispersion (habitat 'matrix' permeability) may also be inferred from a static description of the spatial distribution. Barbu et al. (2013) used data from a single cross-sectional survey to infer the impact of street topology on house infestation. Based on the correlation of infestation between houses of the same block, and controlling for confounding factors (e.g. housing material, presence of non-human animals, imperfect or biased detection), Barbu and co-authors (op. cit.) found that a large proportion of the spatial variation in vector presence could be accounted for by street topology. In this urban context, city block is thus likely to be an appropriate spatial unit to describe the distribution of, and control *T. infestans*.

All these modelling studies have helped to quantify the movement of triatomines by estimating their kernel of dispersal from field data. They represent promising alternatives to standard experimental approaches as they are more tightly connected to the field patterns of vector spatial distribution and dynamics. In addition, the estimates of the kernels of dispersal can be entered into other spatially-explicit models based on integro-difference equations that can make predictions on the spread of triatomine at larger scales (Mahdjoub and Menu, 2008) and could thus be used to model, for instance, the variation of the biogeographical range of vectors with respect to climate change.

2.3 Modelling vector population regulation

As explained in the first part of this chapter, 'basic' (Box 1**(A)**) and 'matrix' (Box 1**(B)**) models are 'linear' mathematical systems that can only lead to ever growing or vanishing population dynamics (Figure 1(a)). Models accounting for dispersal are suited to studying how immigration can sustain 'sink' populations, and are thus an ideal framework to understand the dynamics of 'nondomiciliated' populations (see above). Alternative models deal with the regulation of the size of 'domiciliated' populations, e.g. *T. infestans* (Rabinovich and Himschoot, 1990) and *Rhodnius neglectus* (Rabinovich and Nieves, 2011), by means of density-dependent regulation or seasonal forcing (Figure 1(c) and (d)). Two standard approaches to model density-dependent regulation have been used in the context of triatomine dynamics.

The first kind of density-dependence modelling is 'phenomenological' as the purpose is to represent the 'phenomena' itself, i.e. the observation of a diminishing return on the population growth rate. Most often, the usual phenomenological 'logistic' equation (i.e. $dN/dt=rN(1-N/K)$, Verhulst, 1838) is fitted to temporal series measuring the growth of a triatomine population, providing an estimate of the 'intrinsic rate of increase' (r), i.e. the rate of increase of the population size in the absence of regulation, and the 'carrying capacity' (K), i.e. the level at which the population stabilizes. This was used by Rabinovich and Nieves (2011) to describe the development of an *R. neglectus* colony in seminatural conditions. On one hand, the use of the logistic equation allows for a tight link with theoretical studies that address general principles in a more abstract way since this equation has consistently been used as a convenient tool to prevent population size from unrealistically unbounded dynamics in general models (Figure 1(c)). On the other hand, however, parameters r and K are known not to be independent (Mallet, 2012; Smith, 1998), and they provide only very loose relationships with the vital rates that are commonly measured to describe mechanistically individual life histories in standard eco-evolutionary approaches (Gourbière and Menu, 2009). As a consequence, modelling the impact of conspecific/congeneric competitive interactions or of the effect of human intervention on individual vital rates with a phenomenological framework can be convenient at best, but may well be misleading.

The second approach consists of incorporating a 'mechanistic' description of the density-dependent processes regulating the population. This leads to more complex models that are usually designed for a specific biological system, as the main idea is to include empirical knowledge about the processes of interaction and to predict their effects on the dynamics of the populations. Empirical studies of triatomine populations have mostly focussed on two potential mechanisms of regulation; intraspecific interference on blood-feeding hosts (Schofield, 1980) and competition for refuges inside houses (Cecere et al., 2002). Castañera et al. (2003) provided a perfect example of a mechanistic model, which was designed to study the dynamics of *T. infestans* populations regulated by a density-dependent access to hosts. The regulating mechanism originated from a host defensive response and was assumed to take place after the number of bites per host reached an irritability threshold. This led to a reduction in the per-capita feeding rate of triatomines when triatomine population increased, eventually inducing a decrease in the per-capita fecundity as unfed females were assumed to deposit eggs at a slower rate than fed females. The model simultaneously accounted for differences in house qualities that were assumed to affect egg mortality in a density independent manner and for the temperature-dependent developmental time and fecundity. Interestingly, the paper by Castañera and co-workers (op. cit.) illustrates well the standard steps toward analysing mechanistic models. Authors made a considerable effort to provide estimates of the parameters of the model before using the parameterized model to produce 'reference' predictions to be compared with available field data. The predictions were found to be consistent with the main features of the field *T. infestans* population dynamics; the total number of insects, the seasonal variations in abundance and stage structure, and the proportion of fed/unfed vectors. In such a situation, it is possible to perform sensitivity analyses to investigate the influence of change in parameter values on the reference predictions, which in the case under discussion, allowed identification of three most influential parameters affecting the expected number of insects, namely, the prolonged developmental time induced by low temperatures, the threshold insect density for host irritability, and the mean female fecundity (Castañera et al., 2003). Clearly, such mechanistic models have a strong potential to make helpful predictions about the impact of interaction processes or seasonal variations on specific populations that are monitored in the field (Figure 1(d)). They are currently widely overlooked and future developments should help improving our understanding of the triatomine population dynamics through, for instance, the analysis of the additional effects that host response has on vector survival as well as reproduction. Such developments are not only relevant to the population dynamics of the vectors, but also to the transmission of *T. cruzi* as a better description of vector demography would undoubtedly provide better epidemiological models of the disease (see section 'Host and vector demography').

2.4 Modelling vector control

Vector control remains the main strategy to limit the transmission of Chagas disease, and was estimated to reduce the prevalence in humans by up to 70% in South America (Moncayo and Silveira, 2009). Evaluating and monitoring specific strategies of vector control, schedule schemes and reinfestation are crucial to achieve optimal vector control, and mathematical models can be extremely useful to evaluate the expected efficacy of various potential interventions. They indeed provide opportunities for extensive and cheap virtual 'experiments' when field trials are severely constrained by practical, financial and/or ethical reasons.

2.4.1 Dynamical models

Dynamical models predicting variations in bugs' abundance are especially valuable tools as they can be expanded to include a description of different control strategies. In this way, simulations can be run specifying the control strategies with different frequency, duration, timing and/or effects (on vector survival, reproduction or dispersal) used alone or in combination. The expected efficacy of many interventions can then easily be assessed by comparing vector abundance in simulations with and without control. Such an 'in silico' evaluation of vector control has been intended to address the issue of dispersing vectors that jeopardize control effort based on standard indoor insecticide spraying using general models (Stevens et al., 2013) or models tailored to a specific entomological context (Barbu et al., 2009, 2011). To propose a general approach to the problem, Stevens et al. (2013) expanded their 'patch' model (see above) to account for 'treated' domestic and peridomestic sites, in addition to the 'susceptible' and 'infected' ones. The reproduction number of triatomines (R_0) was shown to increase by an order of magnitude when sylvatic vectors are present. In such case, R_0 was sensitive to the size of the community (ratio of the number of houses to the number of sylvatic burrows) and the mobility of the sylvatic triatomines, but was only weakly affected by house spraying rate, insecticide persistence and spraying efficacy. These results provide general insights into the key parameters of the population dynamics of dispersing vectors and illustrate well the difficulty raised to achieve control. However, such models are not well suited to delivering practical guidelines as this requires models tailored and validated for the specific situation at hand. Barbu et al. (2009) designed a matrix model with seasonal immigration to predict the temporal dynamics of houses infestation by nondomiciliated *T. dimidiata*. This model was fitted to a 2-year longitudinal field study of bug abundance in two villages of the Yucatan peninsula, Mexico and it was shown to predict accurately house infestation in three other villages of the same area. The validated model was then adapted to account for the effect of residual insecticide spraying, insect screens and bednets on vector survival, reproduction and dispersal. Simulations showed that vector control can hardly be achieved by insecticide spraying, because it would require yearly application of 50 mg/m^2 of pyrethroid

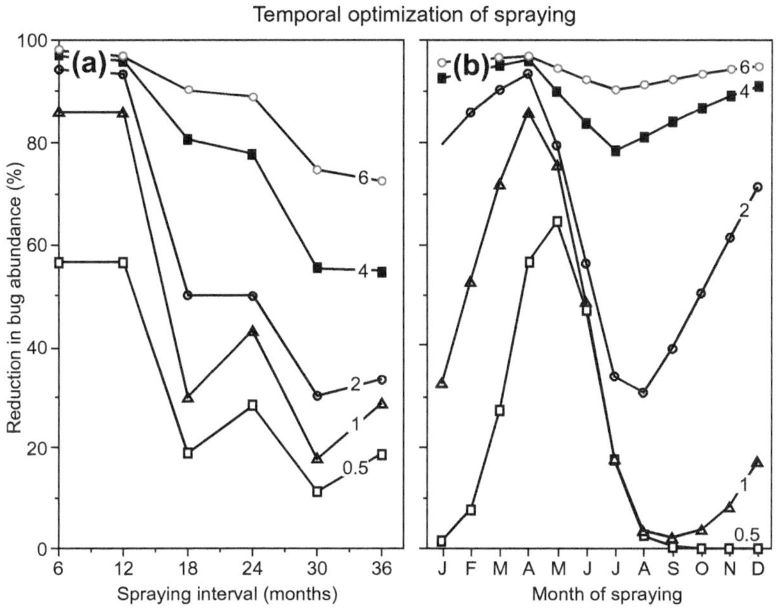

Figure 3 *'In silico'* optimization of insecticide campaigns. Dynamical models can incorporate a mechanistic description of the effect of spraying on triatomine demography. Various strategies can be simulated to identify 'in silico' the optimal frequency (a) and timing (b) of spraying under different assumptions, e.g. about the decay of insecticide in the field (Barbu et al., 2009). Alternative strategies, such as mosquito net or peridomicile management, can be included and compared to standard insecticide spraying.

within a 2-month period matching the immigration season (Figure 3(a)). Larger doses or more frequent interventions would not increase the effectiveness of control, while different timing and lower frequencies of intervention would drastically reduce the impact of interventions (Figure 3(a)). While this study identified the optimal temporal design for insecticide spraying, it also suggested insect screens to be an effective alternative, as their use consistently reduced bug abundance proportionally to the reduction of the vector immigration rate. These modelling results were critical in identifying the set of strategies that were subsequently tested in a comparative field trial (Ferral et al., 2010). The 'in silico' optimization of control strategies was taken a step further to identify spatially-targeted strategies. Barbu et al. (2011) used a spatially-explicit population dynamics model combined with a 2-year field study of infestation dynamics to obtain a validated model predicting both the intra-annual variation and the spatial gradient in *T. dimidiata* abundance typically observed in villages of this area (Pacheco-Tucuch et al., 2012; Ramirez-Sierra et al., 2010). Five different control strategies were then applied in concentric rings to mimic spatial design targeting the periphery of the village, where vectors were most abundant. The most cost-effective intervention combined the use of insect screens in outer zones to reduce infestation by both sylvatic and peridomiciliated vectors, and cleaning of peridomicile in the centre of the village where sylvatic vectors are absent.

2.4.2 Projection models

Projection models have also been used to assess the impact of control strategies on triatomine populations. Levy et al. (2008) expanded their modelling (see 'Modelling dispersal') to include a reduction in the spread of infestation due to deltamethrin-impregnated netting added on guinea pig enclosures. This background was combined with the data on infestation of enclosures with and without impregnated netting to simultaneously estimate the kernel of triatomine dispersal and the impact of control. Levy and colleagues, found evidence for impregnated netting slowing down the rate of vector infestation fourfold, strongly suggesting that such nets would be valuable tools in the control of the disease (Levy et al., 2008). The 'in silico' optimization of vector control in the emblematic city of Arequipa was taken a step further as Levy et al. (2010) used spatially-explicit compartmental models to look at the effect of the sequential spatial ordering in which houses were sprayed on the overall campaign efficacy. They demonstrated that instances of reinfestation can be greatly reduced by an appropriate spatial design. The optimal path to be followed was found using a genetic algorithm, and reduced reinfestation 10-fold relative to the common practice of sequentially spraying contiguous communities. Using similar models of vector (re-) infestation, Zu Dohna et al. (2009) explored the spatial and seasonal dynamics of house (re-) infestation (see 'Modelling dispersal') and the impact of insecticide spraying on such dynamics. A model selection approach was used to identify the covariates/processes that influenced house reinfestation. These authors demonstrated that spraying led to a 60% reduction in the probability of reinfestation during a 6-month season, but found no evidence for a more sustainable effect on triatomine dynamics.

Although projection models have evaluated the impact of control strategies, they do so in a different way from that of dynamical models. They offer a powerful way of analysing data as they are directly embedded in a probabilistic framework that easily allows fitting models to data with and without control interventions. However, in their current form, they are not set up from standard vital rates that would apply recursively to shape vector population dynamics. Accordingly, one cannot include a mechanistic description of control strategies and proceed to in silico control 'experiments', as one can do with dynamical models. Dynamical and projection models are thus two complementary approaches and the choice between them depends on the available data (times series vs. snapshot) and the ultimate goal of the study (exploration/optimization vs description).

2.5 Conclusions on the modelling of vector population dynamics

While entomological studies have been a central part of empirical research on Chagas disease for more than a century, few models have been developed to produce predictions on the dynamics of triatomine populations. Many research groups across Central and South America have been monitoring triatomine infestation in specific ecological settings and undoubtedly have appropriate data sets to adopt a modelling approach. Data from longitudinal (or even cross-sectional) entomological studies could indeed be used to fit simple matrix or spatially-implicit models. Developing validated models to optimize vector control could be more demanding as, ideally, a model should not only fit the data used to estimate its parameters, but also predict well the variation in triatomine abundance in independent datasets, which might require additional specifically designed field studies. However, the cost of such integrative studies might well be balanced by the identification and use of optimal vector control interventions. Another key interest of developing a stronger modelling background would be to connect more tightly studies on triatomines to standard theories and concepts of ecology and evolutionary ecology. The well-established framework of source–sink dynamics and/or metapopulation dynamics could for instance provide valuable conceptual inputs to understand better the importance of dispersal in triatomine population dynamics. This would undoubtedly be beneficial in order to address two key challenges in today's control of the disease; (1) the failure of control due to reinfestation by key domiciliated species and their replacement by sylvatic species (Cecere et al., 2004, 2006) and (2) the transmission of *T. cruzi* due to 'nondomiciliated' species (Ceballos et al., 2011; Noireau et al., 2005).

3. EPIDEMIOLOGICAL MODELS OF *TRYPANOSOMA CRUZI* TRANSMISSION

To study triatomine's vector demography is a clear and necessary step to improve our knowledge of the disease and vector control strategies. However, ultimately, we aim at understanding and limiting the spread of the parasite itself. The modelling of disease transmission can then be very useful as the occurrence of new infections and/or the evolution of infection and immunity can be predicted with respect to host/vector vital rates/abundances and the key features of the parasite. Clearly, such models have a high heuristic value for basic science and can be used to anticipate the effect of various vector as well as nonvector control strategies on the spread of the parasite. Two main kinds of epidemiological models are available for such analyses and are referred to as Force of Infection (FOI) and COMpartmental (COM) models.

3.1 Force of Infection (FOI) models for *T. cruzi*

In this epidemiological approach, the central quantity is the so-called force-of-infection (FOI), i.e. the per unit time probability for a susceptible host to become infected. This probability can be modelled in different ways (see Box 2 and Table 2), but the philosophy always is to provide simple models that can be fitted to prevalence or incidence data in order to unravel historical and/or spatial variations in disease transmission, compare assumptions about their origins and estimate key epidemiological parameters. For the sake of simplicity, FOI models typically do not account for the complex dynamical feedbacks between parasite infection and host/vector population dynamics that are investigated using compartmental models. Instead, they relate the prevalence/incidence of the disease in the focal host population with respect to the ecological and social contexts and try to match predictions with the observed epidemiology of the disease. The flexibility of such models facilitates working at different spatial scales ranging from house to administrative region according to the data available.

3.1.1 The basic FOI model

The simplest FOI model assumes that all susceptible individuals can acquire the pathogen according to a constant force of infection, λ (Box 2(**A**)). The increase in the probability of being infected with age can then be predicted with respect to λ (Figure 4(a)) and the force of infection can be estimated by matching these predictions with cross-sectional data on age-specific seroprevalences. This provides an estimate of the probability for any susceptible individual to acquire the disease in any year of his/her life. Although it can serve as a valuable null hypothesis (e.g. Bowman et al., 2008; Delgado et al., 2011), this basic model has actually never been used to describe Chagas disease transmission in humans as all authors considered or showed that the force of infection varied through time and/or with age or other individual

Box 2 Epidemiological Models of *T. cruzi* Transmission

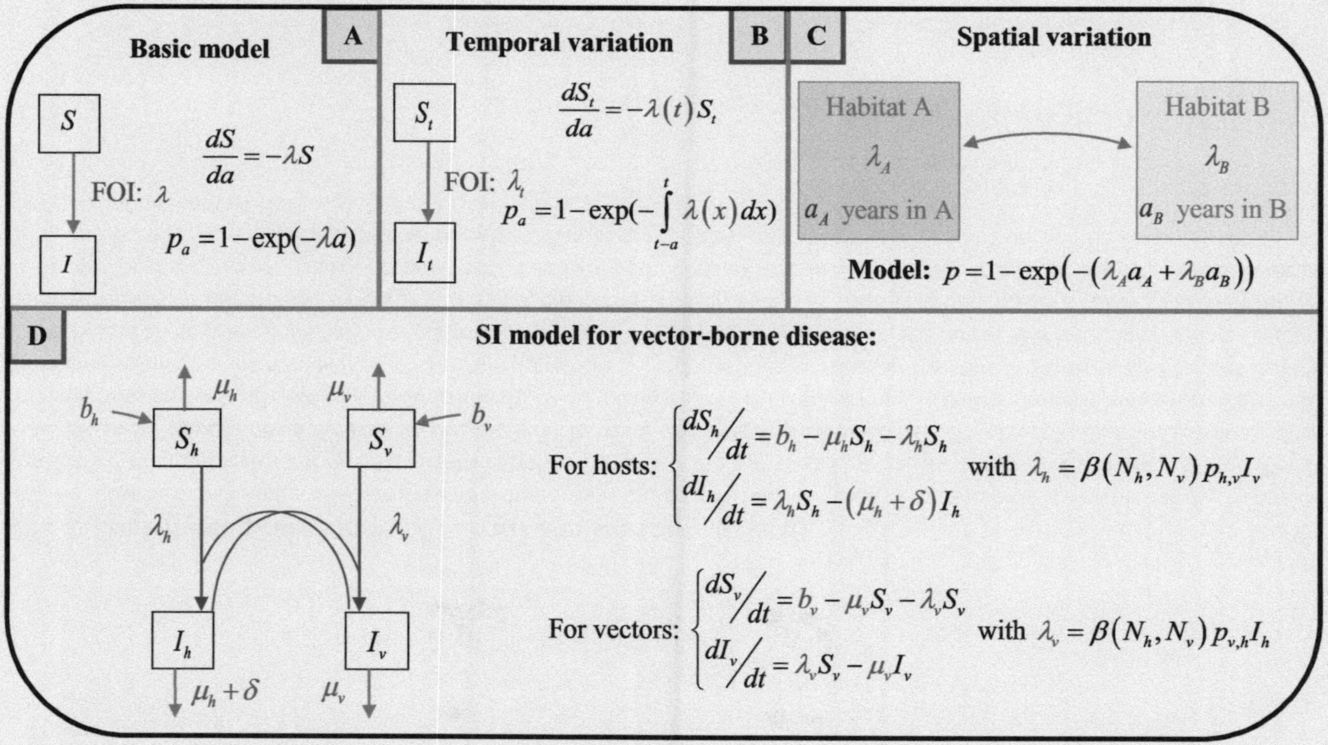

(Continued)

Box 2 Epidemiological Models of *T. cruzi* Transmission (*cont'd*)

(A) The most basic FOI model describes the infection of susceptible individuals at a constant rate (FOI) λ. The probability to remain susceptible then decreases exponentially with exposure time, and the prevalence of infection increases with age. **(B and C)** FOI models can be straightforwardly expanded to account for temporal or spatial variation in λ. The prevalence of infection at age a then depends on the local history of transmission (B), or on the time spent in each habitat (C). **(D)** The simplest compartmental model for vector-borne diseases modelling accounts for susceptible (S) and infectious (I) hosts and vectors. Individuals are born susceptible (b) and die at a rate μ, and infected hosts suffer additional mortality linked to infection (δ). The rate at which susceptible hosts/vectors become infected is determined by the FOIs they impose on each other, λ_h and λ_v. Such FOIs depend on a contact rate defined per vector and per host, $\beta(N_h, N_v)$, the per contact probabilities of pathogen transmission from vector to host ($p_{h,v}$) and from host to vector ($p_{v,h}$), and the density of infected hosts and vectors. Various hypotheses on vector feeding can be accounted for according to the definition of the contact rates (see Table 4).

Table 2 Force of infection models of *T. cruzi* transmission

Temporal variation[1]	Spatial variation[2]	Control measure	Transmission mode[3]	Habitat	and scale	Host	Species[4]	Reference
5 y	Regional	Spraying	Vector	Domestic	Regional	Human	*R. prolixus*	Feliciangeli et al. (2003)
—	—	—	Vector	Domestic	Regional	Dog	*T. infestans*	Gürtler et al. (2005)
20 y	Neighbourhood	—	Vector	Domestic	Village	Human	*T. infestans*	Bowman et al. (2008)
f(risk)	f(risk)	—	Vector, congenital★	Domestic	Village	Human	*T. infestans*	Levy et al. (2011)
Interruption	Migration	Spraying	Vector, congenital★	Domestic	Village	Human	*T. infestans*	Delgado et al. (2011)
B-spline	—	Spraying	Vector, congenital★	Domestic	Village	Human	*T. infestans*	Samuels et al. (2013)
f(Iv)	f(Iv)	—	Vector	Domestic	Household	Dog	*T. infestans*	Rabinovich et al. (1990)
f(Iv)	f(Iv)	—	Vector, oral	—	Lab	Opossum	*T. infestans*	Rabinovich et al. (2001)
f(Iv)	f(Iv)	—	Vector	Domestic Peri-Domestic	Household Village	Human	multiple	Nouvellet et al. (2013)

(1 and 2) Temporal variations include FOI changes between time windows (5 y = 5 years, 20 y = 20 years), according to a flexible time function (B-spline), or more specific form where FOI depend on predefined risk factors, f(risk), including local measure of exposure to infected vectors, f(Iv). (3) ★Congenital transmission via the inclusion of an intercept. (4) T = Triatoma, R = Rhodnius.

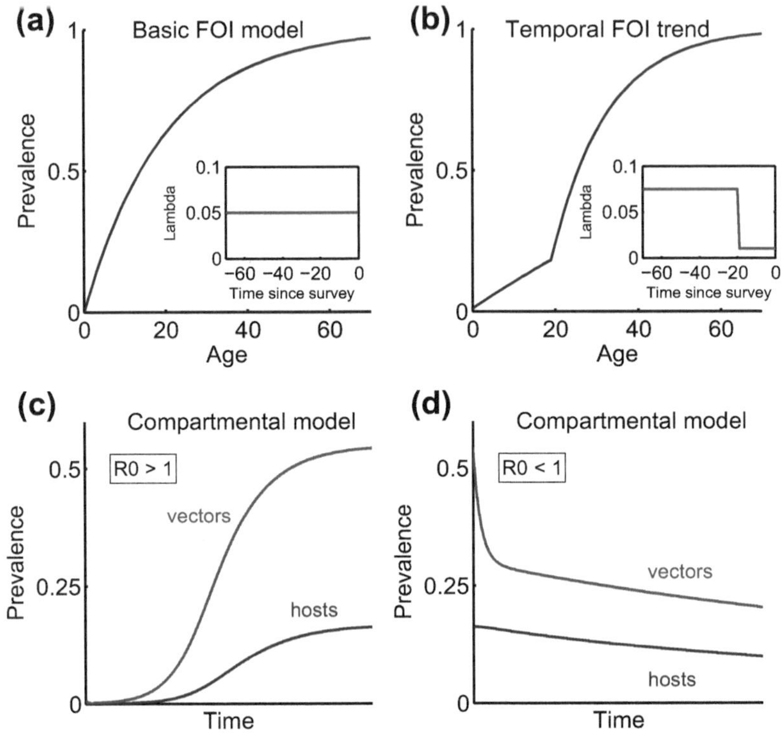

Figure 4 Dynamical behaviour of standard epidemiological models of *T. cruzi.* (a,b) Simple FOI models where the force of infection (λ) is constant or time dependent. The models allow to link the level of exposure in time to the expect prevalence per age. As older individuals are more exposed, their prevalence is expected to increase. The rate in this increase gives information about the past exposure. (c,d) Simple compartmental models for hosts and vectors. When R_0 is greater than 1, the disease spreads in both vectors and hosts until an endemic equilibrium is reached. However, if R_0 is lower than 1, the disease is expected to die out. In these examples, populations are assumed constant (death rate equals birth rate) and transmission is limited by vector feeding rate leading to frequency dependence related to hosts (see Table 4).

risk factors. However, Gürtler et al. (2005) used such a model to estimate the FOI affecting dogs in a village of Argentina.

3.1.2 Describing large-scale spatial and temporal patterns of transmission

The closest alternative to the basic FOI model was provided by Feliciangeli et al. (2003) who accounted for temporal differences in the FOI between 5-year time windows (Box 2(**B**)). Feliciangeli and colleagies fitted their model with seroprevalence data measured among 5-year age groups in highly endemic regions of Venezuela. The resulting estimates of the average annual FOI provided evidence that the national control programme led to a 10-fold decrease in the disease annual incidence, although disease transmission had not been interrupted and seemed to show a re-emerging tendency. Authors subsequently used the same methodology to characterize the spatial heterogeneity in FOI thereby demonstrating that residual transmission was mostly focussed on the foothills of the Andean and coastal mountain ranges. This study can be seen as an emblematic example of how FOI models can be used to provide quantitative figures about spatiotemporal patterns of disease risk and the impact of control programmes at large (regional or national) scales (see 'Cost-effectiveness'). Surprisingly, while many states and national health programmes routinely measure *T. cruzi* seroprevalence, the potentially very informative outcomes of this approach are rarely available in the standard scientific literature. Instead, FOI models of Chagas disease have typically focussed on disease transmission history and the effect of control interventions at much smaller scales such as houses, villages or cities.

3.1.3 Tracing back the local history of transmission

The FOI approach has been used repeatedly and very successfully in the last few years to understand the history of transmission in rapidly evolving peri-urban and rural areas in southern Peru. The simplest FOI model was developed to understand *T. cruzi* emergence at the periphery of the city of Arequipa, where the epidemiological context is changing rapidly due to an important urbanization following immigration from surrounding rural areas. Bowman et al. (2008) compared a basic model with constant FOI (null model) with a model assuming a recent introduction of the pathogen in this community. The period of exposure in this second model depended on the relative values of the age of individuals and the time of introduction of the pathogen. This led to temporal variation in the FOI, e.g. $\lambda = 0$ prior to the parasite introduction (Box 2**(B)**, Figure 4(b)). The two FOI models were fitted to age-specific seroprevalence data and likelihood ratio statistics showed that the second model provided a much better fit to the local epidemiological pattern, suggesting that the pathogen was introduced around 1993 in Guadaloupe, a district on the south-western part of Arequipa. Levy et al. (2011) then inferred not only the timing but also the location of *T. cruzi* introduction in this recent peri-urban settlement. Using a cross-sectional serological survey to characterize the spatial distribution of human infections in the community, they used a Bayesian approach to match these data with different spatially extended FOI models that represented three main hypotheses about the local history of disease transmission; (1) an endemic, (2) a single epidemic and (3) a multiepidemic model. Clearly, these models have the potential to generate qualitatively different spatiotemporal dynamics of the spread of *T. cruzi* in the community and therefore predictions differed regarding the duration of exposure of individuals according to their age and the distance(s) of their household from the site(s) of introduction(s) of the parasite. The FOI and the probability for each individual to be infected were thus calculated for each model using individual measures of exposure that accounted for household heterogeneity in standard risk factors analysis. Risk factors were assessed during an entomological/sociological survey concomitant to the serological survey and they were incorporated in a similar way in the three models. Model fitting and comparison showed that *T. cruzi* was most likely introduced in four locations in the new settlement of Guadaloupe, leading to distinct microepidemics that all started within the last 20 years before the study. The reconstruction of the local history of *T. cruzi* transmission from the parameterized models showed that the prevalence of infection was still increasing sharply at the time (2004) of the surveys. Interestingly, this understanding of the emergence of the disease in the peri-urban area of Arequipa is associated with a similar FOI study in four peri-rural contiguous localities of the La Joya district located 30 km from Arequipa. Delgado et al. (2011) reconstructed the recent history of transmission accounting for human migration in and out of the study area and a postulated insecticide campaign. Sociodemographic and serological cross-sectional surveys were conducted to record the age of the study participants, where they had lived, and their infection status. Again, three models were proposed to explain the observed age-specific patterns of seroprevalence; (1) an age only, (2) a migration and (3) a migration—transmission interruption model. The best fit to the data was provided by the latter, and indicated that transmission was most likely interrupted around 1996 by a forgotten vector control intervention. A posteriori search in past newspapers validated this conclusion by uncovering reports of insecticide sprayings associated with the then-President Fujimori re-election campaign in 1995. Estimates derived from the best model suggested that, before that date, annual incidence was around 1% and that the insecticide campaign brought it down to 0.1%. Epidemiological insights gained from these different FOI studies suggested that perirural communities like La Joya could have been the source of vectors and parasites explaining *T. cruzi* emergence in the 1980—1990s in Guadaloupe. The seasonal migration of inhabitants of Guadaloupe for agricultural employment could indeed be the cause of *Triatoma infestans* and *Trypanosoma cruzi* spread from rural to urban environments. The FOI models developed in this area of southern Peru clearly exemplify the role they can play in improving our understanding of local transmission history by providing theoretical predictions on the impact of different processes on age-specific patterns of infection that can be matched to cross-sectional surveys. A study in Bolivia (Samuels et al., 2013), used a similar approach to investigate the temporal variation in the FOI in relation to an insecticide spraying campaign. Originally, instead of estimating yearly FOIs (or at discrete intervals), the authors assumed a functional form (i.e. a B-spline) allowing them to not only to account for the temporal autocorrelation in the FOI time series but also to limit the number of estimated parameters. The high flexibility of the FOI framework potentially allows for many processes to be accounted for, and thus for many alternative hypotheses to be tested in various eco-epidemiological contexts. There surely are many serological data available across Latin America that, combined with such a modelling approach, would lead to improving our knowledge of disease transmission without extensive additional field work.

3.1.4 Estimating the probability of stercorarian transmission

The last use of FOI models was for a more specific purpose: estimating the per host—vector contact probability of host infection. Because *T. cruzi* is a stercorarian parasite, i.e., it is transmitted by the faeces of its vector rather than by its salivary glands, this quantity is typically small and cannot be estimated from experimental infections (Nouvellet et al., 2013). The only alternative to estimate this quantity is to match the observed incidence of infections in a population with an estimate of the number of potentially infectious contacts (PIC) received by the host individuals, which requires knowledge on the abundance, infection status and biting rate of the triatomine population. The general principle is then to propose an FOI model that links the probability for a host individual to become infected to the number of PIC received during a given time period, and to fit this model to observed incidences as typically measured by seroconversion data between two serosurveys. This approach, pioneered by Rabinovich and his colleagues, was used to obtain estimates of the probability of transmission to humans ($p = 0.0009$, Rabinovich et al., 1990), guinea-pigs ($p = 0.0002$, Basombrío et al., 1999) and opossums ($p = 0.06$ 95% CI (0.023, 0.162), Rabinovich et al., 2001) as well as the probability of oral infection after an infected triatomine has been eaten by an opossum

(0.075 95% CI (0, 0.265), Rabinovich et al., 1990). Recently, the method was refined to derive confidence intervals (instead of point) estimates of the probability of transmission to humans using a likelihood framework (Nouvellet et al., 2013). The modelling proposed by Rabinovich et al. (1990) was expanded to derive estimates when (1) entomological and/or epidemiological data are available at the village rather than at the household scale, and (2) epidemiological data consist of human prevalence rather than incidence. This provided seven additional estimates of the probability of stercorarian transmission of *T. cruzi* to human that all ranged from 10^{-4} to 10^{-3} per contact. Interestingly, these estimates were used to predict accurately human prevalence from the abundance of infected vectors in 7/10 independent data sets that covered various triatomine species and epidemiological situations. Such FOI models could thus provide the quantitative tool required to shift from purely entomological to parasitological indexes for both the assessment of Chagas disease risk and the impact of control interventions.

FOI approaches are very flexible and powerful as many relationships can easily be proposed and fitted to static age-specific epidemiological patterns. However, these relationships do not describe the feedbacks between patterns and processes in a fully dynamical way. They are, with that respect, equivalent of the 'projection models' used to investigate vector population dynamics (see 'Modelling dispersal' and 'Modelling vector control'). Accordingly, they may be less suited for investigation of the potentially complex ecological and/or evolutionary interactions between parasite infection and host/vector population dynamics, which compartmental models may be able to describe better.

3.2 Compartmental models for *T. cruzi*

The general principle behind COMpartmental models (COM) is to define categories of individuals that correspond to compartments and to model the continuous flows of individuals between them, so that the changes in the number of individuals of each category can be described. For most vector-borne diseases, categories at least include susceptible (*S*) and infectious (*I*) hosts and vectors leading to SI models (Box 2**(D)** and see also Chapter 3 by Rock et al. (2015), this volume). The demography of host and vectors is usually modelled by death and birth flows, while transmission is accounted for by flows of susceptible hosts (vectors) to infected hosts (vectors) due to (contact with) infectious vectors (hosts) by virtue of the haematophagic habit of the insect vectors. Further compartments can be added/removed according to the diversity of hosts/vectors or parasites, the inclusion of (e.g. stage, habitat) structures for the host/vector populations or the description of the within-host dynamic of the parasite (Table 3). The typical outcomes of these models are (1) to identify the conditions when the parasite can spread (Figure 4(c)) or dies out (Figure 4(d)) and, in the former case, (2) to investigate the epidemic phases or endemic regimes in terms of incidence or prevalence. The conditions for emergence are typically derived from analytical expressions of the parasite basic reproduction number in a fully susceptible population (R_0) and thus defined with respect to the ecological and epidemiological parameters of the model (Massad, 2008). The epidemic or endemic dynamics are much less mathematically tractable and are usually investigated numerically, which raises an issue of (lower) generality of simulation results. COM models of Chagas disease are no exception, and the intrinsic complexity of the biological system characterized by high levels of host, vector and parasite diversities make this issue even more critical. It is then important to identify modelling assumptions since the theoretical predictions would only apply under those conditions. Here we review COM models of Chagas disease focussing on four key constitutive aspects: (1) hosts, vectors and parasite biodiversity, (2) host and vector demography, (3) the nature of the transmission processes and (4) the within-host dynamic of the parasite.

3.2.1 Host, vector and parasite biodiversity

A significant aspect of the transmission of *T. cruzi* is that it is able to infect a broad diversity of mammalian species (Lent and Wygodzinsky, 1979), and that its triatomine vectors are generalist haematophagous insects feeding on an even larger range of host species distributed across 4 vertebrate classes, namely, Mammalia, Aves, Reptilia, and Amphibia (Gottdenker et al., 2012). Despite this, and while the disease is endemic in areas with high levels of biodiversity, our quantitative understanding of pathogen transmission remains mostly confined to a single-human host or a single-reservoir and a single-vector species (Table 3). These simplified representations of the system have been expanded to account for both a human and a reservoir host (see Table 3) or for small 'communities' of 2–3 domestic and peridomestic species in addition to human hosts (Coffield et al., 2013; Cohen and Gürtler, 2001; Cruz-Pacheco et al., 2012; Rascalou et al., 2012). These latter studies propose mathematical models of *T. cruzi* transmission that account for the structure of the 'community' and the competence of the different species to transmit *T. cruzi*, with the ultimate goal of assessing the effectiveness of control measures. Cohen and Gürtler (2001) developed a simulation model

Table 3 Epidemiology of Chagas disease in compartmental (COMP) models

Mode of transmission	Host			Vector				Parasite			Control	
(1)	Diversity (2)	Structure (3)	Reg. (4)	Div. (2)	Structure (3)	Reg. (4)	VTM (5)	Div. (2)	Within-Host (6)	Vir. (7)	measure (8)	Reference
V	H, R	–	Cst	1	Stage	Dp, Mig	1	1	A, C	Yes	–	Rabinovich and Rossell (1976)
V	H, R	–	Cst	1	Stage	Dm, Mig	1	1	I	–	–	Rabinovich et al. (1990)
V, C, B	H, R	–	Dp	1	–	Cst	1	1	I	Yes	–	Velasco-Hernandez (1991)
C, B	H, R	–	Dp	1	–	Cst	1	1	I	Yes	S	Velasco-Hernandez (1994)
V	H, R, nC	Age	Cst	1	Seas	Cst	1	1	I	–	S, Z	Cohen and Gürtler (2001)
V, B	H	Age	Cst	1	–	Cst	1	1	I	Yes	–	Inaba and Sekine (2004)
V	H	–	Cst	1	–	Cst	1	1	A, C, Res	Yes	–	Das and Mukherjee (2006)
V	H	–	Cst	1	–	Cst	–	2	A, C, C′	Yes	–	Devillers et al. (2008)
V	H	–	Cst	1	–	Cst5	1	1	I	Yes	S, T	Massad (2008)
V, C, B	H	–	Cst	1	–	Cst	1	1	C, c	Yes	–	Das and Mukherjee (2009)
V, O, C	R	–	Cst	1	–	D	1-3	1	I	–	–	Kribs-Zaleta (2010a)
V, O, C	R	–	Dp	1	Stage	D	1-3	1	I	–	–	Kribs-Zaleta et al. (2010b)
C	H	Ch, A, Sex	Cst	–	–	–	–	1	I	Yes	T	Raimundo et al. (2010)

V	H, R, nC	–	Cst	1	Seas	D	1	1	I	Yes	S, Z	Spagnuolo et al. (2011)
V	H, R, nC	–	Cst	1	Delay	Cst	1	1	I	Yes	VC	Clauson et al. (2012)
V, C	H, R, nC	–	Cst	1	–	Dm	1	1	A_H, C_H★, I_R	Yes	S, Z, T	Cruz-Pacheco et al. (2012)
V, O, C	R	–	Dp	1	–	D	1–3	2	I	–	–	Kribs-Zaleta and Mubayi (2012)
V, O, C	R	–	Dp	1	–	D	1–3	2	I	–	–	Pelosse and Kribs-Zaleta (2012)
V	H, R, nC	–	Cst	1	Delay, Seas	D	1	1	I	Yes	S, Z	Spagnuolo et al. (2012)
V	H-R	–	Cst	1	–	Cst, Mig	1–2	1	A, C	Yes	–	Rascalou et al. (2012)
H	H	HIV	Cst	–	–	–	–	1	I_t, I_{th}	Yes	–	Annan and Fisher (2013)
V, O, C	H, R, nC	–	Cst	1	Delay, Seas	D	1	1	I	Yes	–	Coffield et al. (2013)
V	H	–	Dp	1	Stage	Dm, Mig	1–3	1	I	Yes	–	Pelosse et al. (2013)
V, O, C, B, H	H, R	–	Cst	1	–	Cst, T°	1	1	I	–	–	Cordovez et al. (2014)
C, B	H	–	Mig	–	–	–	–	1	A, C, C'★	Yes	Ta, Tc	Fabrizio et al. (2014)
V, O, C	R	–	Dp	1	–	D	1–3	2	I	–	–	Kribs-Zaleta (2014)

(1) Mode of transmission investigated. V = Vectorial, O = Oral, C = Congenital, B = Blood (transfusion), H = horizontal (direct) transmission in opossums. (2) Diversity: in host, H = human, R = reservoir, nC = non-Competent host like chickens; in vector and parasite, number of species/strain considered. (3) Structure in hosts/vector: Ch = children, A = adult, HIV = HIV status of infection. (4) Regulation: Cst = constant population size (constant flow of births or deaths – implies some implicit regulation) Dp = density dependence – phenomenological model, Dm = density dependence – mechanistic model, Mig = migration. (5) Vectorial transmission mode: 1 = typical frequency dependence (relative to hosts) – see one in Table 4 and 2 = Density dependence – see two in Table 4 and 3 = atypical frequency dependence (relative to vectors) – see three in Table 4. (6) Within host dynamics of the parasite: A = acute stage, C = chronic stage, C' = chronic stage with determinate pathology, I_t = *T. cruzi* infections, I_{tH} = *T. cruzi* and HIV coinfection, ★Possibility of returning to the susceptible state. Res = Resistance class to infection in hosts. (9) Vir = virulence – defined as additional host or reservoir mortality induced by the parasite. (10) Control measure included in the model: S = insecticide spraying, Z = zoo-prophylaxis usually using chickens, T = etiological treatment in humans, Ta = treatment in acute phase, Tc = treatment in chronic phase.

of transmission at the scale of a typical household and in close relationship with an impressive set of data collected over long-term field studies in three villages located in the province of Santiago del Estero, northwest of Argentina. They demonstrated a clear 'amplification' of *T. cruzi* transmission to humans when keeping dogs in the household as dogs provide attractive sources of blood meals to vectors (Gürtler et al., 1997) and are highly susceptible to *T. cruzi* (Gürtler et al., 1991). Critically, the first 2 domiciliary dogs contributed strongly to increase the bug population size and the bug and human rates of infection. The model also showed a 'dilution' effect as an increase in the domiciliary chicken population decreased the vector and human infection rates, although in a much smaller proportion. Accordingly, authors suggested that leaving dogs out of the house would reduce human infection rates, whereas having chicken inside houses to divert vectors from humans would only provide a marginal benefit. Cruz-Pacheco et al. (2012) developed a more standard COM model based on a set of differential equations parameterized using the same data. They confirmed that removing dogs from dormitories was the best option to control *T. cruzi* transmission to humans, and further showed that an early diagnosis and treatment of the disease in humans would not substantially reduce the risk of further human infection if animals were kept inside houses. Coffield et al. (2013) built a similar COM model and confirmed that 'domestic' *T. cruzi* would only persist in the presence of dogs even at low density (0.2 dog per house). Interestingly, the risk of *T. cruzi* transmission to humans increased continually with the number of dogs, which contrasts with the damping effect observed by Cohen and Gürtler (2001) when the number of domestic dogs exceeded 2. Rascalou et al. (2015) reached conclusions similar to those of Coffield et al. (2013) in a model including opossums and small rodents, with both contributing substantially to *T. cruzi* transmission. These contrasting predictions are likely explained by differences in the modelling of dog infections. Dogs were set as infected in Cohen and Gürtler (2001) and then served as a constant source of parasites, while in the other frameworks, dogs were introduced in the model as new-born susceptible individuals and had to acquire the infection during their lifetime before contributing to *T. cruzi* spread. This suggests that inferring the effect of such key species requires a dynamical modelling of the eco-epidemiological feedbacks between humans, reservoirs and vectors. (See also Rock et al., 2015; this volume.) Accordingly, there is a clear need to develop further standard COM models to provide more systematic analyses of the direct and indirect effects that typical domestic community assemblages may have on the risk of disease transmission when it is associated with well-known, domiciliated triatomines species. Meanwhile, it is established that sylvatic/peridomestic triatomines can serve as 'bridge' species in bringing wild *T. cruzi* into human dwellings. The risk of transmission is then more likely to be influenced by changes in wild biodiversity and agricultural practices (Briceño-León, 2007). The above modelling studies could thus be expanded to predict the effect of changes in the biodiversity of wild species on disease transmission.

Two other intriguing aspects of *T. cruzi* are that (1) it can be transmitted by a very broad range of vector species (Gourbière et al., 2012), and (2) it shows itself very high levels of genetic diversity (El-Sayed et al., 2005). These two facets of the ecology and the evolution of Chagas disease have only very marginally been included in modelling studies (but see Chapter 7 by Lamberton et al. 2015, this volume, for examples of the importance of incorporating parasite genetics and genomics within transmission dynamics models for other neglected tropical diseases such as schistosomiasis). All current COM (and FOI) models consider only one vector species (Table 3), while in many places the diversity of vectors can range from a few cryptic species (Nouvellet et al., 2011) to very rich assemblages with up to 14 species of triatomines (Gourbière et al., 2012). Some COM models have accounted for more than one type of *T. cruzi* (Table 3). They explored the evolution of both the mode of transmission (Kribs-Zaleta and Mubayi, 2012; Kribs-Zaleta, 2014; Pelosse and Kribs-Zaleta, 2012) and virulence (Pelosse et al., 2013) by building models where susceptibles can become infected by either one of two *T. cruzi* strains. Via invasion analysis, the authors of these papers determined whether a new mutant strain with distinct phenotypic characteristics (e.g. increased virulence, more efficient vertical transmission) can invade an established resident strain. Typically, the outcome depends on underlying assumptions and parameter values controlling the dynamics of the systems. For instance, it was found that the mode of transmission (e.g. oral, stercorarian and congenital) is influenced by the host—vector ratio and that increased stochasticity was associated with increased virulence. Devillers et al. (2008) also use a multistrains model to show that coexistence between 2 strains was possible and required modelling of the reservoir species.

Many challenges remain in order to improve understanding of the impact of biodiversity on *T. cruzi* transmission. Because empirical knowledge is currently lacking about, for instance, the level of susceptibility of the many potential reservoirs or the feeding preferences of many vector species, 'strategic' models that aim at providing simple (hopefully) analytical predictions are likely to be the most appropriate avenue to obtain these new theoretical insights. Standard COM models are very attractive in such context as they can be analysed using mathematical tools (Anderson and May, 1992), and there is a track record of their use in the investigation of the transmission dynamics of other vector-borne diseases such as Lyme disease (Hatermink et al., 2008) or other multi-host disease systems such as zoonotic schistosomiasis (Rudge et al., 2013). Developing such models would provide a more detailed knowledge of the transmission of Chagas disease and help in the identification of its specific features, thereby re-enforcing our general understanding of the impact of biodiversity on human health.

3.2.2 Host and vector demography

By controlling the turnover in populations, and thus the pool of both susceptible host and vector individuals, demography is an important component of transmission. The life-cycle and the variation in abundance of vectors are typically shorter and larger, respectively, than those of humans and reservoirs. This has led many authors to describe vectors' demography in a more detailed way than that of its host and reservoir counterparts.

3.2.2.1 Vector demography

Stage structure or migration is accounted for in almost half of the modelling studies (Table 3). Differences between stages are introduced in fecundity and biting rates (Rabinovich and Rossell, 1976; Rabinovich and Himschoot, 1990), the latter with consequences on both demography and transmission (see '*Trypanosoma cruzi* transmission', Pelosse et al., 2013). While migration is crucial to both vector population dynamics (see 'Modelling dispersal') and the epidemiology of the disease (Rabinovich and Rossell 1976; Rascalou et al., 2012), its treatment in epidemiological models remains surprisingly scarce (Table 3) and in most cases only using 'patch models'. Accordingly, there is very little epidemiological modelling work on the importance of triatomine dispersal in connecting sylvatic, peridomestic and domestic transmission cycles, which contrasts with the growing concerns raised by reinfestation dynamics after the control of domiciliated species (Cecere et al., 2004) and transmission associated with nondomiciliated species (Ceballos et al., 2011). Finally, as illustrated in 'Modelling vector regulation', triatomine density is likely controlled by the availability of food resources, resting sites and seasonal variations of environmental factors (Table 3). As specified above (see 'Modelling vector regulation'), density dependence can be modelled either (1) in a phenomenological way, with a defined carrying capacity at which population size stabilizes (Rabinovich and Rossell,

1976) or (2) mechanistically, with an explicit link to host abundance and search for blood meals (Rascalou et al., 2012; Pelosse et al., 2013) or to the predation by host/reservoir. The importance of explicit regulation goes beyond demography as transmission processes are linked to the frequency of contact (see '*T. cruzi* transmission' below). Interestingly, the description of density dependence easily accounts for host/reservoir diversity: each host is assumed to support π vectors, leading to a carrying capacity of $K_v = \pi N_h$ vectors. This is straightforward to expand to multiple host species. Considering one host and one reservoir, the carrying capacity of vectors then becomes $K_v = \pi_h N_h + \pi_r N_r$ where π_h and π_r can be interpreted in the same way as above (Cruz-Pacheco et al., 2012), or as vector preference for alternative food sources. The complexity of regulation can lead to unintuitive dynamics, with the emergence of cyclic dynamics (Rabinovich and Himschoot, 1990) even in the absence of environmental forcing. However, such cyclic patterns are most often generated by environmental forcing (see 'Modelling vector regulation') such as seasonal variation in fecundity (Cohen and Gürtler, 2001), development time or biting frequency (Spagnuolo et al., 2012).

An accurate description of vector demography is especially crucial when considering vector control (see 'Modelling vector control'). For instance, Spagnuolo et al. (2011) modelled both the dynamics of domiciliary transmission of *T. cruzi* by *T. infestans* and its control by assuming a given impact of insecticide spraying on survival. These authors considered that some domiciliated vectors survived insecticide spraying by resting within cracks, from where they re-emerged once the effects of the insecticide had waned. Spagnuolo and co-workers concluded that the abundance of infected vectors and human incidence returned to their original values within 5–10 years following interruption of insecticide spraying, even after 30 years of uninterrupted yearly spraying (Spagnuolo et al., 2011). In a follow-up study (Spagnuolo et al., 2012), the authors showed that skipping spraying every third year did not influence much the control outcome, although this was highly dependent on the reservoir population. Typically, skipping spraying became ineffective when the number of chickens and dogs was increased from one dog/one chicken per house to 2 dogs or 5 chickens (see 'Host, vector and parasite biodiversity').

3.2.2.2 Host and reservoir demography

Such demographic processes are usually modelled in a simple way because host and reservoir dynamics are typically more stable, i.e. with no cycle, and, most importantly, unrelated to vector demography. Population size is then considered to be a constant, which results from modelling reproduction as a fixed flow (see Box 1**(B)**, Das and Mukherjee, 2009) or using a standard phenomenological density-dependent function of regulation (Coffield et al., 2013). In addition, it is normally assumed that the course of the disease is related to infection-age (e.g. time since infection) rather than the host/reservoir age. However, Inaba and Sekine (2004) constructed a full continuous age-structured compartmental model where both transmission and demography might become age/stage dependent (Inaba and Sekine, 2004). Here again (host and reservoir) migration is seldom incorporated (but see Fabrizio et al., 2014, Table 3), despite its well known importance, especially in urban settings (Delgado et al., 2011).

3.2.3 T. cruzi *transmission*

Describing pathogen transmission is a cornerstone of epidemiological modelling as the way pathogen, host and vector populations are linked together strongly influences the dynamic of the disease (McCallum et al., 2001). For *T. cruzi*, this is challenging as there are multiple ways hosts can become infected. While most Chagas disease models primarily focus on vector transmission, some also include congenital or oral transmission and infection from blood transfusion (Table 3).

3.2.3.1 Vector transmission

A critical aspect in modelling vectorial transmission is the description of the contact patterns between host(s) and vector(s), which requires clear assumptions on the insect feeding behaviour. The majority of Chagas disease models assume that triatomines are not limited in the number of bloodmeals they take (Table 4), which relies on evidences that triatomines have developed adaptations to efficiently locate (Guerenstein et al., 1995) and almost inconspicuously feed (Lent and Wygodzinsky, 1979). In such a case, a fixed number of bites per vector are distributed among the hosts and the number of contacts per individual host is directly proportional to vector density and inversely proportional to host density. The rate of new host (vector) infections then depends on the frequency of susceptible (infectious) hosts, which leads to the so-called 'frequency-dependent' transmission. The basic reproduction number (R_0) then increases with the (square root of the) number of vectors per host. Accordingly, the larger the vector population, the higher the probability for an individual host to get infected and to transmit the parasite back to a vector. On the contrary, in a host

Table 4 Modelling the transmission: biting rate and saturation mechanisms

	1	2	3
Limiting factor	**Vector feeding rate**	**Encounter probability**	**Host irritability**
Interpretation	All vectors feed to satiety	Vectors struggle to find hosts	Too many vectors per host, which trigger hosts response
Contact rate per vector and per host: $\beta(N_v, N_h)$	$\frac{b^{max}}{N_h}$	r	$\frac{b^{max}}{N_v}$
Rate of new host infection:	$p_{h,v}\frac{b^{max}}{N_h}I_vS_h$	$p_{h,v}rI_vS_h$	$p_{h,v}\frac{b^{max}}{N_v}I_vS_h$
Rate of new vector infection: λ_vS_v	$p_{v,h}\frac{b^{max}}{N_h}I_hS_v$	$p_{v,h}rI_hS_v$	$p_{v,h}\frac{b^{max}}{N_v}I_hS_v$
Nature of transmission	Frequency dependence on host	Density dependence	Frequency dependence on vector
Reproductive number	$R_0 = \sqrt{\frac{(b^{max})^2 p_{h,v}p_{v,h}\frac{N_v}{N_h}}{\mu_v\mu_h}}$	$R_0 = \sqrt{\frac{r^2 p_{h,v}p_{v,h}N_vN_h}{\mu_v\mu_h}}$	$R_0 = \sqrt{\frac{(b^{max})^2 p_{h,v}p_{v,h}\frac{N_h}{N_v}}{\mu_v\mu_h}}$
Relevant when	Vectors: + Hosts: ++	Vectors: + Hosts: +	Vectors: ++ Hosts: +

Modelling transmission with respect to vector feeding ecology and host behavioural response. b^{max} represents the maximum number of vector bites, N_h the contact rate per host, and N_v the contact rate per vector. r is the finding rate and $p_{h,v}$ the per contact probabilities of pathogen transmission from vector to host and $p_{v,h}$ from host to vector, μ is the mortality rate for vectors (μ_v) and humans (μ_h). N, I and S stand for the size of the total population and the number of susceptible and infectious individuals, respectively.

population of large size, each individual host receives less bites and the spread of the pathogen is harder. This may be particularly relevant when considering noncompetent (NC) hosts, i.e. hosts species that do not permit *T. cruzi* development, such as poultry. Increasing the number of NC hosts can impede transmission by reducing the average number of bites per host and by making some of those bites nonefficient with respect to parasite transmission. However, it is also likely to increase vector abundance (see section 'Host and vector demography') leading to counterintutitive outcomes that strongly depend on assumptions and parameter estimates (see 'Host, vector and parasite biodiversity'). Although frequency-dependent transmission is the standard assumption in most vector-borne studies, it is biologically sensible to consider that vectors may struggle to encounter hosts. In such situation, the contact per vector and per host is constant, and corresponds to a 'finding' rate (Table 4). The rate of new host (vector) infection then increases proportionally with the density of susceptible (infectious) hosts, which lead to the so-called 'density-dependent' transmission. Rascalou et al. (2012) used an original function of transmission that links the biting rate to two standard quantities characterizing the vector feeding behaviour; its 'finding' rate and 'handling time', i.e. the time separating two consecutive blood meals. This function allows for a gradual switching from frequency to density dependence as the frequency of blood meals and the searching efficacy of the vector increase. In this context, the emergence and prevalence of *T. cruzi* in humans were both shown to be very sensitive to the frequency of vector feeding, as compared to other key human vector-borne pathogens. A third type of *T. cruzi* transmission function has been proposed that would permit modelling of the host behavioural response when the density of triatomines per individual host is increasing. Host irritability typically leads to vector-avoidance (or predation) that starts when the number of triatomines per host exceeds some threshold (see 'Modelling vector regulation'). This was explored using a threshold function whereby the number of bites per host increased with vector density until it reached a constant level. At that point the total number of bites is fixed by the number of hosts and the biting rate per host and per vector becomes inversely proportional to vector density (Table 4). This leads to another kind of 'frequency-dependence' transmission, where the rate of new infection depends on the frequency of susceptible and infectious vectors (instead of hosts). Importantly, the sizes of the host and vector populations then have the opposite effect on R_0 compared to the situation where vectors fed to satiety. As the number of host increases, there are more opportunities for vectors to bite, so that the rate of transmission and R_0 increases. Accordingly, lowering vector density provides more opportunity for each vector to become infected and transmit back to the host, which favours transmission. Kribs-Zaleta (2010a,b) found good evidence for this type of frequency dependence (relative to vector, type 3 in Table 4) when opossums or racoons are the principal hosts as typically vectors populations are large and can trigger a host response. One the contrary, when wood-rats are the main hosts, vectors are likely to feed to satiety and transmission would be the standard type of frequency dependence (relative to host, type 1 in Table 4). Analogously to Rascalou et al. (2012), Pelosse et al. (2013) provided a formulation that allows for a gradual switching between the 2 forms of frequency dependence. These various studies strongly suggest that the type of vector transmission of *T. cruzi* is variable in the field and that its variations can drastically change the consequences of vector and/or host variation on *T. cruzi* transmission.

3.2.3.2 Other forms of transmission

Additional routes of transmission have been modelled by adding further components to the standard force of infection (Box 2**(C)**, Table 2). Congenital transmission is typically modelled as an additional force of infection applied on new borns from infected women, the latter giving birth to a proportion of infected children that has been estimated to range between 2.5% and 11% (Howard et al., 2014). To describe transmission via blood transfusion, the added force of infection has been considered as a constant (Cordovez et al., 2014) or as a frequency-dependent process between susceptible and infected hosts (Velasco-Hernandez, 1991). While congenital and blood transfusion contribute only weakly to R_0 (Das and Mukherjee, 2009, 2006; Velasco-Hernandez, 1991; Cordovez et al., 2014; Coffield et al., 2013), these routes of transmission can maintain the disease for decades in the absence of vector transmission (Raimundo et al., 2010; Fabrizio et al., 2014). Oral transmission has also usually been modelled as a constant force of infection (Cordovez et al., 2014) or using density/frequency-dependent functions (Coffield et al., 2013). In this last contribution the authors used a Holling type II functional response to describe oral transmission to dogs, which switches from density dependence (at low vector density) to frequency dependence (at high vector density). While oral transmission remains a secondary road of Chagas disease infection in humans (Coffield et al., 2013), it is epidemiologically very important given its high fatality rate (Bastos et al., 2010) and the substantial size of related outbreaks (Alarcón de Noya et al., 2010). In addition, for reservoirs such as opossum and dogs, oral transmission plays a more important role because of a substantial rate of triatomine ingestion (Coffield et al., 2013; Rabinovich, 1972) that leads to a substantially higher probability of infection than pure stercorarian transmission (Rabinovich et al., 2001).

While COM models have traditionally been focussed on vector transmission of *T. cruzi*, recent studies tend to account for alternative routes. This may reflect the decrease in active vector transmission due to large-scale insecticide spraying, and the concomitant inclination to bring forward new targets. Also this brings original insights; the above studies and the various failures of vector control clearly demonstrate that further knowledge on triatomine feeding and host ecology are clear pre-requisites to accurately understand and model Chagas disease epidemiology. Potentially, high quality epidemiological data could then facilitate the selection among various alternative COM models comparing vector, congenital and/or oral routes of transmission, thereby improving understanding of their quantitative contributions to human infections.

3.2.4 Modelling within-host dynamics of T. cruzi

In the typical course of many diseases, individuals go from being exposed (infected but not yet infectious) to infections, then recovered (and immune) until eventually losing immunity. However, for Chagas disease, humans and animals hosts become rapidly infectious and never clear the parasites, which explains why almost no model includes an 'exposed' stage nor allows for infecteds to recover to a 'susceptible' state (Table 3). While this lack of parasite clearance simplifies the interpretation of seroprevalence data (as no seroreversion occurs), it implies that the infectiousness period in human is very long. This period can be divided into an acute and a chronic period, and chronically infected individuals may develop into a symptomatic/pathogenic phase which contributes most to the fatality burden of the disease (Rassi and Marin-Neto, 2010). Individuals may spend different amounts of time in these various compartments where they suffer distinct mortality rates and show various levels of infectiousness (Figure 5). Such details, in addition to providing better insights into the dynamics of the course of the disease, are critical to predict the burden associated with Chagas disease (Lee et al., 2012, 2013). Despite this, the majority of Chagas disease models include a single infectious stage; some accounted for both the acute and chronic phases of the disease, and only a few consider the chronic phase with determinate pathology (Table 3). In these latter models, the mortality associated with the disease, i.e. the virulence of the parasite, is assumed to be low and is

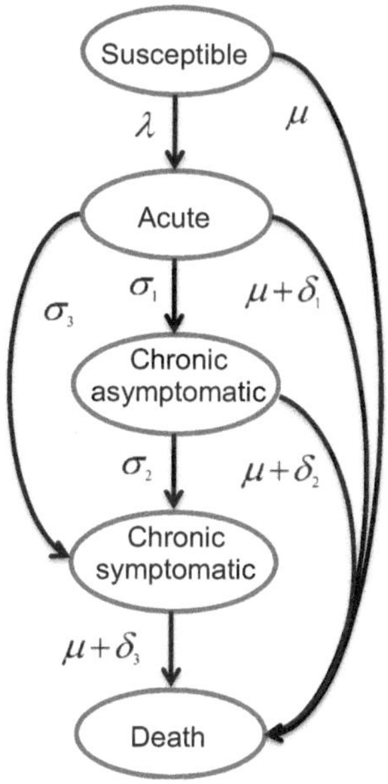

Figure 5 Basic Markov model for Chagas disease. The ovals represent health states and are linked with transition probabilities, σ_x: transition probabilities between states. Additionally λ is the force of infection; μ, general mortality; $\delta_{1,2,3}$ additional mortality in (1) acute, (2) chronic asymptomatic and (3) chronic symptomatic phase (usually $\delta_2 \sim 0$).

sometimes even omitted (Table 3). When acute/chronic stages are incorporated, the highest mortality occurs during the chronic with determinate pathology phase, followed by the acute and chronic asymptomatic phases (Bastos et al., 2010; Devillers et al., 2008; Ramírez et al., 2013). There are empirical evidences backing such assumptions (Sanches et al., 2014), although mortality estimates remain uncertain and heterogeneous (Carrasco et al., 1994; Maguire et al., 1987). In particular, mortality associated with oral transmission during the acute phase is increased, most likely because of higher parasitaemia and the presence of sylvatic strain of *T. cruzi* (Shikanai-Yasuda and Carvalho, 2012). Finally, while Chagas disease is thought as a lifelong infection, some models have nonetheless considered a probability of recovering (Velasco-Hernandez, 1991) and have shown important consequences when aetiological treatment is involved as a control strategy (Cruz-Pacheco et al., 2012; Fabrizio et al., 2014). We anticipate that the emergence of the modelling of the interaction between *T. cruzi* and the immune system in the intra- and extra-cellular environment will contribute to our understanding of the course of infection and its variability at the individual scale (Galvão and Miranda, 2010; Isasi et al., 2001; Sibona et al., 2007), which may help refining its description in COM models.

3.3 Cost–benefit and cost-effectiveness analysis

Mathematical models have the potential to influence health policy makers aiming at best controlling, preventing and treating the disease. Models, therefore, need to account for the economic values of health status and estimators of the quality of life. These, and the social or health system perspective used will determine the outcomes of those economic analyses (see Lee et al., 2015, Chapter 8 this volume, for a discussion of economic evaluations in NTDs and the various perspectives that can be taken). As such, these models are different from the dynamical models presented above that primarily deal with the eco-epidemiology of the disease.

Two different approaches have been used for the economic evaluation of Chagas disease control and treatment measures: cost–benefit and cost-effectiveness analyses. While the focal point of a cost–benefit analysis is the monetary benefit of life saved or cases averted, cost-effectiveness analysis focuses on the health benefit of interventions typically measured by the Quality Adjusted or Disability Adjusted Life Years (QALYs and DALYs, respectively) that assign specific disability weights according to disease stages (Lozano et al., 2012; Murray et al., 2012).

Because of their complexity, cost–benefit and effectiveness studies of Chagas disease do not link the force of infection with the number of vectors or infectious individual hosts. Given their public health importance, we briefly review some of these contributions.

Schofield and Dias (1991) conducted the first study to assess the cost–benefit (monetary value of healthy lives gained) of insecticide spraying to control vectors in Latin America. The model only included the cost incurred by death in the acute phase and the cost of supportive treatment in the chronic phase, and did not consider other treatments, opportunity costs or incapacity for work. Schofield and Dias (op. cit.) found a high, over 14%, internal rate of return, which measures the profitability of investments, and highlighted that without a surveillance phase of the control programme the cost would only decrease by 5.4% while the overall benefits would be reduced by 33%. In another study, Basombrío et al. (1998) also estimated the cost–benefit through a societal perspective in Argentina, i.e. the cost of medical care plus incapacity for work. These authors estimated the number of averted cases by the vector control strategy and found an internal rate of return in excess of 60%. Both studies assumed a constant incidence in absence of control, and a reduced and constant incidence once the programme was implemented. These studies were very important to support vector control strategies in the South Cone countries over the last three decades.

In recent years, with the advent of the Burden of Disease Studies (Hotez et al., 2014), cost-effectiveness analyses have become more widely accepted to inform decision making in health systems. As mentioned above, in the context of Chagas disease, these studies did not consider the dynamical feedbacks between infectious and susceptibles directly. They were based on FOI or COM models with constant incidence or 'Markov models' and 'decision trees', which can be viewed as discretized COM with fixed incidence (Kim and Sue, 2008, Turner et al., 2014a). These models provided economic evaluations of vector control (Wilson et al., 2005), congenital transmission (Billot et al., 2005; Sicuri et al., 2011), transmission associated with transfusion (Wilson et al., 2008), aetiological treatment (Ramsey et al., 2014), impact of potential vaccines (Lee et al., 2010) and assessment of the disease burden (Lee et al., 2013).

Markov models are particularly useful for inclusion of several exclusive stages of the disease linked by transition probabilities that can be constant or vary over time (Briggs and Sculpher, 1998). In the case of Chagas disease, they typically include acute, chronic asymptomatic, chronic symptomatic and death stages (see Figure 5). Interestingly, Castillo-Riquelme et al. (2008) built a model where the incidence, while constant, depended on the state of the house (e.g. infested or not). They then overlaid another ('Markov type') model to specify the number of infested houses as a function of spraying schedule and monitoring (e.g. sensitivity and specificity of surveillance). This allowed some degree of freedom into the FOI and could be interpreted as somehow linking vector populations to incidence in humans. Contrastingly, the recent projections of Chagas disease by the Global Burden of Disease Study 2010 have not used dynamical or force of infection models, but rather estimates of the fraction of the heart disease burden which can be attributed to Chagas disease (Murray et al., 2012).

Some promising results combining more standard COM models and health economics have better informed control strategies not only for directly-transmitted infections like influenza or HPV (Vanni et al., 2012; Chen and Liao, 2013), but also for vector borne diseases such as onchocerciasis (Turner et al., 2014b, 2014c) or malaria (Briët and Penny, 2013). However, for Chagas disease, as for many NTDs (Turner et al., 2014a), combining dynamical models with economic evaluations is still lacking for Chagas disease. While such a combination would help in achieving a more comprehensive understanding of the best strategies to be implemented, they rely on a robust and data-driven understanding of the dynamics of the disease that may not be available yet in various regions of Latin America where the importance of this public health concern is still underrecognized.

3.4 Conclusion on epidemiological models of *T. cruzi* infection

As compared with COM models, FOI models require less knowledge to estimate parameters and provide arguably more robust results. They can easily be adapted to combine multiple paths of transmission and integrated within a statistical framework to permit both estimation of parameters and selection of models. Their use is not restricted to human hosts, and much could be learned from adapting those models to serological data from animal reservoirs. While they omit the dynamical link between the force of infection and the number of infectious individuals, some contributions have tried to link the force of infection to the vector population (Levy et al., 2011; Nouvellet et al., 2013; Rabinovich et al., 2001). These offer interesting perspectives as these links can be seen as an equilibrium state of COM models, therefore bridging gaps between the two standard approaches. Statistical models conceptually related to those associated with FOI models are becoming increasingly available (Cauchemez et al., 2009; Churcher et al., 2014; Ferguson et al., 2001). Combined with epidemiological and ecological data they could thus make valuable contributions to our understanding of Chagas dynamics, especially by reducing uncertainty surrounding key parameters. COM models have brought interesting insights on the key determinants of incidence and how to reduce it, although a systematic analysis of the role of each modelled component is still largely lacking. Such an orderly analysis would help to identify the key dynamical specificities of Chagas disease as compared to other vector-borne diseases, and it would undoubtedly offer insights on the failures and successes of control strategies by providing the opportunity for rigorous quantitative comparisons. Finally, developing COM models of Chagas disease in a harmonized manner would help building the robust ecological modelling basis on which theories of adaptive evolution can be built. Not only this would benefit our general understanding of the evolution of disease transmission, but also could be used in a practical context to tailor evolutionary-proof control strategies.

4. GENERAL CONCLUSION

Mathematics is biology's 'next microscope' (Cohen, 2004) as they 'can reveal otherwise invisible worlds in all kinds of data'. Complementary to more common field and experimental studies, this 'third road' provides a powerful way of grasping knowledge and understanding complex systems when observations and experiments are severely constrained by logistic, financial or ethical reasons, which is typically the situation faced by scientists working on neglected tropical diseases. Additionally, it is now well established that tackling Chagas disease and other NTDs requires a multidisciplinary approach to comprehend its ecology, evolution and epidemiology together with the surrounding socioeconomic context. Mathematical models are unique in their ability to intermingle knowledge within and between these various fields of research, and they can thus serve as the *agora* where an efficient quantitative integration of those empirical pieces of knowledge can be achieved.

While one should thus expect dynamical models to be a common practice in the field of Chagas disease, as it is for other human infectious diseases such as, e.g., malaria (Mandal et al., 2011), dengue (Arnauld et al., 2012) or helminthiasis (Anderson and May 1992), this approach has been overlooked in the last 100 years of research on *T. cruzi* and its triatomine vectors (see Rascalou and Gourbière, unpublished data for a quantitative comparison of the theoretical literature across all neglected tropical diseases). Although several important and inspirational contributions were made as far back as the 1970s, the use of modelling remains even today confined to a few research groups. Accordingly, while modelling studies have clearly demonstrated their potential in the Chaco (Cohen and Gürtler, 2001; Fabrizio et al., 2014), in Peru (Levy et al. 2011., Delgado et al., 2011) and in the Yucatan peninsula (Barbu et al., 2009, 2010, 2011, Gourbière et al., 2008), many excellent empirical research efforts have not benefited yet from the integration of collected data into the quantitative conceptual frameworks that models can provide. The multiplication of such attempts, which could in part rely on already available data, would facilitate a more general understanding of the local specificities of *T. cruzi* transmission as well as its differences with other vector-borne diseases. This would greatly contribute to the scientific positioning of the disease in the general ecological and evolutionary literature.

It can be anticipated that attracting the interest of biologists outside of the field will also require the development of simpler and more general approaches that can be connected to theoretical ecology and epidemiology.

This would indeed help to emphasize research outcomes toward the key features of the system rather than developing more complex models. Efforts have already been made in this direction by applying key concepts of theoretical ecology such as source/sink (Rascalou et al., 2012) or metapopulation (Lardeux, 2013) dynamics to explain the spatiotemporal distribution of Chagas disease vectors. Another prospective example relates to the assessment of the impact of biodiversity loss on human well-being. Many recent and stimulating studies have been undertaken with the inviting idea that biodiversity could dilute transmission and thus protect human society against the (re-) emergence of infectious diseases (e.g. Cardinale et al., 2012; Levy et al. 2011). Chagas disease is an excellent model system to test such 'dilution' effect as it presents all four key features to promote dilution: (1) a generalist vector that (2) principally acquire infection horizontally and (3) can feed on reservoir species with heterogeneous competences with (4) the most competent host being dominant in the community of interest (Johnson and Thieltges, 2010; Keesing et al., 2010). Without dismissing the complexity of the system, we believe that more harmonization and comparison Chagas disease models and a better integration into the general literature could help to improve our understanding of its transmission dynamics. Topics such as dispersion, environmental stochasticity, heterogeneity in exposure (especially linked to socioeconomic context), local adaptation, resistance to insecticides are among key issues that are shared with many other diseases and across areas other than epidemiology, and for which a robust analytical understanding is critical.

Asides from their contribution to the development of a theoretical understanding of the complex eco-epidemiological interactions between *T. cruzi* and its multiple hosts and vectors, dynamical models can also be used to test hypotheses about the specific mechanisms of interaction that could be responsible for observed patterns. Mathematical models can indeed be used to represent a hypothetical mechanism of interaction and to identify what patterns should be expected under the specific hypothesis. Predictions can then be tailored to the type of data that can be or have been collected, so that tests are easier. If several hypotheses are competing, one can confront the data with the predictions of several models looking for the model that best fit the observations. The fit of each model to data can be made using standard methods based on basic likelihood statistics and, when combined with theoretical information criteria, the support received by each fitted model (and thus by each hypothesis), can be compared (Johnson and Omland, 2004). This not only allows for estimating the values of the parameters describing the potential mode of interactions in the various models, but also comparing the support that the data give to each of the models. This 'model comparison' approach can be seen as a valuable alternative to standard statistical tests as it allows the testing of several hypothesis at a time (Anderson et al., 2000). Such a coupling between mathematical and statistical models has already been used successfully, for example, to infer from the spatiotemporal dynamics of vector populations, the dispersal mechanisms of triatomines (Barbu et al., 2010, 2011), which is a component of their life-history that remains hard to study in the field. Mathematical models embedded into theoretical information frameworks are more than simple microscopes as they help to dissect the interaction processes at work. We believe these scientific 'scalpels' constitute a promising avenue for a better integration of dynamical modelling into the practice of primarily empirical research groups in the near future. They will ease the virtuous circle between data collection and model building that benefit both empirical and theoretical studies. This concurs well with the ever-increasing requirement for the research on Chagas disease to produce robust and data-supported results if it claims to influence policy makers and stakeholders through an evidence-based approach.

Hopefully, this review of the last century of the modelling of *T. cruzi* and Chagas disease transmission will provide the opportunity for more research groups or public health agencies to appreciate the potential of this 'third road' to study biology. As we truly wish this effort to be instrumental and to serve as a turning point toward an improved use of mathematical models in the community, we will propose regular updates on the modelling in the field and facilitate tools that will be made available through the Website of the last author of this contribution.

ACKNOWLEDGMENTS

This work has benefited from an "Investissements d'Avenir" grant managed by Agence Nationale de la Recherche (CEBA, ref. ANR-10-LABX-25-01). This work was performed within the framework of the LABEX ECOFECT (ANR-11-LABX-0048) of Université de Lyon, within the program "Investissements d'Avenir" (ANR-11-IDEX-0007) operated by the French National Research Agency (ANR). PN thank the Medical Research Council Centre for Outbreak Analysis and Modelling for support and received funding from the European Union Seventh Framework Programme [FP7/2007–2013] under Grant Agreement n°278433-PREDEMICS and from the Health Protection Research Units of the National Institute for Health Research. ZMC is PhD student funded by the Departamento Administrativo Nacional de Ciencia y Tecnología de Colombia – 'COLCIENCIAS' and RED CHAGAS COLOMBIA, grant number 380-2011, code 5014-537-30398.

REFERENCES

Alarcón de Noya, B., et al., 2010. Large urban outbreak of orally acquired acute Chagas disease at a school in Caracas, Venezuela. J. Infect. Dis. 201 (9), 1308–1315.

Annan, K., Fisher, M., 2013. Stability conditions of chagas-HIV co-infection disease model using the next generation method. Appl. Math. Sci. 7 (57), 2815–2832.

Anderson, R., May, R., 1992. Infectious Diseases of Humans: Dynamics and Control. OUP Oxford (New Ed edition).

Anderson, D.R., et al., 2000. Null hypothesis testing: problems, prevalence, and an alternative. J. Wildl. Manage. 64, 912–923.

Andraud, M., Hens, N., Marais, C., Beutels, P., 2012. Dynamic epidemiological models for dengue transmission: a systematic review of structural approaches. PLoS One 7 (11), e49085.

Aparicio, J.P., Castillo-Chavez, C., 2009. Mathematical modelling of tuberculosis epidemics. Math. Biosci. Eng. MBE 6 (2), 209–237.

Barbu, C., Dumonteil, E., Gourbière, S., 2009. Optimization of control strategies for non-domiciliated *Triatoma dimidiata*, Chagas disease vector in the Yucatán Peninsula, Mexico. PLoS Negl. Trop. Dis. 3 (4), e416.

Barbu, C., Dumonteil, E., Gourbière, S., 2010. Characterization of the dispersal of non-domiciliated *Triatoma dimidiata* through the selection of spatially explicit models. PLoS Negl. Trop. Dis. 4 (8), e777.

Barbu, C., Dumonteil, E., Gourbière, S., 2011. Evaluation of spatially targeted strategies to control non-domiciliated *Triatoma dimidiata* vector of Chagas disease. PLoS Negl. Trop. Dis. 5 (5), e1045.

Barbu, C.M., et al., 2013. The effects of city streets on an urban disease vector. PLoS Comput. Biol. 9 (1), e1002801.

Basombrío, M.A., Schofield, C.J., Rojas, C.L., del Rey, E.C., 1998. A cost-benefit analysis of Chagas disease control in north-western Argentina. Trans. R. Soc. Trop. Med. Hyg. 92 (2), 137–143.

Basombrío, M.A., et al., 1999. The transmission de Chagas disease in salta and the detection of congenital cases. Medicina 59 (Suppl. 2), 143–146.

Bastos, C.J.C., et al., 2010. Clinical outcomes of thirteen patients with acute chagas disease acquired through oral transmission from two urban outbreaks in northeastern Brazil. In: Sinnis, P. (Ed.), PLoS Neglected Tropical Diseases, vol. 4(6), p. e711.

Billot, C., Torrico, F., Carlier, Y., 2005. Cost effectiveness study of a control program of congenital Chagas disease in Bolivia. Rev. Soc. Bras. Med. Trop. 38 (Suppl. 2), 108–113.

Boischio, A., et al., 2009. Health and sustainable development: challenges and opportunities of ecosystem approaches in the prevention and control of dengue and Chagas disease. Cad. Saúde Pública 25 (Suppl. 1), S149–S154.

Bowman, N.M., et al., 2008. Chagas disease transmission in periurban communities of Arequipa, Peru. Clin. Infect. Dis. Off. Publ. Infect. Dis. Soc. Am. 46 (12), 1822–1828.

Briceño-León, R., 2007. Chagas disease and globalization of the Amazon. Cad. Saúde Pública 23, S33–S40.

Briceño-León, R., 2009. Chagas disease in the Americas: an ecohealth perspective. Cad. Saúde Pública 25 (Suppl. 1), S71–S82.

Briët, O.J., Penny, M.a, 2013. Repeated mass distributions and continuous distribution of long-lasting insecticidal nets: modelling sustainability of health benefits from mosquito nets, depending on case management. Malar. J. 12, 401.

Briggs, A., Sculpher, M., 1998. An introduction to Markov modelling for economic evaluation. Pharmacoeconomics 13 (4), 397–409.

Cardinale, B.J., et al., 2012. Biodiversity loss and its impact on humanity. Nature 486 (7401), 59–67.

Carrasco, H., Parada, H., Guerrero, L., 1994. Prognostic implications of clinical, electrocardiographic and hemodynamic findings in chronic Chagas' disease. Int. J. Cardiol. 43 (1), 27–38.
Castañera, M.B., Aparicio, J.P., Gürtler, R.E., 2003. A stage-structured stochastic model of the population dynamics of *Triatoma infestans*, the main vector of Chagas disease. Ecol. Model. 162 (1–2), 33–53.
Castillo-Riquelme, M., et al., 2008. The costs of preventing and treating chagas disease in Colombia. PLoS Negl. Trop. Dis. 2 (11), e336.
Caswell, H., 2000. Matrix Population Models: Construction, Analysis and Interpretation, second ed. Sinauer Associates Inc, U.S. 2nd Revised edition.
Cauchemez, S., et al., 2009. Household transmission of 2009 Pandemic influenza a (H1N1) virus in the United States. N. Engl. J. Med. 361 (27), 2619–2627.
Ceballos, L.A., Piccinali, R.V., Marcet, P.L., Vazquez-Prokopec, G.M., Cardinal, M.V., Schachter-Broide, J., Dujardin, J.P., Dotsonm, E.M., Kitron, U., Gürtler, R.E., 2011. Hidden sylvatic foci of the main vector of Chagas disease *Triatoma infestans*: threats to the vector elimination campaign? PLoS Negl. Trop. Dis. 5 (10), e1365.
Cecere, M.C., Canale, D.M., Gürtler, R.E., 2002. Effects of refuge availability on the population dynamics of *Triatoma infestans* in experimental huts under natural climatic conditions in central Argentina. J. Appl. Ecol. 40 (4), 742–756.
Cecere, M.C., Vazquez-Prokopec, G.M., Gürtler, R.E., Kitron, U., 2004. Spatio-temporal analysis of reinfestation by *Triatoma infestans* (Hemiptera: reduviidae) following insecticide spraying in a rural community in northwestern Argentina. Am. J. Trop. Med. Hyg. 71 (6), 803–810.
Cecere, M.C., Vazquez-Prokopec, G.M., Gürtler, R.E., Kitron, U., 2006. Reinfestation sources for chagas disease vector, *Triatoma infestans*, Argentina. Emerg. Infect. Dis. 12 (7), 1096–1102.
Chagas, C., 1909. Nova tripanozomiaze humana. Estudos sobre a morfolojia e o ciclo evolutivo do Schizotrypanum cruzi n. gen, n. sp., ajente etiolojico de nova entidade morbida do homem. Mem Inst Oswaldo Cruz. 1, 11–80.
Chaves, L.F., et al., 2004. Mortality profiles of *Rhodnius prolixus* (Heteroptera: reduviidae), vector of Chagas disease. Acta Trop. 92 (2), 119–125.
Chen, S.-C., Liao, C.-M., 2013. Cost-effectiveness of influenza control measures: a dynamic transmission model-based analysis. Epidemiol. Infect. 141, 1–14.
Churcher, T.S., et al., 2014. Measuring the path toward malaria elimination. Science 344 (6189), 1230–1232.
Clauson, M., et al., 2012. Analysis of the steady states of a mathematical model for Chagas disease. Involve 5 (3), 237–246.
Coffield, D.J., et al., 2013. A model for Chagas disease with oral and congenital transmission. In: Tanowitz, H.B. (Ed.), PloS One, vol. 8(6), p. e67267.
Cohen, J.E., 2004. Mathematics is biology's next microscope, only better; biology is mathematics' next physics, only better. PLoS Biol. 2 (12), e439.
Cohen, J.E., Gürtler, R.E., 2001. Modeling household transmission of American trypanosomiasis. Sci. (New York, N.Y.) 293 (5530), 694–698.
Colbert, J., et al., 2001. Dispersal. Oxford University Press, New York, NY.
Cordovez, J.M., et al., 2014. Using the basic reproduction number to assess the effects of climate change in the risk of Chagas disease transmission in Colombia. Acta trop. 129, 74–82.
Costa, J., Peterson, A.T., 2012. Ecological niche modeling as a tool for understanding distributions and interactions of vectors, hosts, and etiologic agents of Chagas disease. Adv. Exp. Med. Biol. 710, 59–70.
Costa, J., Peterson, A.T., Beard, C. Ben, 2002. Ecologic niche modeling and differentiation of populations of *Triatoma brasiliensis neiva*, 1911, the most important Chagas' disease

vector in northeastern Brazil (hemiptera, reduviidae, triatominae). Am. J. Trop. Med. Hyg. 67, 516–520.

Cruz-Pacheco, G., Esteva, L., Vargas, C., 2012. Control measures for Chagas disease. Math. Biosci. 237 (1), 49–60.

Das, Mukherjee, 2009. Aspects of an epidemic model of chagas disease. Tamsui Oxf. J. Math. Sci. s25 (3), 301–311.

Das, P., Mukherjee, D., 2006. Qualitative study of a model of Chagas' disease. Math. Comput. Model. 43 (3), 413–422.

Delgado, S., et al., 2011. A history of chagas disease transmission, control, and re-emergence in peri-rural La Joya, Peru. PLoS Negl. Trop. Dis. 5 (2), e970.

Devillers, H., Lobry, J.J.R., Menu, F., 2008. An agent-based model for predicting the prevalence of *Trypanosoma cruzi* I and II in their host and vector populations. J. Theor. Biol. 255 (3), 307–315.

Dias, J.C.P., Silveira, A.C., Schofield, C.J., 2002. The impact of Chagas disease control in Latin America: a review. Memórias do Inst. Oswaldo Cruz 97 (5), 603–612.

Dumonteil, E., et al., 2013. Eco-bio-social determinants for house infestation by non-domiciliated *Triatoma dimidiata* in the Yucatan Peninsula, Mexico. PLoS Negl. Trop. Dis. 7 (9), e2466.

El-Sayed, N.M., et al., 2005. The genome sequence of *Trypanosoma cruzi*, etiologic agent of Chagas disease. Science 309, 409–415.

Fabrizio, M.C., Schweigmann, N.J., Bartoloni, N.J., 2014. Modelling inter-human transmission dynamics of Chagas disease: analysis and application. Parasitology 141 (6), 837–848.

Feliciangeli, M.D., Rabinovich, J., 1985. Vital statistics of triatominae (hemiptera, reduviidae) under laboratory conditions .2. *Triatoma-maculata*. J. of Med. Entomol. 22 (1), 43–48.

Feliciangeli, M.D., et al., 2003. Chagas disease control in Venezuela: lessons for the Andean region and beyond. Trends Parasitol. 19 (1), 44–49.

Ferguson, N.M., Donnelly, C.A., Anderson, R.M., 2001. The foot-and-mouth epidemic in Great Britain: pattern of spread and impact of interventions. Science 292 (5519), 1155–1160.

Ferral, J., et al., 2010. Comparative field trial of alternative vector control strategies for non-domiciliated *Triatoma dimidiata*. Am. J. Trop. Med. Hyg. 82, 60–66.

Galvão, V., Miranda, J.G.V., 2010. A three-dimensional multi-agent-based model for the evolution of Chagas' disease. Bio Syst. 100 (3), 225–230.

Gilpin, M.E., Hasnky, I., 1991. Metapopulation Dynamics: Empirical and Theoretical Investigations. Academic Press.

Goldbaum, M., Barreto, M.L., 2008. Commentary: the contribution and example of Carlos Chagas. Int. J. Epidemiol. 37 (4), 697–698.

Gottdenker, N.L., Chaves, L.F., Calzada, J.E., Saldaña, A., Carroll, C.R., 2012. Host life history strategy, species diversity, and habitat influence *Trypanosoma cruzi* vector infection in changing landscapes. PLoS Negl. Trop. Dis. 6 (11), e1884.

Gourbiere, S., Gourbiere, F., 2002. Competition between unit-restricted fungi: a metapopulation model. J. Theor. Biol. 217 (3), 351–368.

Gourbière, S., et al., 2008. Demographic and dispersal constraints for domestic infestation by non-domicilated chagas disease vectors in the Yucatan Peninsula, Mexico. Am. J. Trop. Med. Hyg. 78 (1), 133–139.

Gourbière, S., et al., 2012. Genetics and evolution of triatomines: from phylogeny to vector control. Heredity 108 (3), 190–202.

Gourbière, S., Menu, F., 2009. Adaptive dynamics of dormancy duration variability: evolutionary trade-off and priority effect lead to suboptimal adaptation. Evolution 63, 1879–1892.

Guerenstein, P.G., Lorenzo, M.G., Núñez, J.A., Lazzari, C.R., 1995. Baker's yeast, an attractant for baiting traps for Chagas' disease vectors. Experientia 51, 834–837.
Guhl, F., et al., 2005. Lessons from a national survey of Chagas disease transmission risk in Colombia. Trends Parasitol. 21 (6), 259–262.
Gurgel-Gonçalves, R., et al., 2012. Geographic distribution of chagas disease vectors in Brazil based on ecological niche modeling. J. Trop. Med. 2012, 705326.
Gürtler, R.E., et al., 1991. Chagas disease in north-west Argentina: infected dogs as a risk factor for the domestic transmission of *Trypanosoma cruzi*. Trans. R. Soc. Trop. Med. Hyg. 85 (6), 741–745.
Gürtler, R.E., et al., 1997. Shifting host choices of the vector of Chagas disease, *Triatoma infestans*, in relation to the availability of hosts in houses in north-west Argentina. J. Appl. Ecol. 34 (3), 699–715.
Gürtler, R.E., et al., 2005. Incidence of *Trypanosoma cruzi* infection among children following domestic reinfestation after insecticide spraying in rural northwestern Argentina. Am. J. Trop. Med. Hyg. 73 (1), 95–103.
Hartemink, N.A., Randolph, S.E., Davis, S.A., Heesterbeek, J.A., 2008. The basic reproduction number for complex disease systems: defining R(0) for tick-borne infections. Am. Nat. 171 (6), 743–754.
Hanski, I., 1991. Single-species metapopulation dynamics – concepts, models and observations. Biol. J. Linn. Soc. 42 (1), 17–38.
Hotez, P.J., et al., 2014. The Global Burden of Disease Study 2010: interpretation and implications for the Neglected Tropical Diseases. PLoS Negl Trop Dis. 8 (7), e2865.
Howard, E.J., et al., 2014. Frequency of the congenital transmission of *Trypanosoma cruzi*: a systematic review and meta-analysis. BJOG: Int. J. Obstet. Gynaecol. 121 (1), 22–33.
Inaba, H., Sekine, H., 2004. A mathematical model for Chagas disease with infection-age-dependent infectivity. Math. Biosci. 190 (1), 39–69.
Isasi, S.C., Sibona, G.J., Condat, C.A., 2001. A simple model for the interaction between *T. cruzi* and its antibodies during Chagas infection. J. Theor. Biol. 208 (1), 1–13.
Johnson, P.T., Thieltges, D.W., 2010. Diversity, decoys and the dilution effect: how ecological communities affect disease risk. J. Exp. Biol. 213 (6), 961–970.
Johnson, J.B., Omland, K.S., 2004. Model selection in ecology and evolution. TREE 19 (2), 101–108.
Keesing, F., et al., 2010. Impacts of biodiversity on the emergence and transmission of infectious diseases. Nature 468 (7324), 647–652.
Kim, S.Y., Goldie, S.J., 2008. Cost-effectiveness analyses of vaccination programmes: a focused review of modelling approaches. PharmacoEconomics 26 (3), 191–215.
Kribs-Zaleta, C.M., 2010a. Alternative transmission modes for *Trypanosoma cruzi*. Math. Biosci. Eng. MBE 7 (3), 657–673.
Kribs-Zaleta, C.M., 2010b. Estimating contact process saturation in sylvatic transmission of *Trypanosoma cruzi* in the United States. PLoS Negl. Trop. Dis. 4 (4), e656.
Kribs-Zaleta, C.M., 2014. Graphical analysis of evolutionary trade-off in sylvatic *Trypanosoma cruzi* transmission modes. J. Theor. Biol. 353, 34–43.
Kribs-Zaleta, C.M., Mubayi, A., 2012. The role of adaptations in two-strain competition for sylvatic *Trypanosoma cruzi* transmission. J. Biol. Dyn. 6 (2), 813–835.
Lamberton, P.H.L., Crellen, T., Cotton, J.A., Webster, J.P., 2015. Modelling the effects of mass drug administration on the molecular epidemiology of schistosomes. Adv. Parasitol. 87, 293–327.
Lardeux, F., 2013. Niche invasion, competition and coexistence amongst wild and domestic Bolivian populations of Chagas vector *Triatoma infestans* (Hemiptera, Reduviidae, Triatominae). Comptes Rendus Biol. 336 (4), 183–193.

Lee, B.Y., et al., 2010. The potential economic value of a *Trypanosoma cruzi* (Chagas disease) vaccine in Latin America. PLoS Negl. Trop. Dis. 4 (12), e916.
Lee, B.Y., et al., 2012. Modeling the economic value of a Chagas' disease therapeutic vaccine. Hum. Vaccines Immunother. 8 (9), 1293–1301.
Lee, B.Y., et al., 2013. Global economic burden of Chagas disease: a computational simulation model. Lancet Infect. Dis. 13 (4), 342–348.
Lee, B.Y., Bartsch, S.M., Gorham, K.M., 2015. Economic and financial evaluation of Neglected Tropical Diseases. Adv. Parasitol. 87, 329–417.
Lent, H., Wygodzinsky, P., 1979. Revision of the triatominae hemiptera heteroptera reduviidae and their significance as vectors of chagas disease. Bull. Am. Mus. Nat. Hist. 163 (3), 123–520.
Levy, M.Z., et al., 2008. Impregnated netting slows infestation by *Triatoma infestans*. Am. J. Trop. Med. Hyg. 79 (4), 528–534.
Levy, M.Z., et al., 2010. Rational spatio-temporal strategies for controlling a Chagas disease vector in urban environments. J. R. Soc. Interface/R. Soc. 7 (48), 1061–1070.
Levy, M.Z., et al., 2011. Retracing Micro-Epidemics of Chagas Disease Using Epicenter Regression. PLoS Comput Biol 7 (9), e1002146.
Lozano, R., et al., 2012. Global and regional mortality from 235 causes of death for 20 age groups in 1990 and 2010: a systematic analysis for the Global Burden of Disease Study 2010. Lancet 380 (9859), 2095–2128.
Lucero, D.E., et al., 2013. Ecohealth interventions limit triatomine reinfestation following insecticide spraying in La Brea, Guatemala. Am. J. Trop. Med. Hyg. 88 (4), 630–637.
Maguire, J.H., et al., 1987. Cardiac morbidity and mortality due to Chagas' disease: prospective electrocardiographic study of a Brazilian community. Circulation 75 (6), 1140–1145.
Mahdjoub, T., Menu, F., 2008. Prolonged diapause: a trait increasing invasion speed? J. Theor. Biol. 251 (2), 317–330.
Mallet, J., 2012. The struggle for existence: how the notion of carrying capacity, K, obscures the links between demography, Darwinian evolution, and speciation. Evol. Ecol. Res. 14, 627–665.
Mandal, S., Sarkar, R.R., Sinha, S., 2011. Mathematical models of malaria—a review. Malar. J. 10 (1), 202.
Martinez-Ibarra, J.A., et al., 2013. Comparative bionomics of four populations of Meccus longipennis (Hemiptera: reduviidae: Triatominae) under laboratory conditions. Memorias do Inst. Oswaldo Cruz 108 (2), 239–242.
Massad, E., 2008. The elimination of Chagas' disease from Brazil. Epidemiol. Infect. 136 (9), 1153–1164.
McCallum, H., Barlow, N., Hone, J., 2001. How should pathogen transmission be modelled? Trends Ecol. Evol. 16 (6), 295–300.
Moncayo, A., Silveira, A.C., 2009. Current epidemiological trends for Chagas disease in Latin America and future challenges in epidemiology, surveillance and health policy. Memórias do Inst. Oswaldo Cruz 104 (Suppl. (May)), 17–30.
Murray, C.J.L., et al., 2012. GBD 2010: design, definitions, and metrics. Lancet 380 (9859), 2063–2066.
Noireau, F., Cortez, M.G., Monteiro, F.A., Jansen, A.M., Torrico, F., 2005. Can wild *Triatoma infestans* foci in Bolivia jeopardize Chagas disease control efforts? Trends Parasitol. 21 (1), 7–10.
Nouvellet, P., Ramirez-Sierra, M.J., Dumonteil, E., Gourbière, S., 2011. Effects of genetic factors and infection status on wing morphology of *Triatoma dimidiata* species complex in the Yucatán peninsula, Mexico. Infect. Genet. Evol. 11 (6), 1243–1249.

Nouvellet, P., Dumonteil, E., Gourbière, S., 2013. The improbable transmission of *Trypanosoma cruzi* to human: the missing link in the dynamics and control of Chagas disease. PLoS Negl. Trop. Dis. 7 (11), e2505.

Orozco, M.M., et al., 2013. New sylvatic hosts of *Trypanosoma cruzi* and their reservoir competence in the humid Chaco of Argentina: a longitudinal study. Am. J. Trop. Med. Hyg. 88 (5), 872–882.

Otto, S.P., Day, T., 2011. A Biologist's Guide to Mathematical Modeling in Ecology and Evolution. Princeton University Press.

Pacheco-Tucuch, F.S., Ramirez-Sierra, M.J., Gourbière, S., Dumonteil, E., 2012. Public street lights increase house infestation by *Triatoma dimidiata*, vector of Chagas disease in the Yucatan peninsula. PLoS One 7 (4), e36207.

Pelosse, P., et al., 2013. Influence of vectors' risk-spreading strategies and environmental stochasticity on the epidemiology and evolution of vector-borne diseases: the example of Chagas' disease. PloS One 8 (8), e70830.

Pelosse, P., Kribs-Zaleta, C.M., 2012. The role of the ratio of vector and host densities in the evolution of transmission modes in vector-borne diseases. The example of sylvatic *Trypanosoma cruzi*. J. Theor. Biol. 312C, 133–142.

Rabinovich, J., 1972. Vital statistics of triatominae (hemiptera: reduviidae) under laboratory conditions. 9(4), 351–370.

Rabinovich, J.E., Himschoot, P., 1990. A population-dynamics simulation model of the main vectors of Chagas' disease transmission, *Rhodnius prolixus* and *Triatoma infestans*. Ecol. Model. 52, 249–266.

Rabinovich, J., et al., 2001. Probability of *Trypanosoma cruzi* transmission by *Triatoma infestans* (Hemiptera: reduviidae) to the opossum *Didelphis albiventris* (Marsupialia: Didelphidae). Am. J. Trop. Med. Hyg. 65 (2), 125–130.

Rabinovich, J., Rossell, O., 1976. Mathematical models and ecology of Chagas disease. PAHO Sci. Pub. 245–250.

Rabinovich, J.E., et al., 1990. Probability of transmission of Chagas disease by *Triatoma infestans* (Hemiptera: reduviidae) in an endemic area of Santiago del Estero, Argentina. Bull. World Health Organ. 68 (6), 737–746.

Rabinovich, J.E., Nieves, E.L., Chaves, L.F., 2010. Age-specific mortality analysis of the dry forest kissing bug, *Rhodnius neglectus*. Entomol. Exp. Appl. 135 (3), 252–262.

Rabinovich, J.E., Nieves, E.L., 2011. Vital statistics of triatominae (Hemiptera: reduviidae) under laboratory Conditions: III. *Rhodnius neglectus*. J. Med. Entomol. 48 (4), 775–787.

Raimundo, S.M., Massad, E., Yang, H.M., 2010. Modelling congenital transmission of Chagas' disease. Bio Syst. 99 (3), 215–222.

Ramírez, J.D., et al., 2013. Molecular epidemiology of human oral Chagas disease outbreaks in Colombia. PLoS Negl. Trop. Dis. 7 (2), e2041.

Ramirez-Sierra, M.J., et al., 2010. Patterns of house infestation dynamics by non-domiciliated *Triatoma dimidiata* reveal a spatial gradient of infestation in rural villages and potential insect manipulation by *Trypanosoma cruzi*. Trop. Med. Int. Health 15, 77–86.

Ramsey, J.M., et al., 2014. Opportunity cost for early treatment of Chagas disease in Mexico. PLoS Negl. Trop. Dis. 8 (4), e2776.

Rascalou, G., et al., 2012. Emergence and prevalence of human vector-borne diseases in sink vector populations. PloS One 7 (5), e36858.

Rascalou, G., Dumonteil, E., Gourbiere, S., 2015. Transmission of Trypanosoma cruzi within a host community. Submited manuscrit.

Rascalou, G., Gourbière, S. How neglected are the NTD? unpublished data.

Rassi, A., Marin-Neto, J.A., 2010. Chagas disease. Lancet 375 (9723), 1388–1402.

Rocha, F.L., et al., 2013. *Trypanosoma cruzi* infection in neotropical wild carnivores (Mammalia: Carnivora): at the top of the *T. cruzi* transmission chain. PloS One 8 (7), e67463.

Rock, K., Stone, C.M., Hastings, I.M., Keeling, M.J., Torr, S.J., Chitnis, N., 2015. Mathematical models of Human African Trypanosomiasis epidemiology. Adv. Parasitol. 87, 53–133.
Rudge, J.W., Webster, J.P., Lu, D.B., Wang, T.P., Fang, G.R., Basáñez, M.G., 2013. Identifying host species driving transmission of schistosomiasis japonica, a multihost parasite system, in China. Proc. Natl. Acad. Sci. U. S. A. 110 (28), 11457–11462.
Samuels, A.M., et al., 2013. Epidemiology of and Impact of Insecticide Spraying on Chagas Disease in Communities in the Bolivian Chaco. PLoS Negl Trop Dis 7 (8), e2358.
Sanches, T.L.M., et al., 2014. The use of a heterogeneously controlled mouse population reveals a significant correlation of acute phase parasitemia with mortality in Chagas disease. In: Marinho, C.R.F. (Ed.), PloS One, vol. 9(3), p. e91640.
Schofield, C.J., 1980. Density regulation of domestic populations of *Triatoma infestans* in Brazil. Trans. R. Soc. Trop. Med. Hyg. 74, 761–769.
Schofield, C.J., Dias, J.C.P., 1991. A cost-benefit-analysis of Chagas-disease control. Mem. Inst. Oswaldo Cruz 86 (3), 285–295.
Schofield, C.J., Jannin, J., Salvatella, R., 2006. The future of Chagas disease control. Trends Parasitol. 22 (12), 583–588.
Shikanai-Yasuda, M.A., Carvalho, N.B., 2012. Oral transmission of Chagas disease. Clin. Infect. Dis. Off. Publ. Infect. Dis. Soc. Am. 54 (6), 845–852.
Sibona, G.J., Condat, C.A., Isasi, S.C., 2007. Relevance of intracellular replication to the evolution of chagas' disease. Math. Model. Biol. Syst. I, 289–297.
Sicuri, E., et al., 2011. Economic evaluation of Chagas disease screening of pregnant Latin American women and of their infants in a non endemic area. Acta Trop. 118 (2), 110–117.
Slimi, R., et al., 2009. A cellular automata model for Chagas disease. Appl. Math. Model. 33 (2), 1072–1085.
Smith, J.M., 1998. Evolutionary Genetics. Oxford University Press.
Spagnuolo, A.M., et al., 2012. A logistic delay differential equation model for Chagas disease with interrupted spraying schedules. J. Biol. Dyn. 6 (2), 377–394.
Spagnuolo, A.M., Shillor, M., Stryker, G.A., 2011. A model for Chagas disease with controlled spraying. J. Biol. Dyn. 5 (4), 299–317.
Stevens, L., et al., 2013. Household model of Chagas disease vectors (Hemiptera: Reduviidae) considering domestic, peridomestic, and sylvatic vector populations. J. Med. Entomol. 50 (4), 907–915.
Turner, H.C., Walker, M., French, M.D., Blake, I.M., Churcher, T.S., Basáñez, M.G., 2014a. Neglected tools for neglected diseases: mathematical models in economic evaluations. Trends Parasitol. 30 (12), 562–570.
Turner, H.C., et al., 2014b. Modelling the impact of ivermectin on River Blindness and its burden of morbidity and mortality in African Savannah: EpiOncho projections. Parasites Vectors 7 (1), 241.
Turner, H.C., Walker, M., Churcher, T.S., Osei-Atweneboana, M.Y., Biritwum, N.K., Hopkins, A., Prichard, R.K., Basáñez, M.G., 2014c. Reaching the London Declaration on Neglected Tropical Diseases goals for onchocerciasis: an economic evaluation of increasing the frequency of ivermectin treatment in Africa. Clin. Infect. Dis. 59 (7), 923–932.
Vanni, T., et al., 2012. Economic modelling assessment of the HPV quadrivalent vaccine in Brazil: a dynamic individual-based approach. Vaccine 30, 4866–4871.
Velasco-Hernandez, J.X., 1994. A model for Chagas disease involving transmission by vectors and blood transfusion. Theor. Popul. Biol. 46 (1), 1–31.
Velasco-Hernandez, J.X., 1991. An epidemiologic model for the dynamics of chagas-disease. Biosystems 26 (2), 127–134.

Verhulst, P.F., 1838. Notice sur la loi que la population poursuit dans son accroissement. Corresp. Math. Phys. 10, 113–121.
Wendel, S., Gonzaga, A.L., 1993. Chagas' disease and blood transfusion: a new world problem? Vox Sang. 64 (1), 1–12.
WHO, 2014a. Accelerating Work to Overcome the Global Impact of Neglected Tropical Diseases a Roadmap for Implementation (France).
WHO, 2014b. Chagas Disease (American Trypanosomiasis). WHO.
Wilson, L.S., et al., 2008. Cost-effectiveness of implementation methods for ELISA serology testing of *Trypanosoma cruzi* in California blood banks. Am. J. Trop. Med. Hyg. 79 (1), 53–68.
Wilson, L.S., Strosberg, A.M., Barrio, K., 2005. Cost-effectiveness of Chagas disease interventions in latin america and the Caribbean: Markov models. Am. J. Trop. Med. Hyg. 73 (5), 901–910.
Wood, S., 1951. Importance of feeding and defecation times of insect vectors in transmission of Chagas disease. J. of Eco. Entomol. 44 (1), 52–54.
Zu Dohna, H., et al., 2009. Spatial re-establishment dynamics of local populations of vectors of Chagas disease J. In: Valenzuela, G. (Ed.), PLoS Negl. Trop. Dis., vol. 3(7), p. e490.

CHAPTER FIVE

Mathematical Inference on Helminth Egg Counts in Stool and Its Applications in Mass Drug Administration Programmes to Control Soil-Transmitted Helminthiasis in Public Health

Bruno Levecke[*,1], Roy M. Anderson[§], Dirk Berkvens[¶], Johannes Charlier[*], Brecht Devleesschauwer[*,||], Niko Speybroeck[||], Jozef Vercruysse[*], Stefan Van Aelst[#,**]**

*Department of Virology, Parasitology and Immunology, Faculty of Veterinary Medicine, Ghent University, Merelbeke, Belgium
§Imperial College London, Department of Infectious Disease Epidemiology, Faculty of Medicine, School of Public Health St Mary's Campus, Norfolk Place, London, UK
¶Department of Biomedical Sciences, Institute of Tropical Medicine, Antwerp, Belgium
||Institute of Health and Society (IRSS), Université catholique de Louvain, Brussels, Belgium
#Department of Applied Mathematics, Computer Science and Statistics, Faculty of Sciences, Ghent University, Ghent, Belgium
**Department of Mathematics, Faculty of Sciences, KU Leuven University, Leuven, Belgium
[1]Corresponding author: E-mail: bruno.levecke@ugent.be

Contents

Advances in Parasitology, Volume 87
ISSN 0065-308X
http://dx.doi.org/10.1016/bs.apar.2015.01.001

Abstract

In the present study, we present a hierarchical model based on faecal egg counts (FECs; expressed in eggs per 1 g of stool) in which we first describe the variation in FECs between individuals in a particular population, followed by describing the variance due to counting eggs under a microscope separately for each stool sample. From this general framework, we discuss how to calculate a sample size for assessing a population mean FEC and the impact of an intervention, measured as reduction in FECs, for any scenario of soil-transmitted helminth (STH) epidemiology (the intensity and aggregation of FECs within a population) and diagnostic strategy (amount of stool examined (~sensitivity of the diagnostic technique) and examination of individual/pooled stool samples) and on how to estimate prevalence of STH in the absence of a gold standard. To give these applications the most wide relevance as possible, we illustrate each of them with hypothetical examples.

1. INTRODUCTION

The soil-transmitted helminths (STHs, *Ascaris lumbricoides*, *Trichuris trichiura* and the two hookworm species, *Necator americanus* and *Ancylostoma duodenale*) cause the highest burden among all neglected tropical diseases

(NTDs) (Murray et al., 2012). Recent global numbers indicate that in 2010 more than 1.4 billion people were infected with at least one of the four STH species, resulting in a global burden of approximately 5.2 million disability-adjusted life years (DALYs) (~20% of total DALYs attributable to NTDs) (Pullan et al., 2014). Mass drug administration (MDA) programmes in which a single oral dose of albendazole (ABZ) or mebendazole (MBZ) — the drugs of choice for STHs — are periodically administered to pre-school and school-aged children are the main strategy to control the morbidity caused by STHs (WHO, 2011), and these programmes have recently received increased political and scientific attention. The World Health Organization (WHO) has devised a roadmap to guide implementation of the policies and strategies set out in a global plan to combat NTDs (period 2008–2015), and more than 70 pharmaceutical companies, governments and global health organizations committed to supporting this roadmap in the London Declaration on NTDs in January 2012 by sustaining or expanding drug donation programmes (WHO, 2012a; NTD Partner Website, 2013). With this growing attention, WHO aims to increase the coverage of the pre-school and school-aged children in need of drug administration from 30% (estimated coverage in 2010; WHO, 2012b) to at least 75% by 2020, and to ultimately eliminate soil-transmitted helminthiasis as a public health problem in children (WHO, 2012c). These pledges of drug donations are at place, but this global upscale of MDA programmes also creates the need for a monitoring system that allows programme managers, policymakers and donors of the drugs to assess whether the objectives are being met and, if necessary, to adjust the implemented strategy (WHO, 2011). Thus, it will be imperative to periodically assess STH infections by means of prevalence and infection intensity to determine whether the MDA programme progresses as anticipated.

MDA programmes are currently poorly monitored, and one of the main reasons for this lack of monitoring systems is the absence of a framework that guides healthcare decision-makers in designing surveys. Development of such a framework, however, is not straightforward. First, due to the heterogeneity of STH infections, it will be impossible to apply one survey design to all implementation units of MDA, both within and between countries (Brooker et al., 2010; Hürlimann et al., 2011; Pullan and Brooker, 2012; Pullan et al., 2014). Moreover, as discussed by Bergquist et al. (2009), survey designs applied in an early stage of MDA programmes may not guarantee a reliable assessment of STH infections in a later stage, as both prevalence and intensity of infections will decrease over consecutive rounds of MDA.

Second, demonstration and quantification of eggs in stool is the current standard means to diagnose STH infections; however, this diagnosis is thwarted by the absence of a gold-standard technique, day-to-day variation in egg excretion and heterogenous distribution of eggs within stool samples (Sinniah, 1982; Ye et al., 1997; Krauth et al., 2012). Finally, MDA programmes typically operate in resource-constrained settings, and hence it is indispensable that healthcare decision-makers have some pliancy to minimize both financial and technical resources, while assuring a reliable assessment of the progress made.

The overall aim of this chapter is to develop a mathematical framework based on helminth egg counts in stool that allows healthcare decision-makers to adapt their survey design according to both local STH epidemiology and resources. Specifically, we will first list the most important sources of variability in egg counts in stool, and how they affect the design of studies. Next, we will outline a general mathematical framework for helminth egg counts in stool. From this general framework, we will continue by working out a selected number of applications for surveys designed to monitor MDA programmes to control STHs. To give these applications the most wide relevance as possible, we will illustrate each of them with hypothetical examples.

2. SOURCES OF VARIABILITY IN EGG COUNTS

In principle, the presence of STH eggs in stool is the result of at least one adult female in the gastrointestinal tract that is laying eggs. However, the number of eggs that is excreted in stool and which is eventually counted under the microscope is affected by a variety of factors. These can be classified in two groups of variation, including variation due to the egg excretion and due to the egg counting procedure. The variation in egg excretion is mainly due to biological factors, whereas the variation in the egg counting procedure is due to technical factors. We will address both sources of variability separately.

2.1 Sources of variability in egg excretion

Important sources that affect the number of eggs excreted in stool include fecundity of adult female worms and host-parasite-environment interactions. We will illustrate each of the two aspects by field data obtained during an epidemiological survey conducted in three countries in East

Africa (Ethiopia, Kenya and Uganda). The main objective of this survey was to investigate the distribution and heterogeneity of co-infection with *Plasmodium falciparum* and helminths, including STHs and *Schistosoma mansoni*. The presence and intensity of these helminth infections were determined by examination of one stool sample per child with duplicate Kato-Katz thick smears. Brooker et al. (2012) describe this epidemiological survey more in detail. We will use data obtained in Kenya which are made publicly available at http://www.thiswormyworld.org and are composed of 17,871 children across 178 schools (median number of children per school = 104).

There is a manifest difference in fecundity of adult female worms between the different STH species. An adult female *A. lumbricoides* worm produces approximately 200,000 eggs per day, whereas this ranges from 25,000 to 30,000, and from 9,000 to 10,000 eggs per day for *A. duodenale* and *N. americanus*, respectively. The daily egg output is the lowest for *T. trichiura*, with a daily egg output ranging from 3,000 to 5,000 eggs (Bethony et al., 2006). These differences in fecundity translate into differences in mean faecal egg counts (FECs) per gram of stool (EPG) across the STH species. For example, the mean FECs for *A. lumbricoides* in the Kenyan survey equalled 5,672 EPG ($N = 582$), whereas this was 332 EPG ($N = 2,086$) and 241 ($N = 1,672$) for hookworm and *T. trichiura*, respectively. As a consequence of this difference in fecundity, a survey designed for assessing *A. lumbricodes*, may not always allow assessing *T. trichiura* or hookworm infections with an equal level of precision (Levecke et al., 2011a). This is challenging, particularly because mixed STH infections are very common. In the Kenyan survey, 20% of the subjects infected with STH were excreting eggs of more than one STH species.

Although it is difficult to unravel the contribution of host, parasite and environment factors to the variation in egg excretion separately, their impact on egg excretion is pertinent. The most important consequences of this complex interplay of host-parasite-environment are a heterogenous distribution of FECs both within and between populations (e.g. school and community), a day-to-day variation in egg excretion, and a heterogeneous distribution in stool. Generally, a minority of the individuals excrete the majority of the eggs, and as illustrated in Figure 1 for school ID1175 of the survey in Kenya; this typically results in a skewed distribution of FECs. In this school 20% of children contributed to the total number of eggs excreted for *A. lumbricoides* and hookworm, and 96.7% for *T. trichiura*. This skewed distribution of FECs expressed in EPG can be modelled using a negative binomial distribution or

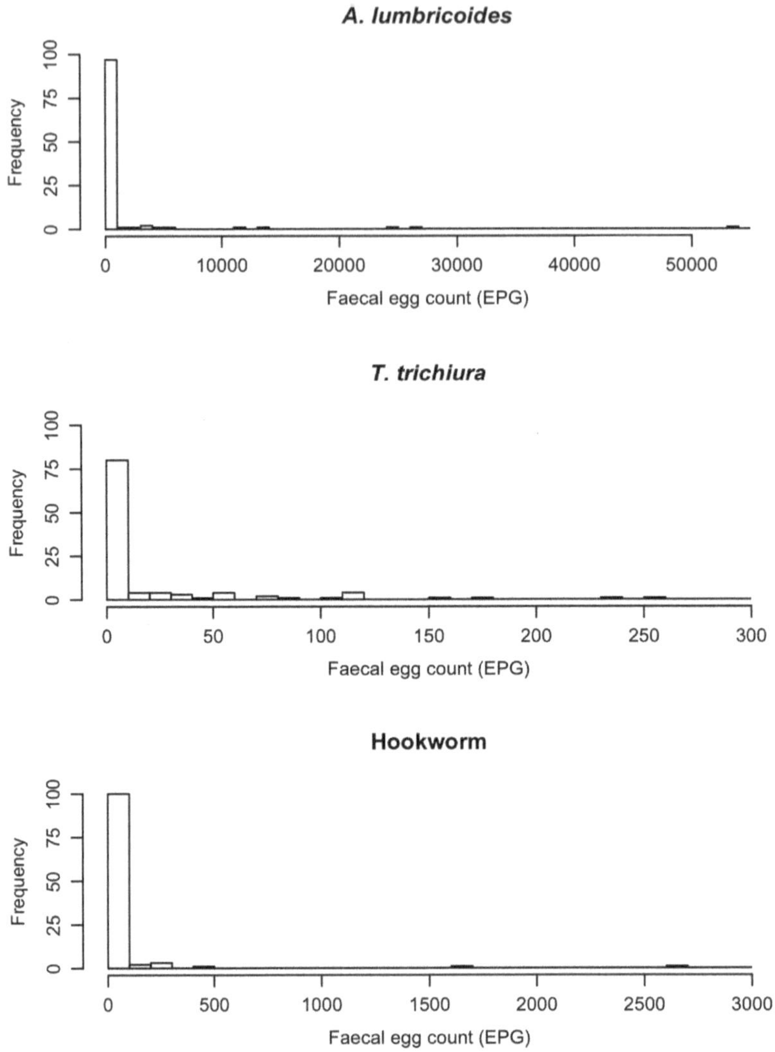

Figure 1 Histogram of individual faecal egg counts (expressed in eggs per gram of stool (EPG)) of *Ascaris lumbricoides, Trichuris trichiura* and hookworm of a random sample of 108 subjects from school ID1175 included in the epidemiological survey conducted in Kenya (Brooker et al., 2012). Note the differential skewness across the three soil-transmitted helminth species.

zero-inflated count distributions. In our framework, we will assume that FECs within a population follow a negative binomial distribution, which may be defined by a mean population FEC μ and an aggregation parameter k (Dobson et al., 2009; Torgerson et al., 2005). The parameter k can be calculated as a function of the mean and variation of the individual FECs through the moment estimator (k = mean FEC^2/(variance FEC − mean FEC)), and is inversely correlated with aggregation: low values indicate a high level of aggregation and high values indicate a more random distribution of FECs. In school ID1175, the mean FEC and k equalled 1,400 EPG and 0.05 for *A. lumbricoides*, 20 EPG and 0.18 for *T. trichiura* and 56 EPG and 0.03 for hookworm. However, these values for mean FEC and k vary considerably among populations. The mean FEC across the 178 schools included in the survey in Kenya, ranged from 0 EPG up to 6,173 EPG for *A. lumbricoides*, up to 499 EPG for *T. trichiura* and up to 482 EPG for hookworm infections. As illustrated in Figure 2, infections are also aggregated between schools, the minority of the schools accounting for the majority of the egg excretion (20% of the schools cover 97.2% of total number of *A. lumbricoides* eggs excreted; for *T. trichiura* and hookworm this was 60.9% and 83.0%, respectively). The values for k were up to 0.32 for *A. lumbricoides*, 0.45 for *T. trichiura* and 0.28 for hookworm infections. These values for k increased as a function of increasing mean FEC (Figure 2). These skewed and species-specific FEC distributions have three important implications. First, they highlight that surveys will have to be designed for each STH species and population separately. Second, they pose a serious risk of bias, as the mean of a small subsample of individual FECs is very likely to underestimate the mean population FEC (Gregory and Woolhouse, 1993). Finally, current formulae to calculate sample sizes are based on a normal distribution of the mean FEC (central limit theorem), an approximation which may be very poor when sample size is small, especially in some scenarios of mean population FEC and k (see Figure 3).

The consequences of day-to-day variation in egg excretion and heterogeneous distribution of eggs within stool on FEC results are well known and documented (Sinniah, 1982; Ye et al., 1997; Krauth et al., 2012). To minimize the day-to-day variation in individual FECs, it has been suggested to examine several samples per subject, collected over consecutive days (Booth et al., 2003; Knopp et al., 2008). To minimize the heterogeneous distribution within a stool sample, one can either examine multiple Kato-Katz thick smears per stool sample or use a diagnostic technique that allows examining a larger amount of stool such as the FLOTAC technique (up to 0.5 g) (Knopp et al., 2009; Cringoli et al., 2010). It is expected that these measures will decrease the variation between individual FECs, and hence increase the precision of population

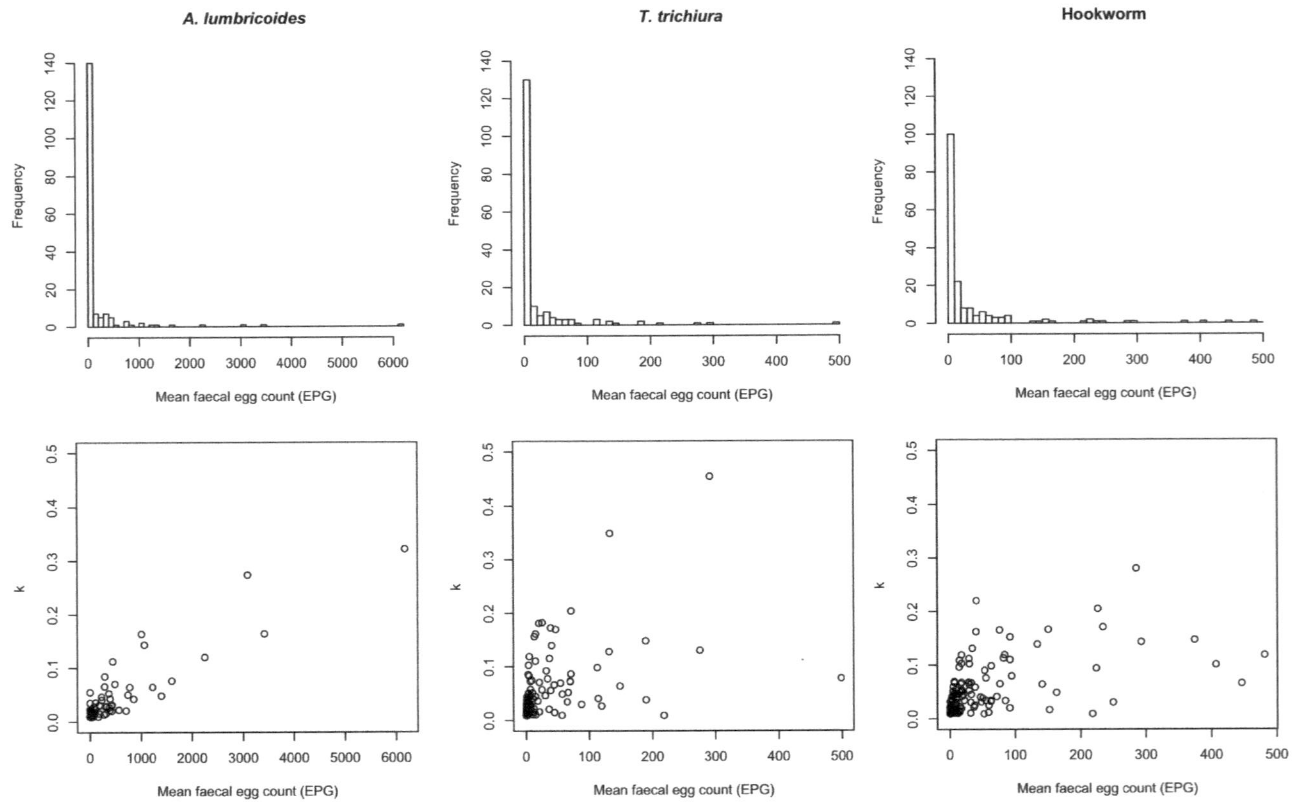

Figure 2 Histogram of mean school faecal egg counts (FECs, expressed in eggs per gram of stool (EPG)) of *Ascaris lumbricoides*, *Trichuris trichiura* and hookworm of all 178 schools included in the epidemiological survey conducted in Kenya (Brooker et al., 2012) and scatter plots of aggregation parameter *k* as a function of mean school FECs (bottom graphs). Note the differential skewness across the three soil-transmitted helminth species.

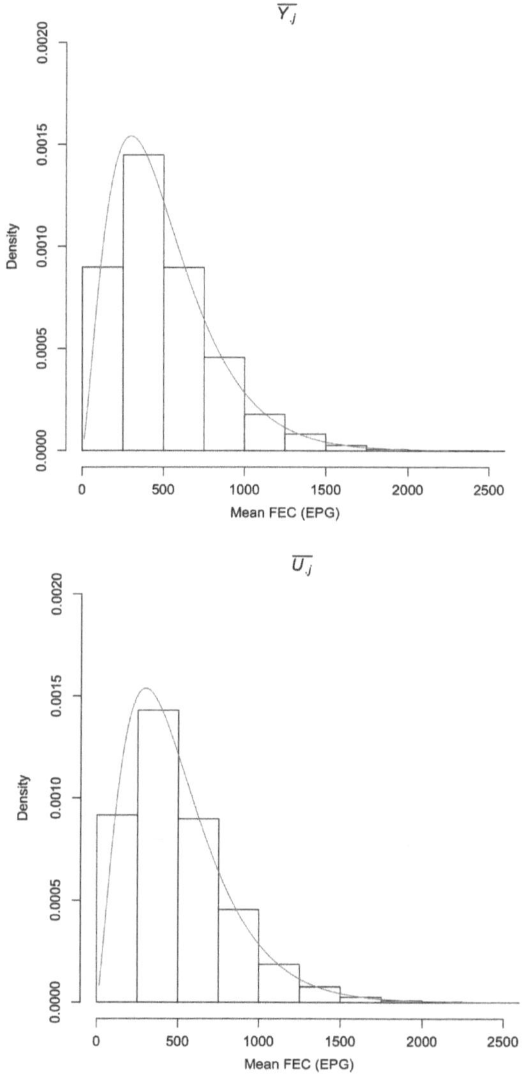

Figure 3 The histograms illustrate the probability density of the mean of observed individual faecal egg counts (FECs; expressed in eggs per g of stool (EPG)) $(=\overline{Y_{\cdot j}})$ and mean of observed pooled FECs $(=\overline{U_{\cdot j}})$ for *Ascaris lumbricoides* generated from 10,000 theoretical surveys in which 50 subjects were sampled from an infinitive population j ($\mu_j = 500$ EPG and $k_j = 0.05$). Both individual $(N_{j_{ind}} = 50)$(top graph) and pooled stool samples $(m_j = 5, N_{j_{pool}} = 10)$ (bottom graph) were examined using a single Kato-Katz thick smear ($f_j = 0.0417$ g). The red (light gray in print versions) lines describe the corresponding probability density of a gamma distribution with $\gamma_{j_{ind}} = \frac{\mu_j \cdot N_{j_{ind}}}{1/f_j + 1 + \mu_j/k_j}$ and $\theta_{j_{ind}} = \frac{1/f_j + 1 + \mu_j/k_j}{N_{j_{ind}}}$ (top graph), and $\gamma_{j_{pool}} = \frac{\mu_j \cdot N_{j_{pool}}}{1/f_j + (1 + \mu_j/k_j)/m_j}$ and $\theta_{j_{pool}} = \frac{1/f_j + (1 + \mu_j/k_j)/m_j}{N_{j_{pool}}}$ (bottom graph).

mean FEC estimates. As consequence of this, a smaller sample size will be required to estimate population mean FEC based on a duplicate Kato-Katz thick smear with an equal level of precision. These measures, however, will have less impact on the accuracy of these estimates, because the variation in egg excretion across days and within stool is random (subjects will not systematically excrete more eggs at the first sample collection and eggs are not systematically found in one particular part of stool). To verify this, we have summarized the mean *A. lumbricoides* FECs and their corresponding variance based on the examination of a single Kato-Katz thick smear and a duplicate Kato-Katz thick smear on the same stool sample in 10 schools from the epidemiological survey in Kenya (Table 1). Compared to a duplicate Kato-Katz thick smear, a single Kato-Katz resulted in a higher variance in eight out of 10 schools. In the two remaining schools (ID1020 and ID1023), a lower variance was found. Mean FEC based on a single and duplicate Kato-Katz thick smear were comparable, and were not systematically under- or overestimated, suggesting that indeed an increase in sampling and diagnostic effort increases the precision of population mean FECs estimates, but not the accuracy. For a more detailed analysis we refer the reader to a recent study by Levecke et al. (2014a), in which a comparison of FECs was made based on collection of one or two stool samples that were processed with single or duplicate Kato-Katz thick smears.

Table 1 The mean and variance in *Ascaris lumbricoides* faecal egg counts (FECs, expressed in eggs per gram of stool (EPG)) in 10 schools for a single and a duplicate Kato-Katz thick smear. The data were collected during an epidemiological survey conducted in Kenya (Brooker et al., 2012)

		Single Kato-Katz		Duplicate Kato-Katz	
School ID	***N***	**Mean FEC (EPG)**	**Variance**	**Mean FEC (EPG)**	**Variance**
1002	102	44	$2.02 \cdot 10^5$	43	$1.88 \cdot 10^5$
1003	100	345	$3.87 \cdot 10^6$	306	$3.30 \cdot 10^6$
1005	100	423	$6.34 \cdot 10^6$	427	$5.75 \cdot 10^6$
1006	100	1	51.8	<1	12.9
1007	104	<1	11.0	1	6.65
1010	108	184	$1.58 \cdot 10^6$	112	$5.37 \cdot 10^5$
1020	107	3	$7.79 \cdot 10^2$	4	$1.47 \cdot 10^3$
1023	104	175	$6.33 \cdot 10^5$	225	$1.08 \cdot 10^6$
1026	103	850	$3.51 \cdot 10^7$	564	$1.44 \cdot 10^7$
1027	103	339	$5.07 \cdot 10^6$	329	$3.75 \cdot 10^6$

2.2 Sources of variability in egg counting procedure

The egg counting procedure provides an estimate of the number of eggs in 1 g of stool by examining a small amount of stool (<1 g). Generally, it consists of three consecutive steps, including (1) preservation of stool, (2) processing the samples with a diagnostic technique that allows counting eggs under the microscope and (3) microscopically counting the eggs. Each of these steps can potentially affect the observed number of eggs excreted per g of stool, resulting in imprecise or inaccurate FEC estimates. Traditionally, samples are prepared fresh within 24 h after stool collection, but when this is not possible samples are often preserved in formalin. Although this allows to postpone the examination of samples, it thwarts the FECs. Indeed, examination of preserved samples resulted in lower FEC compared to FEC obtained after examining fresh samples (Albonico et al., 2013). Currently, a variety of diagnostic techniques have been applied to process human stool (e.g. Kato-Katz thick smear (Katz et al., 1972; WHO, 1991), McMaster (MAFF, 1986; Levecke et al., 2011b), Mini-FLOTAC (Cringoli et al., 2013; Barda et al., 2013) and FLOTAC techniques (Cringoli et al., 2010)), but they rarely result in identical FECs. These techniques differ in many aspects, but we will focus on the five most important differences that contribute to the variation in FECs between these techniques, including mode of action, amount of stool that is sampled, the way this amount of stool is determined, the amount of stool that is examined and the use of a counting apparatus (Table 2).

When using the Kato-Katz technique a simple stool smear is made, whereas for McMaster and (Mini-)FLOTAC stool is suspended in a solution of which the density is higher than that of STH eggs, allowing eggs to float to the surface. Although techniques based on flotation result in more clean microscopic views (debris will not always float), and hence reduce the inter-rater variability in FECs, they may result in lower FECs compared to Kato-Katz thick smear. For example, Levecke et al. (2011b) noticed that unfertilized *A. lumbricoides* eggs did often not float to the surface, contributing to discrepancies in FECs observed between flotation-based techniques and Kato-Katz technique (Knopp et al., 2009). Sampling a larger proportion of the stool will result in more precise estimates of FECs, as it will overcome the heterogeneous distribution of eggs in stool (see also Table 1). Of the four currently applied diagnostic techniques, Kato-Katz processes the least stool (~0.0417 g), FLOTAC the most (in theory the entire stool sample can be processed). To assess FECs by means of EPG, it is essential that a known mass of stool is processed. For the flotation-based techniques, a

Table 2 Mode of action, amount of stool that is sampled, the way this amount of stool is determined, the amount of stool that is examined and the use of a counting apparatus for four commonly diagnostic techniques used for the detection and quantification of soil-transmitted helminth eggs in human stool

	Kato-Katz	McMaster	Mini-FLOTAC	FLOTAC
Mode of action	Smear	Flotation	Flotation	Flotation
Mass of stool sampled (g)	0.0417	2	2	Entire stool sample
Determination of amount of stool	Volume	Mass	Mass	Mass
Sieving stool prior/after the determination of amount of stool	Prior	After	After	After
Mass of stool examined (g)	0.0417	0.02	0.1	0.5
Multiplication factor (=1/mass of stool examined)	24	50	10	2
Counting apparatus	No	Yes	Yes	Yes

known mass of stool is weighed, after which the stool is suspended in flotation solution and sieved to withhold the large debris. This is in contrast with the Kato-Katz technique. For this technique, the stool samples are first sieved, after which a known volume of sieved stool is sampled. The difference in sequence of sieving, contributes to the higher FECs observed for Kato-Katz technique compared to the flotation-based techniques (Knopp et al., 2009; Levecke et al., 2011b). Measuring stool volumetrically introduces additional variation across FECs, as the density of stool is not fixed across individuals (Levecke et al., 2011b). Examination of a larger proportion of the stool will result in more precise estimates of FECs. Of the four currently applied diagnostic techniques, the least stool is examined with McMaster (= 0.02 g), the most with FLOTAC (up to 0.5 g). Finally, in contrast to the Kato-Katz technique, the flotation-based techniques make use of a counting apparatus in which the number of eggs is counted within grids. These grids are designed to guide the laboratory technicians in examining the samples, and hence minimize the probability of missing eggs or counting eggs twice. Despite these specific measures to increase the agreement in FECs between laboratory technicians and laboratories, variation in FEC due to human error remains considerable (Bogoch et al., 2006; Levecke et al., 2011b). Therefore, it has been recommended to re-examine a subset of the samples by a senior laboratory technician to further minimize the inter-rater variation in FECs.

Currently, we lack detailed studies that can provide estimates of bias and variation introduced by each of the aforementioned aspects across a large number of laboratories for STH diagnosis. Moreover, keeping the final application of the mathematical framework in mind, we would like to minimize the number of parameters that needs to be defined by the end user. In the mathematical framework described in this chapter, we will assume that the number of eggs observed by microscopy follows a Poisson distribution which is defined by a parameter λ. This parameter, however, depends on how much stool is examined, and hence which diagnostic technique is applied. For example, if the true FEC in a stool sample equals 500 EPG, than the expected number of eggs λ to be counted under the microscope in 0.02 g of stool equals 10 (=true FEC·amount of stool examined (in g) = 500·0.02) and 250 when 0.5 g of stool is examined (=500·0.5). To obtain the FECs expressed in EPG, the observed egg counts are subsequently multiplied by 50 and 2 (= 1/amount of stool examined (in g)), respectively. However, due to random variation the FECs obtained after re-processing the same sample will not be identical. This assumption that the number of eggs observed by microscopy follows a Poisson distribution has been mainly studied in veterinary parasitology (Morgan et al., 2005; Torgerson et al., 2012), and as we will illustrate, it already explains most of the variation in FECs reported in field studies. Figure 4 illustrates the variation in FECs obtained by theoretically processing 0.02 (McMaster) and 0.5 g (FLOTAC) of the same stool sample 100 times. To explore the change in variation as a function of FEC, we show the variation in a stool sample in which the true underlying FEC equals 5, 50 and 500 EPG. From this figure, we can deduce that (1) on average the observed FECs approaches the true FEC, (2) the variance (~1/precision) increases as a function of decreasing amount of stool that is examined and increasing true FECs and (3) the sensitivity increases as a function of increasing amount of stool examined and the true FEC. Each of these aspects has been observed in field data (see Table 1, Levecke et al., 2011b, 2014a), highlighting that the assumption that the number of eggs counted by microscopy follows a Poisson distribution is justified.

3. GENERAL MATHEMATICAL FRAMEWORK FOR FECs

In this section we will describe a general mathematical framework based on FECs. Given that the current diagnostic techniques only

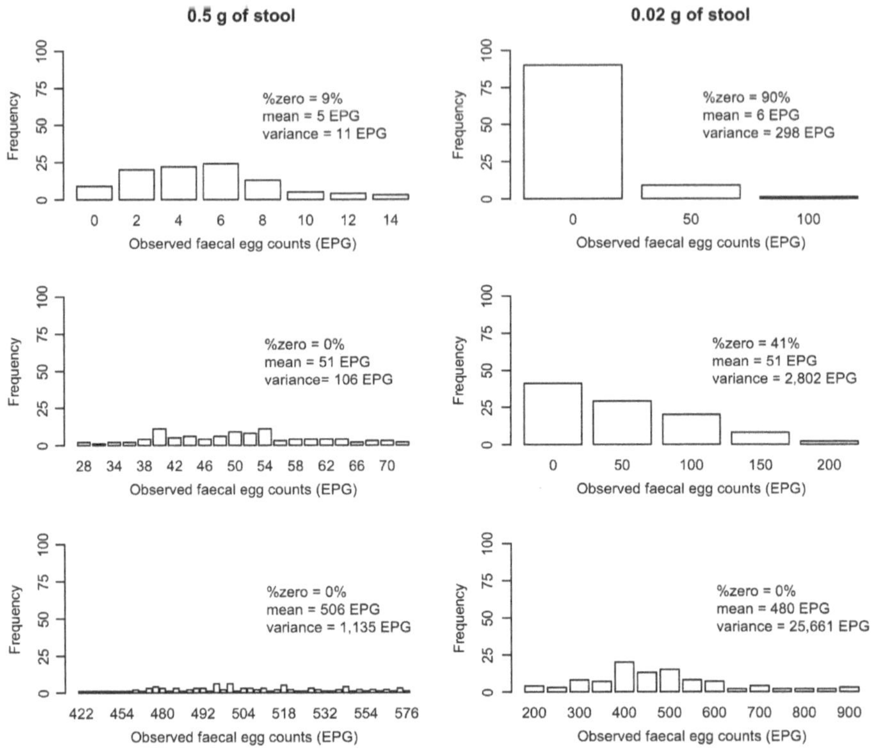

Figure 4 Histogram of 100 theoretical faecal egg counts (FECs, expressed in eggs per gram of stool (EPG)) obtained by examining 0.02 and 0.5 g of stool. It was assumed that the number of eggs observed by microscopy follows a Poisson distribution with λ equal to true FEC·amount of stool examined (in g). The true FEC was set at 5 (top graphs), 50 (middle graphs) and 500 EPG (bottom graphs). We provide the proportion of zero FECs (*%zero*), mean and variance of FECs.

provide estimates of the number of eggs in 1 g of stool as a proxy of egg excretion, we will express these FECs in EPG. We will assume that individual FECs within a population follow a negative binomial distribution, that there is no day-to-day variation in FECs, and that variation in egg counts from the same stool sample under the microscope is completely explained by a Poisson distribution. To give healthcare decision-makers more flexibility to adapt their survey design, the framework will be described for the examination of both individual and pooled stool samples. A list of all variables in this general framework is given in Table 3.

Table 3 The variables included in the hierarchical models describing faecal egg counts (FECs, expressed in eggs per gram of stool (EPG)) prior and after an intervention, and for examination of individual and pooled stool samples

	Definition
e_j	True efficacy of an intervention in a population *j*
f_j	Amount of stool (in g) collected from a population *j* that is microscopically examined
μ_j	True population mean FEC (in EPG) in a population *j* prior to an intervention
λ_{ij}	True number of eggs in f_j g of stool of an individual *i* of population *j* prior to an intervention
λ'_{ij}	True number of eggs in f_j g of stool of an individual *i* of population *j* after an intervention
λ_{lj}	True number of eggs in f_j g of a pooled stool sample *l* from a population *j* prior to an intervention
λ'_{lj}	True number of eggs in f_j g of a pooled stool sample *l* from a population *j* after an intervention
k_j	True aggregation of FEC (in EPG) in a population *j* prior to an intervention
m_j	The number of stool samples collected from individuals of population *j* that are pooled into one sample
$n_{j_{ind}}$	The number of individuals in a population *j*
$n_{j_{pool}}$	The number pools in a population *j*
U_{lj}	Observed FEC (in EPG) obtained by examining f_j g of a pooled stool sample *l* from a population *j* prior to an intervention
U'_{lj}	Observed FEC (in EPG) obtained by examining f_j g of a pooled stool sample *l* from a population *j* after an intervention
V_{lj}	True FEC (in EPG) in a pooled stool sample *l* from a population *j* prior to an intervention
V'_{lj}	True FEC (in EPG) in a pooled stool sample *l* from a population *j* after an intervention
X_{ij}	True FEC (in EPG) of an individual *i* of population *j* prior to an intervention
X'_{ij}	True FEC (in EPG) of an individual *i* of population *j* after an intervention
Y_{ij}	Observed FEC (in EPG) obtained by examining f_j g of stool of an individual *i* of a population *j* prior to an intervention
Y'_{ij}	Observed FEC (in EPG) obtained by examining f_j g of stool of an individual *i* of a population *j* after an intervention
Z_{ij}	Observed number of eggs counted under the microscope in f_j g of stool from an individual *i* of a population *j* prior to an intervention
Z'_{ij}	Observed number of eggs counted under the microscope in f_j g of stool from an individual *i* of a population *j* after an intervention
Z_{lj}	Observed number of eggs counted under the microscope in f_j g of a pooled stool sample *l* from a population *j* prior to an intervention
Z'_{lj}	Observed number of eggs counted under the microscope in f_j g of a pooled stool sample *l* from a population *j* after an intervention

3.1 Faecal egg counts

3.1.1 FECs based on individual stool samples

Generally, the framework describes a hierarchical model based on FECs in which we first describe the distribution of FECs between individuals, followed by describing the process of counting eggs under a microscope separately for each stool sample. For the distribution of FECs between individuals, we assume that FECs X_{ij} (in EPG) between individuals $i = 1, 2, \ldots, n_{j_{ind}}$ of a population j follow a negative binomial distribution with a population mean μ_j and aggregation parameter k_j. For the process of counting eggs, we assume that the number of eggs observed by microscopy in stool of an individual i of population j is Poisson distributed with λ_{ij} as the true underlying number of eggs in f_j g of stool equal to $f_j \cdot X_{ij}$, where f_j is the amount of stool examined (in g) and where X_{ij} is the true FEC expressed in EPG for an individual i. Because f_j is always less than 1 g we multiply these observed egg counts under the microscope Z_{ij} by a factor $1/f_j$ to obtain the observed FEC Y_{ij} in EPG for an individual i in a population j. This results in the following hierarchical model for observed FECs Y_{ij},

$$\begin{aligned} X_{ij} &\sim NB\left(\mu_j,\ k_j\right) \\ Z_{ij} \middle| X_{ij} &\sim Poiss\left(\lambda_{ij}\right) \\ &\sim Poiss\left(f_j \cdot X_{ij}\right) \\ Y_{ij} &= \frac{Z_{ij}}{f_j} \end{aligned} \tag{1}$$

It can be derived that the expected value of Y_{ij} equals μ_j, the mean of the negative binomial distribution describing the FECs between individuals in a population j. Due to the additional variation introduced by the egg counting process, the variance of the observed individual FECs Y_{ij} will be larger than those of true individual FECs $X_{ij}(= \mu_j + \mu_j^2/k_j)$. The variance of Y_{ij} given Eqn (1) is described below in Eqn (2). The variance of Y_{ij} will increase with increasing levels of egg excretion (μ_j) and aggregation ($\sim 1/k_j$), and with decreasing quantity of stool examined (f_j). The quantity of stool that is examined can be increased by changing diagnostic technique (Kato-Katz thick smear ($f_j = 0.0417$ g) vs FLOTAC ($f_j = 0.5$ g)) or by repeatedly processing the same sample (single ($f_j = 0.0417$ g) vs duplicate ($f_j = 0.0834$ g) Kato-Katz thick smears).

$$E\left[Y_{ij}\right] = \mu_j$$
$$Var\left[Y_{ij}\right] = \mu_j \cdot \left(\frac{1}{f_j} + 1 + \frac{\mu_j}{k_j}\right) \tag{2}$$

when an intervention with a fixed efficacy e_j is implemented to a population j with population mean FEC μ_j and aggregation parameter k_j we can write the hierarchical model for the observed FECs Y'_{ij} after an intervention as,

$$\begin{aligned} X_{ij} &\sim NB\left(\mu_j,\ k_j\right) \\ X'_{ij} &= \left(1 - e_j\right) \cdot X_{ij} \\ Z'_{ij}\left|X'_{ij}\right. &\sim Poiss\left(\lambda'_{ij}\right) \\ &\sim Poiss\left(f_j \cdot X'_{ij}\right) \\ Y'_{ij} &= \frac{Z'_{ij}}{f_j} \end{aligned} \tag{3}$$

By analogy with Eqn (2) the expected value and variance of Y'_{ij} are equal to,

$$E\left[Y'_{ij}\right] = \left(1 - e_j\right) \cdot \mu_j$$
$$Var\left[Y'_{ij}\right] = \left(1 - e_j\right) \cdot \mu_j \cdot \left(\frac{1}{f_j} + \left(1 - e_j\right) \cdot \left(1 + \mu_j / k_j\right)\right) \tag{4}$$

The derivation of the expected valued and variance of Y_{ij} and Y'_{ij} can be found in Appendix A.

3.1.2 FECs based on pooled stool samples

A hierarchical model for FECs based on pooled stool samples can be described based on the models for FECs obtained by examining individual stool samples (see Eqns (1) and (3)). Compared to the model for individual stool samples, it is also necessary to define the number of samples pooled and the true FECs in these pools. For this model, we assume (1) that the pool size m_j (= number of individual samples pooled) is the same for all pools and (2) that the true FEC V_{lj} of a pool $l = 1, 2, \ldots, n_{j_{pool}}$ equals the mean of the true FECs of m_j individual stool samples from individuals $i = 1, 2, \ldots, n_{j_{ind}}$ from a population j ($n_{j_{ind}} = m_j \cdot n_{j_{pool}}$). The latter implies that we ignore any variation in FEC V_{lj} due to the pooling process. In practice, this means that for each individual stool sample an equal amount of stool is pooled and that pools are well homogenized to avoid any heterogeneous distribution of eggs in the pool. The observed FEC U_{lj} prior to an intervention can be described as the hierarchical model below,

$$\begin{aligned} X_{ij} &\sim NB\left(\mu_j,\ k_j\right) \\ V_{lj} &= \frac{\sum_{i=1}^{m_j} X_{ij}}{m_j} \\ Z_{lj}\middle| V_{lj} &\sim Poiss\left(\lambda_{lj}\right) \\ &\sim Poiss\left(f_j \cdot V_{lj}\right) \\ U_{lj} &= \frac{Z_{lj}}{f_j} \end{aligned} \tag{5}$$

From this model and by analogy with Eqns (1) and (2), it can be shown that in a population j the expected value and the variance of observed FECs U_{lj} are equal to,

$$\begin{aligned} E\left[U_{lj}\right] &= \mu_j \\ Var\left[U_{lj}\right] &= \mu_j \cdot \left(\frac{1}{f_j} + \frac{1 + \mu_j / k_j}{m_j}\right) \end{aligned} \tag{6}$$

After an intervention with a fixed efficacy e_j, we can write the hierarchical model for the observed FECs U'_{lj} in a pool of m_j stool samples from a population j as,

$$\begin{aligned} X_{ij} &\sim NB\left(\mu_j,\ k_j\right) \\ X'_{ij} &= \left(1 - e_j\right) \cdot X_{ij} \\ V'_{lj} &= \frac{\sum_{i=1}^{m_j} X'_{ij}}{m_j} \\ Z'_{lj}\middle| V'_{lj} &\sim Poiss\left(\lambda'_{lj}\right) \\ &\sim Poiss\left(f_j \cdot V'_{lj}\right) \\ U'_{lj} &= \frac{Z'_{lj}}{f_j} \end{aligned} \tag{7}$$

The expected value and the variance of observed FECs U'_{lj} after an intervention are equal to,

$$E\left[U'_{lj}\right] = (1 - e_j) \cdot \mu_j$$
$$Var\left[U'_{lj}\right] = (1 - e_j) \cdot \mu_j \cdot \left(\frac{1}{f_j} + (1 - e_j) \cdot \frac{1 + \mu_j / k_j}{m_j}\right) \quad (8)$$

The derivations of the expected value and the variance of U_{lj} and U'_{lj} are provided in Appendix B.

3.1.3 Distribution of $\overline{Y_{\cdot j}}$ and $\overline{U_{\cdot j}}$

Given the central limit theorem, it is expected that the mean $\overline{Y_{\cdot j}}\left(= \frac{\sum_{i=1}^{N_{j_{ind}}} Y_{ij}}{N_{j_{ind}}}\right)$ and $\overline{U_{\cdot j}}\left(= \frac{\sum_{l=1}^{N_{j_{pool}}} U_{lj}}{N_{j_{pool}}}\right)$ would approach a normal distribution when sample size is large. However, as illustrated in Figure 3, this assumption is violated when sample size is small. This figure illustrates the distribution of estimated population mean FEC for *A. lumbricoides* generated from 10,000 theoretical surveys in which 50 subjects $(= N_{j_{ind}})$ from a population j ($\mu_j = 500$ EPG and $k_j = 0.1$) are randomly screened. The mean FECs are obtained by examining both individual and pooled ($m_j = 5$, $N_{j_{pool}} = 10$) stool samples using a single Kato-Katz thick smear ($f_j = 0.0417$ g). For both examination of individual and pooled samples the distribution of $\overline{Y_{\cdot j}}$ and $\overline{U_{\cdot j}}$ is skewed, and this will become more pronounced when the sample size, μ_j and f_j decrease, and the level of aggregation increases ($\sim 1/k_j$) (data not shown). For the described hierarchical models we will assume that both $\overline{Y_{\cdot j}}$ and $\overline{U_{\cdot j}}$ follow a gamma distribution. This distribution is defined by two parameters, i.e. the shape parameter γ and the scale parameter θ. The expected value for gamma distributed variables equals $\gamma \cdot \theta$, the variance equals $\gamma \cdot \theta^2$. From these equations and given Eqns (2) and (6) we can derive γ_j and θ_j of the gamma distribution for $\overline{Y_{\cdot j}}$ Eqn (9) and $\overline{U_{\cdot j}}$ Eqn (10) for a population j given a sample size $N_{j_{ind}}$ and $N_{j_{pool}}$, respectively. As illustrated in Figure 3, these gamma distributions approach those of the above described theoretical surveys.

$$\begin{aligned}\gamma_{j_{ind}} &= \frac{\mu_j \cdot N_{j_{ind}}}{1/f_j + 1 + \mu_j/k_j} \\ \theta_{j_{ind}} &= \frac{1/f_j + 1 + \mu_j/k_j}{N_{j_{ind}}}\end{aligned} \tag{9}$$

$$\begin{aligned}\gamma_{j_{pool}} &= \frac{\mu_j \cdot N_{j_{pool}}}{1/f_j + \left(1 + \mu_j/k_j\right)/m_j} \\ \theta_{j_{pool}} &= \frac{1/f_j + \left(1 + \mu_j/k_j\right)/m_j}{N_{j_{pool}}}\end{aligned} \tag{10}$$

To verify whether these theoretical gamma distributions also fit empirical data, we have plotted the probability density of both $\overline{Y_{\cdot j}}$ generated by bootstrap analysis (10,000 iterations) of 108 individual data obtained by a single Kato-Katz thick smear in school ID1175 and the expected gamma distribution in Figure 5 for each of the STH species. Overall, the plots suggest that the theoretically derived gamma distributions for $\overline{Y_{\cdot j}}$ fit the empirical data. The distribution of $\overline{Y'_{\cdot j}}$ and $\overline{U'_{\cdot j}}$ will not be discussed here, but will be incorporated in Section 3.2.2, Distribution of different scenarios of $FECR_j$.

3.2 Reduction in FECs

There is an ongoing debate as to whether reduction in prevalence is an appropriate metric to monitor the impact of an intervention, as opposed to reduction in FECs (FECR syn. egg reduction rate), respectively (Humphries et al., 2011; Montresor, 2011; Montresor et al., 2011). Anderson and colleagues, highlighted that a drop in FEC may not always be reflected in a drop in prevalence, with changes in the latter hence underestimating the impact of control interventions (Anderson et al., 2012). Analogously, an intervention may fail to cure helminth infections (reduction in prevalence = 0%), but may result in an FECR of 99% which is satisfactory. Additionally, it has been shown that estimates of reduction in prevalence are highly affected by both sampling and diagnostic effort, being underestimated when the sampling and diagnostic effort are minimized. This is in sharp contrast with FECR estimates, which remain unchanged regardless of both sampling and diagnostic effort (Levecke et al., 2014a). To date, a wide range of formulae has been used to calculate FECR, each differing in terms of the statistical unit (individual vs group) and how the mean

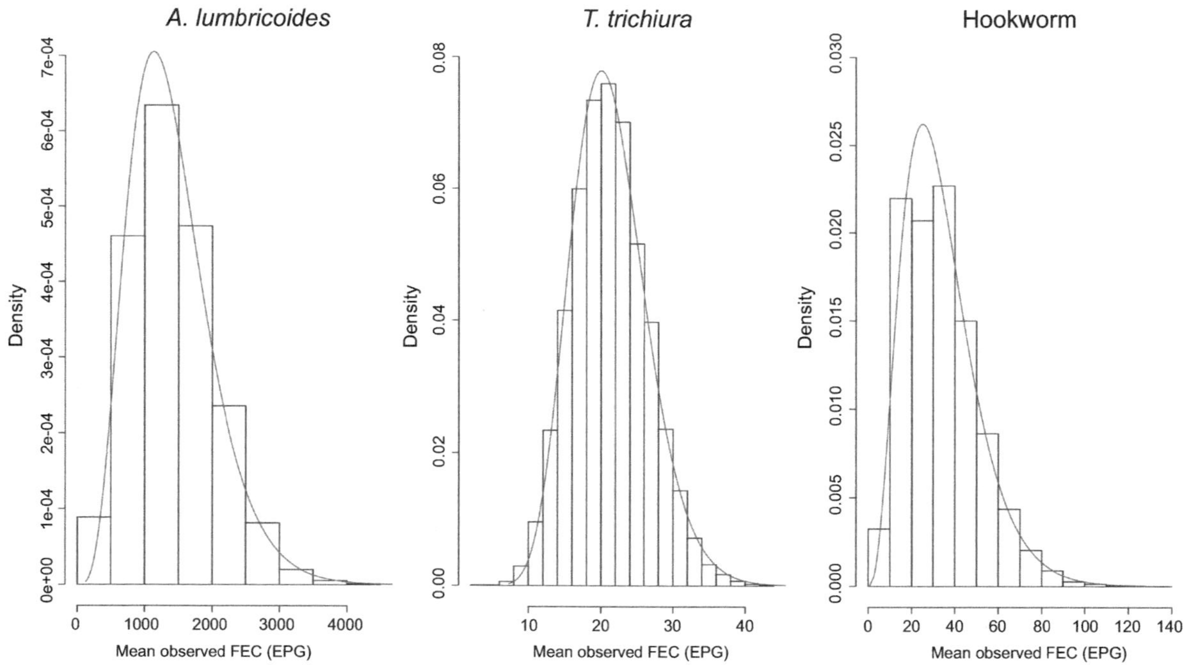

Figure 5 The histograms illustrate the probability density of mean observed faecal egg counts (FECs; expressed in eggs per gram of stool (EPG)) $(= \overline{Y_{\cdot j}})$ for *Ascaris lumbricoides*, *Trichuris trichiura* and hookworm generated by bootstrap analysis (10,000 iterations) of 108 individual data obtained by a single Kato-Katz thick smear in school ID1175. The red (light gray in print versions) lines describe the corresponding probability density of the expected gamma distribution. For these gamma distributions $\gamma_{j_{ind}}$ equalled $\frac{\mu_j \cdot N_{j_{ind}}}{1/f_j+1+\mu_j/k_j}$, $\theta_{j_{ind}}$ equalled $\frac{1/f_j+1+\mu_j/k_j}{N_{j_{ind}}}$. For each of the three helminth species, μ_j and k_j were substituted by their corresponding $\overline{Y_{\cdot j}}$ and $\overline{Y_{\cdot j}}/(Var[Y_{ij}]/\overline{Y_{\cdot j}} - 1/f_j - 1)$ (derived from Eqn (2)), respectively. For all helminth species, $N_{j_{ind}}$ was substituted by 108 and f_j by 0.0417.

FEC is calculated (arithmetic vs geometric mean). However, recent studies suggest that the group-based formula using the arithmetic mean, as described below, is a suitable metric for evaluating an intervention. Compared to the other formulae, it represents a robust indicator (vs individual-based formula) (Vercruysse et al., 2011) that provides accurate efficacy of an intervention (vs group-based formula using the geometric mean (Dobson et al., 2009; Vercruysse et al., 2011)). We will work out the applications for the assessment of FECR based on the group-based formula using the arithmetic mean.

$$FECR = 1 - \frac{\textit{arithmetic mean of FECs after an intervention}}{\textit{arithmetic mean of FECs prior an intervention}} \tag{11}$$

3.2.1 Expected value and variance of different FECR scenarios

When an intervention with efficacy e_j is assessed in a population j (population mean FEC of μ_j and aggregation of k_j) by means of $FECR_j$ based on the individual examination of f_j g of stool of the same individuals, we can deduce that the expected value asymptotically equals,

$$\begin{aligned} E\left[FECR_{j_{ind}}\right] &\simeq 1 - \frac{E\left[Y'_{ij}\right]}{E\left[Y_{ij}\right]} \\ &\simeq 1 - \frac{\left(1 - e_j\right) \cdot \mu_j}{\mu_j} \\ &\simeq e_j \end{aligned} \tag{12}$$

Given that two variables are dependent (examination of the same subjects both prior and after an intervention) and that both their expected value is nonzero, one can demonstrate, based on Casella and Berger (2002), that the variance of the ratio of these variables asymptotically equals,

$$\begin{aligned} Var\left[FECR_{j_{ind}}\right] \simeq \left(\frac{E\left[Y'_{ij}\right]}{E\left[Y_{ij}\right]}\right)^2 \cdot \left(\frac{Var\left[Y_{ij}\right]}{E\left[Y_{ij}\right]^2} + \frac{Var\left[Y'_{ij}\right]}{E\left[Y'_{ij}\right]^2}\right. \\ \left. - 2 \cdot corr\left(Y_{ij}, Y'_{ij}\right) \cdot \frac{\sqrt{Var\left[Y_{ij}\right] \cdot Var\left[Y'_{ij}\right]}}{E\left[Y_{ij}\right] \cdot E\left[Y'_{ij}\right]}\right) \end{aligned} \tag{13}$$

In the scenario that different subjects are examined after the intervention, the expected value remains unchanged, but the variance will change into,

$$Var\left[FECR_{j_{ind\Delta}}\right] \simeq \left(\frac{E\left[Y'_{ij}\right]}{E\left[Y_{ij}\right]}\right)^2 \cdot \left(\frac{Var\left[Y_{ij}\right]}{E\left[Y_{ij}\right]^2} + \frac{Var\left[Y'_{ij}\right]}{E\left[Y'_{ij}\right]^2}\right) \tag{14}$$

In analogy with Eqns (12)—(14), it is possible to write the expected value and the variance of $FECR_j$ based on the examination of f_j g of m_j pooled stool samples as,

$$E\left[FECR_{j_{pool}}\right] \simeq e_j \tag{15}$$

$$Var\left[FECR_{j_{pool}}\right] \simeq \left(\frac{E\left[U'_{lj}\right]}{E\left[U_{lj}\right]}\right)^2 \cdot \left(\frac{Var\left[U_{lj}\right]}{E\left[U_{lj}\right]^2} + \frac{Var\left[U'_{lj}\right]}{E\left[U'_{lj}\right]^2} - 2 \cdot corr\left(U_{lj}, U'_{lj}\right) \cdot \frac{\sqrt{Var\left[U_{lj}\right] \cdot Var\left[U'_{lj}\right]}}{E\left[U_{lj}\right] \cdot E\left[U'_{lj}\right]}\right) \tag{16}$$

$$Var\left[FECR_{j_{pool\Delta}}\right] \simeq \left(\frac{E\left[U'_{lj}\right]}{E\left[U_{lj}\right]}\right)^2 \cdot \left(\frac{Var\left[U_{lj}\right]}{E\left[U_{lj}\right]^2} + \frac{Var\left[U'_{lj}\right]}{E\left[U'_{lj}\right]^2}\right) \tag{17}$$

For each of the variances of $FECR_j$ described above it is possible to substitute the expected values and variances of Y_{ij}, Y'_{ij}, U_{lj}, U'_{lj} by the equations described in (2), (4) and (6) and (8), respectively. It was not possible to write an analytical solution for the correlation between Y_{ij} and Y'_{ij}, and between U_{lj} and U'_{lj}. Henceforth, we will estimate these correlations based on simulations.

3.2.2 Distribution of different FECR scenarios

Given the central limit theorem, it is to be expected that $FECR_j$ estimates would approach a normal distribution when sample size is large. However, as illustrated in Figure 6, this assumption is violated when sample size is small. This figure illustrates the probability density of $FECR_j$ generated from 10,000 theoretical surveys in which 50 subjects ($= N_{j_{ind}}$) from a population j ($\mu_j = 500$ EPG and $k_j = 0.1$) are included to assess the impact of three interventions e_j with three different levels of efficacy (0.50, 0.80 and 0.99) against *A. lumbricoides*. The $FECR_j$ are obtained by examining both individual and pooled stool samples ($m_j = 5, N_{j_{pool}} = 10$) using a single Kato-Katz

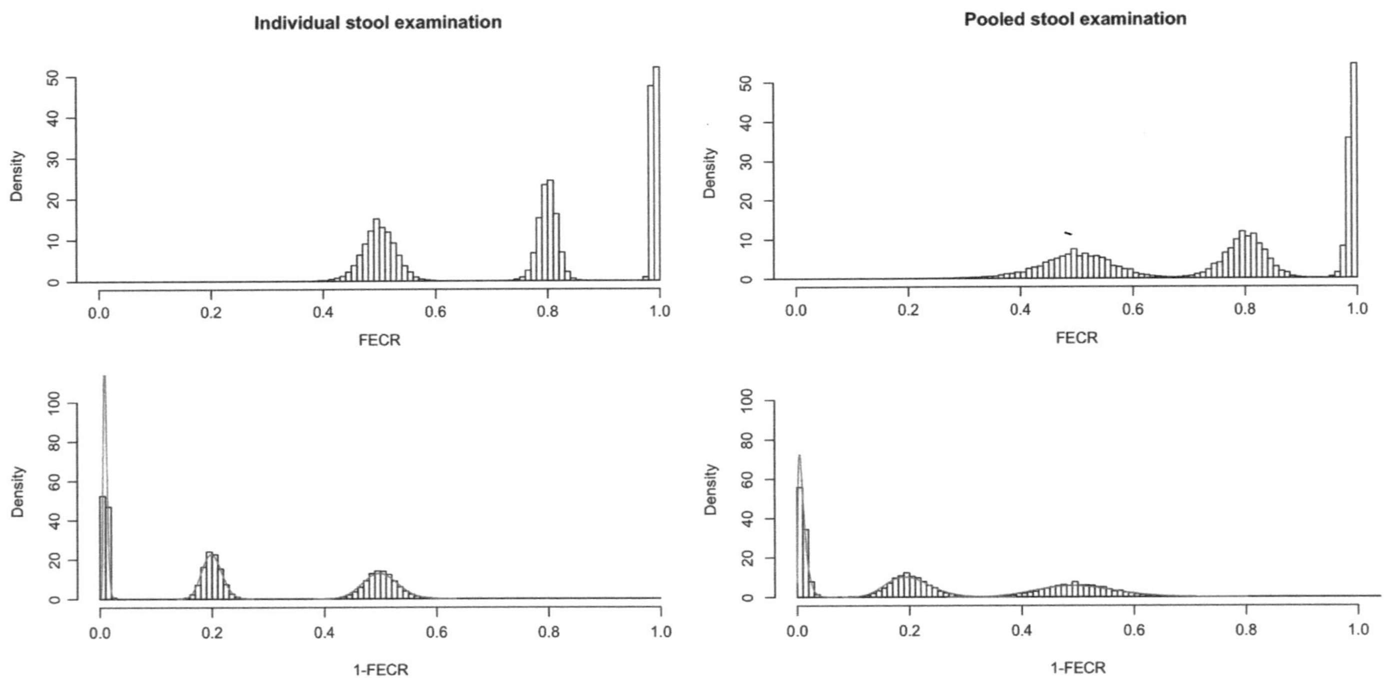

Figure 6 The top graphs illustrate the probability density of $FECR_j$ (faecal egg count reduction) estimates generated from 10,000 theoretical surveys in which 50 subjects ($= N_{j_{ind}}$) from a population *j* ($\mu_j = 500$ EPG and $k_j = 0.1$) are included to assess the impact of three interventions e_j with three different levels of efficacy (0.50, 0.80 and 0.99) against *Ascaris lumbricoides*. The $FECR_j$ are obtained by examining both individual (left top graph) and pooled ($m_j = 5, N_{j_{pool}} = 10$) stool samples (right top graph) using a single Kato-Katz thick smear ($f_j = 0.0417$ g). The same individuals are screened both prior and after the intervention. The graphs at the bottom illustrate the probability density of $1 - FECR_j$ from the 10,000 theoretical surveys described above (histograms) and the probability density of the corresponding expected gamma distribution (red (light gray in print versions) line). For these gamma distributions $\gamma_{j_{ind}} = (1 - e_j)^2 \cdot N_{j_{ind}} / Var[FECR_{j_{ind}}]$ and $\theta_{j_{ind}} = Var[FECR_{j_{ind}}]/(1 - e_j) \cdot N_{j_{ind}}$ (left bottom graph), and $\gamma_{j_{pool}} = (1 - e_j)^2 \cdot N_{j_{pool}} / Var[FECR_{j_{pool}}]$ and$\theta_{j_{pool}} = Var[FECR_{j_{pool}}]/(1 - e_j) \cdot N_{j_{pool}}$.

thick smear ($f_j = 0.0417$ g). For both examination of individual and pooled samples the distribution of $FECR_j$ becomes skewed when e_j approaches 1.

$FECR_j$ can take any value from $-\infty$ (mean FEC at follow-up > mean FEC at baseline) up to 1 (mean FEC at follow-up = 0), which makes it not trivial to identify the most appropriate distribution. Instead we will focus on the function $1 - FECR_j$. This function can take any value between zero and $+\infty$, and hence we will assume that the function $1 - FECR_j$ follows a gamma distribution. The expected value of this function equals $1 - e_j$, the variance equals that of $FECR_j$ (Eqns (13), (14), (16) and (17)). Based on these equations, the general format to calculate γ_{1-FECR_j} and θ_{1-FECR_j} equals,

$$\gamma_{1-FECR_j} = \frac{(1 - e_j)^2 \cdot N_j}{Var[FECR_j]}$$
$$\theta_{1-FECR_j} = \frac{Var[FECR_j]}{(1 - e_j) \cdot N_j} \tag{18}$$

To obtain γ_{1-FECR_j} and θ_{1-FECR_j} for the different $FECR_j$ scenarios, it is necessary to substitute $Var[FECR_j]$ by Eqn (13) for $FECR_{j_{ind}}$, by Eqn (14) for $FECR_{j_{ind\Delta}}$, by Eqn (16) for $FECR_{j_{pool}}$, and by Eqn (17) for $FECR_{j_{pool\Delta}}$. As illustrated in Figure 6, these gamma distributions approach those of the above described theoretical surveys.

To verify whether these theoretical gamma distributions also fit empirical data, we have plotted the probability density of both the $1 - FECR_j$ generated by bootstrap analysis (10,000 iterations) of individual data obtained in a drug efficacy trial in Ethiopia and the expected gamma distributions in Figure 7 for each of the STH species. The drug efficacy trial was part of a multinational study designed to assess the efficacy of a single oral dose of albendazole (ABZ) against STH infections in school children in endemic countries (Vercruysse et al., 2011). For this validation all 410 ($= N_{j_{ind}}$) Ethiopian subjects screened at baseline were included. However, 154 subjects with a baseline FEC of 0 EPG were not treated nor re-examined at follow-up. To include these subjects it was assumed that the FEC at follow-up of these noninfected subjects (falsely/truly) also equalled zero after drug administration. In addition to this, seven infected subjects did not provide a stool sample at follow-up. These subjects were replaced by a random sample of subjects for which complete data were available. The values for e_j, μ_j and k_j required to estimate the parameters of the gamma distribution were estimated from the observed data. The amount of stool

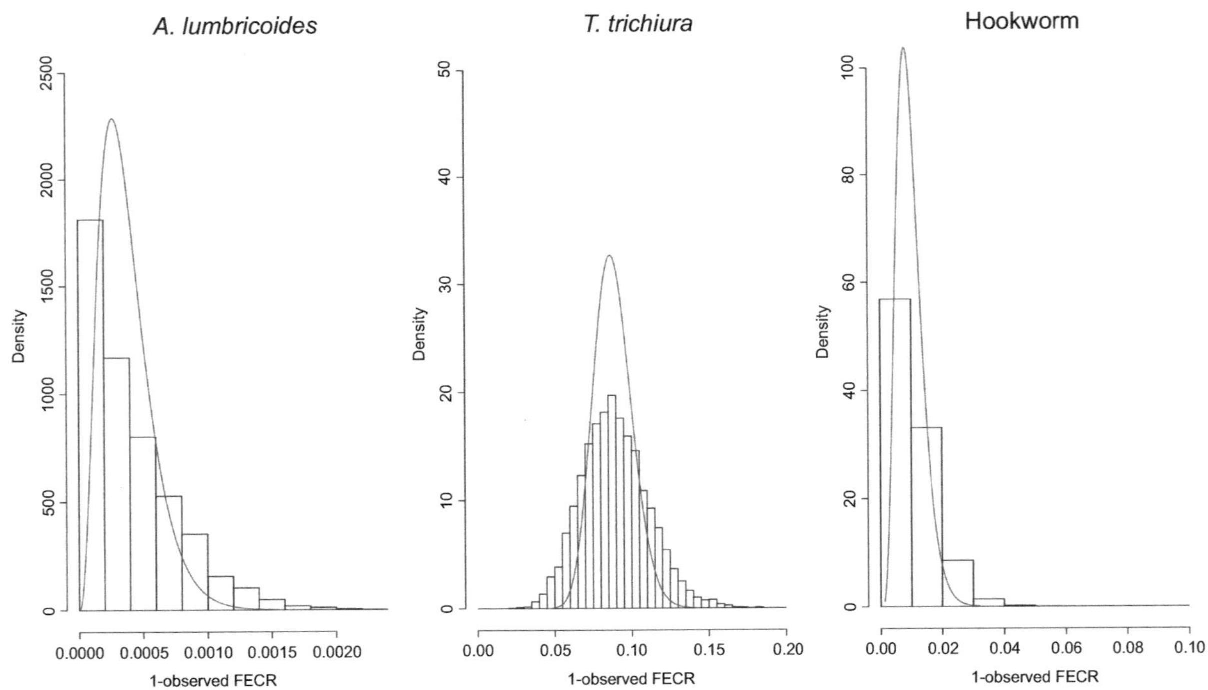

Figure 7 The probability density of $1 - FECR_j$ (faecal egg count reduction) estimates of a single oral dose of albendazole against *Ascaris lumbricoides*, *Trichuris trichiura* and hookworm infections generated by bootstrap analysis (10,000 iterations) of 410 individual data using McMaster during a drug efficacy trial in Ethiopia and the probability density of the expected gamma distribution (red (light gray in print versions) lines). In these Gamma distributions $\gamma_{j_{ind}}$ equalled $(1 - e_j)^2 \cdot N_{j_{ind}} / Var[FECR_{j_{ind}}]$ and $\theta_{j_{ind}}$ equalled $Var[FECR_{j_{ind}}]/(1 - e_j) \cdot N_{j_{ind}}$. For each of the three helminth species e_j, μ_j and k_j were separately substituted by $FECR_{j_{ind}}$, $\overline{Y_{\cdot j}}$ and $\overline{Y_{\cdot j}}/(Var[Y_{ij}]/\overline{Y_{\cdot j}} - 1/f_j - 1)$ (derived from Eqn (2)), respectively. For all helminth species, $N_{j_{ind}}$ was substituted by 410, and f_j by 0.02.

examined f_j was set at 0.02 g as all samples were processed with McMaster. Overall, the plots indicate that there is no perfect match between the empirical data and the theoretical gamma distributions, the theoretical distribution underestimating the variation observed in the empirical data.

4. APPLICATIONS OF THE GENERAL MATHEMATICAL FRAMEWORK

4.1 Sample size calculation for the assessment of FECs

Depending on the objectives of the survey, there are various ways to determine the samples size for the assessment of population mean FECs. We will discuss how to calculate a sample size for assessing population mean FEC with a predefined level of precision, whether population mean FEC exceeds a predefined level of infection intensity using a lot quality assurance sampling strategy (LQAS), and the absence of STH with a predefined level of probability.

4.1.1 Assessment of population mean FEC with a predefined precision

Given the variance derived from two hierarchical models (see Eqns (2) and (6)), and the distributions of $\overline{Y_{\cdot j}}$ Eqn (9) and $\overline{U_{\cdot j}}$ Eqn (10), it is possible to determine the sample size N_j to assess the population mean FEC μ_j with a precision W (= width of the $1 - \alpha$ confidence interval (CI)). This can be obtained by taking the $\alpha/2th$ and $1 - \alpha/2th$ percentile of the gamma distribution and determining W for a wide range of values of N_j. The required sample size is that N_j for which W does not exceed a predefined value. Figure 8 illustrates the increase in sample size as a function of W to assess the population mean of *A. lumbricoides* FEC in a population j ($\mu_j = 500$ EPG and $k_j = 0.1$). Both individual and pooled stool samples were examined using the Kato-Katz thick smear ($f_j = 0.0417$ g). In this example α was set at 0.05. When individual samples are examined, a minimum of 37 individuals ($= N_{j_{ind}}$) will need to be screened to assess the population mean FEC with a precision of at least 1,000 EPG, when pools of 5 are examined 40 individuals will need to be screened ($= N_{j_{pool}} \cdot m_j = 8 \cdot 5$).

4.1.2 Assessment of FECs using an LQAS strategy

LQAS was initially developed to assure the quality in industrial processes at minimum cost and involves three consecutive steps. First, a random sample ('lot') of items is taken, subsequently it is verified whether the lot meets a

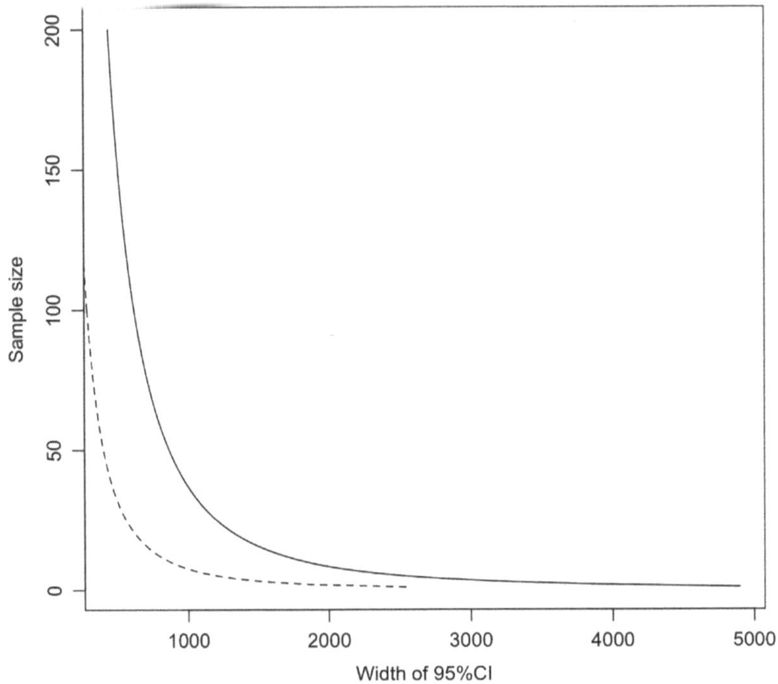

Figure 8 The number of individuals ($= N_{j_{ind}}$, straight line) and pools of 5 ($N_{j_{pool}}$, dashed line) as a function of the width of the 95% confidence interval ($\alpha = 0.05$) for the assessment of population mean *Ascaris lumbricoides* FEC in a population j ($\mu_j = 500$ EPG and $k_j = 0.1$). Both individual and pooled stool samples were examined using the Kato-Katz thick smear ($f_j = 0.0417$ g).

predefined standard of quality (e.g. number of items with production faults). Finally, further actions are taken when the quality is revealed to be unacceptable. Since its first description in the late 1920s, LQAS has also found various applications in the field of public health. For helminthiasis, it can be used to verify whether the observed number of infected subjects in a random sample exceeds a predefined decision threshold (Addiss et al., 2003; Brooker et al., 2005), followed by administration of drugs if this is the case. The mathematical underpinnings of LQAS based on helminth prevalence have been described by Olives et al. (2012). At present, LQAS has not yet been developed for FECs.

For the application of LQAS based on FECs, we want to identify those populations j for which μ_j equals or exceeds a decision threshold t, and hence those populations j that are at the highest need of drug

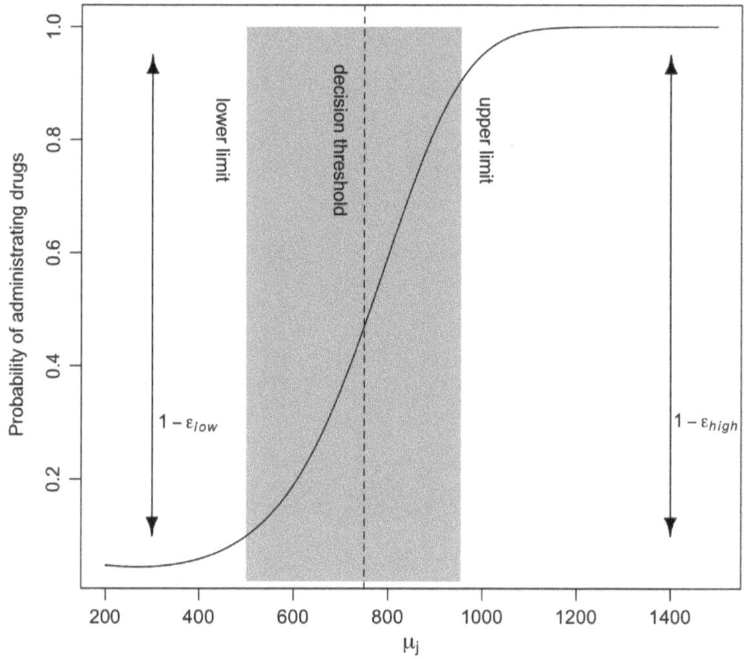

Figure 9 The probability of administering drugs to a population j as a function of population mean faecal egg count μ_j (expressed in eggs per g of stool). The dashed line indicates the decision threshold t. The grey zone between the lower and the upper limit represents the range of μ_j values for which the probability of correctly withdrawing drugs ($= 1 - \varepsilon_{low_j}$) or correctly administering drugs ($= 1 - \varepsilon_{high_j}$) is less than 0.90 ($\varepsilon_{low_j} = \varepsilon_{high_j} = 0.1$).

administration. However, we want to reduce the probability (ε_{low_j}) that populations are unnecessarily receiving drugs and that populations in need of drugs are not covered (ε_{high_j}). Therefore, we want to be confident that infected populations in no need of drugs with $\mu_j \leq ll$ (lower limit) will not be covered with a probability of at least $1 - \varepsilon_{low_j}$ and that heavily infected populations with $\mu_j \geq ul$ (upper limit) will receive drugs with a probability of at least $1 - \varepsilon_{high_j}$. We illustrated the concept of LQAS in Figure 9. In this figure, the sample size N_j allowed to correctly withhold and provide drugs to a population j with a probability of at least 0.90 ($\varepsilon_{low_j} = \varepsilon_{high_j} = 0.1$) as long as $\mu_j \leq 500$ EPG, and $\mu_j \geq 955$ EPG, respectively.

Given that $\overline{Y_{\cdot j}}$ and $\overline{U_{\cdot j}}$ are gamma distributed, we can write these probabilities for the examination of individual stool samples as,

$$\begin{aligned} \varepsilon_{low_{j_{ind}}} &\leq P\left(\overline{Y_{\cdot j}} \geq t \middle| \mu_j = ll\right) = 1 - P\left(\overline{Y_{\cdot j}} < t \middle| \mu_j = ll\right) \\ &\leq 1 - F\left(t; \gamma_{j_{ind}}, \theta_{j_{ind}}, N_{j_{ind}} \middle| \mu_j = ll\right) \\ \varepsilon_{high_{j_{ind}}} &\leq P\left(\overline{Y_{\cdot j}} < t \middle| \mu_j = ul\right) \\ &\leq F\left(t; \gamma_{j_{ind}}, \theta_{j_{ind}}, N_{j_{ind}} \middle| \mu_j = ul\right) \end{aligned} \tag{19}$$

and for the examination of pooled stool samples as

$$\begin{aligned} \varepsilon_{low_{j_{pool}}} &\leq P\left(\overline{U_{\cdot j}} \geq t \middle| \mu_j = ll\right) = 1 - P\left(\overline{U_{\cdot j}} < t \middle| \mu_j = ll\right) \\ &\leq 1 - F\left(t; \gamma_{j_{pool}}, \theta_{j_{pool}}, N_{j_{pool}} \middle| \mu_j = ll\right) \\ \varepsilon_{high_{j_{pool}}} &\leq P\left(\overline{U_{\cdot j}} < t \middle| \mu_j = ul\right) \\ &\leq F\left(t; \gamma_{j_{pool}}, \theta_{j_{pool}}, N_{j_{pool}} \middle| \mu_j = ul\right) \end{aligned} \tag{20}$$

F in these equations represents the lower tail of the cumulative distribution function of the gamma distribution. For the parameterization of $\gamma_{j_{ind}}$, $\theta_{j_{ind}}$, $\gamma_{j_{pool}}$and $\theta_{j_{pool}}$, we refer the reader to Eqns (9) and (10). Subsequently, the required sample size N_j can be determined for which ε_{low_j} at $\mu_j = ll$ and ε_{high_j} at $\mu_j = ul$ do not exceed a predefined value. In the following hypothetical example, we will determine the sample size N_j to correctly assign drugs against *A. lumbricoides* infections in a population j. The threshold t to administer drugs is set at 750 EPG. We would like to have 90% confidence that populations with $\mu_j = 500$ EPG are withhold from treatment, and that populations with $\mu_j = 1000$ EPG receive treatment, implying that both ε_{low_j} and ε_{high_j} equal 0.1. Given that μ_j and k_j are correlated with each other (see Figure 2), we will assume that k_j equals 0.1 and 0.6 for a population j for which μ_j equals 500 EPG and 1,000 EPG, respectively. We will screen both individual and pooled samples ($m_j = 5$) using a single Kato-Katz thick smear ($f_j = 0.0417$ g). Figure 10 illustrates the sample size as a function of ε_{low_j} and ε_{high_j}. When samples are examined individually at least 72 individuals are required (N_{ind} for $\varepsilon_{low_{j_{ind}}} \leq 0.1$ equals 72, $N_{j_{ind}}$ for $\varepsilon_{high_{j_{ind}}} \leq 0.1$ equals 41). When samples are pooled into pools of 5, 75 individuals are required ($N_{j_{pool}}$ for $\varepsilon_{low_{j_{pool}}} \leq 0.1$ equals 15, $N_{j_{pool}}$ for $\varepsilon_{high_{j_{pool}}} \leq 0.1$ equals 9).

4.1.3 Assessment of absence of STHs

When assessing the absence of STH infections, we would like to minimize the probability of falsely declaring the absence of eggs based on a subset of

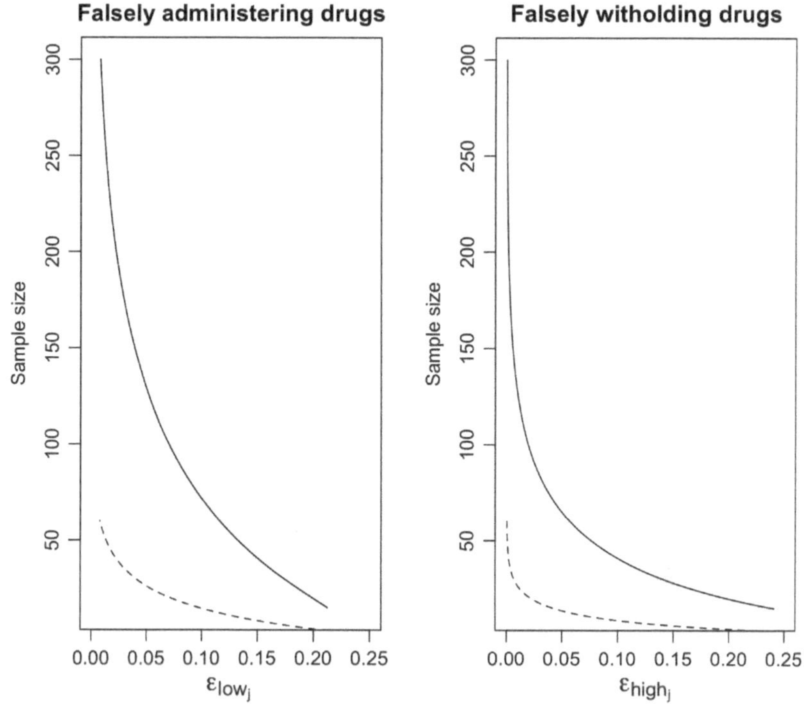

Figure 10 The number of individuals ($= N_{j_{ind}}$, straight line) and pools of 5 ($N_{j_{pool}}$, dashed line) as a function of the probability of falsely distributing drugs ε_{low_j} (left graph) in a population j ($\mu_j = 500$ EPG and $k_j = 0.1$), and falsely withholding drugs ε_{high_j} in a population j ($\mu_j = 1000$ EPG and $k_j = 0.6$) (right graph).

$N_{j_{ind}}$ individuals ($\overline{Y_{\cdot j}} = 0$) or $N_{j_{pool}}$ pools ($\overline{U}_{\cdot j} = 0$), while population mean FEC μ_j of a population j exceeds zero. Given that $\overline{Y_{\cdot j}}$ and $\overline{U_{\cdot j}}$ are gamma distributed, we can deduce this probability ε_{zero_j} for examination of individual stool samples as described in Eqns (21) and (22). The terms $1/(f \cdot N_{j_{ind}})$ and $1/(f \cdot N_{j_{pool}})$ in these equation correspond to the least nonzero mean FEC possible. For example, if 25 subjects ($= N_{j_{ind}}$) are screened individually for the presence of eggs using a single Kato-Katz thick smear ($f_j = 0.0417$ g), then the least nonzero possible number of eggs detected is one single egg in one subject, or a mean FEC of 1 egg/(0.0417 g · 25) = 0.96 EPG.

$$\begin{aligned}\varepsilon_{zero_{j_{ind}}} &= P\left(\overline{Y_{\cdot j}} = 0 \middle| \mu_j > 0\right) = P\left(\overline{Y_{\cdot j}} < 1\middle/\left(f_j \cdot N_{j_{ind}}\right) \middle| \mu_j > 0\right) \\ &= F\left(1\middle/\left(f_j \cdot N_{j_{ind}}\right); \gamma_{j_{ind}}, \theta_{j_{ind}}, N_{j_{ind}} \middle| \mu_j > 0\right)\end{aligned} \tag{21}$$

$$\begin{aligned}\varepsilon_{zero_{j_{pool}}} &= P\left(\overline{U_{\cdot j}} = 0 \middle| \mu_j > 0\right) = P\left(\overline{U_{\cdot j}} < 1 \middle/ \left(f_j \cdot N_{j_{pool}}\right) \middle| \mu_j > 0\right) \\ &= F\left(1 \middle/ \left(f_j \cdot N_{j_{pool}}\right); \gamma_{j_{pool}}, \theta_{j_{pool}}, N_{j_{pool}} \middle| \mu_j > 0\right)\end{aligned} \tag{22}$$

For the parameterization of $\gamma_{j_{ind}}$, $\theta_{j_{ind}}$, $\gamma_{j_{pool}}$, $\theta_{j_{pool}}$ we refer the reader to Eqns (9) and (10). Figure 11 illustrates the probability ε_{zero_j} to falsely declare a population j free of *A. lumbricoides* infections as a function of sample size N_j. In this theoretical example, the population mean FEC μ_j and aggregation parameter k_j equalled 100 EPG and 0.001, respectively. Both individual and pooled ($m_j = 5$) samples were examined using a Kato-Katz thick smear ($f_j = 0.0417$ g). If we would like to minimize ε_{zero_j} to 0.1, 290 individuals need to be screened when samples are examined individually. When pools of 5 are examined, 360 individuals need to be screened ($= N_{j_{pool}} \cdot m_j = 72 \cdot 5$).

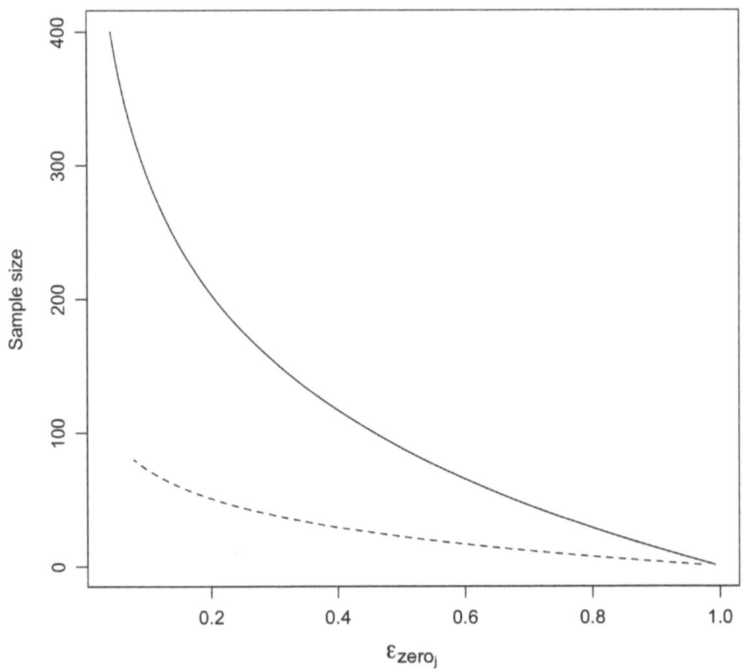

Figure 11 The number of individuals ($= N_{j_{ind}}$, straight line) and pools of 5 ($N_{j_{pool}}$, dashed line) as a function of the probability ε_{zero_j} of falsely declaring that a population j is free of *Ascaris lumbricoides* infections, while the population mean FEC μ_j and aggregation parameter k_j equals 100 eggs per gram of stool and 0.001, respectively. Both individual samples and pooled samples were examined with a Kato-Katz thick smear ($f_j = 0.0417$ g).

4.2 Sample size calculation for the assessment of FECR

4.2.1 Assessment of FECR with a predefined precision

We will determine the sample size that allows to assess e_j with a precision W (= width of the $1 - \alpha$ CI). This can be determined by taking the $\alpha/2$th and $1-\alpha/2$th percentile of the gamma distribution and determine W for a wide range of values of N_j. The required sample size is that N_j for which W does not exceed a predefined value. Figure 12 illustrates the increase in sample size as a function of W to assess the efficacy of an intervention e_j (= 98%) in a population j ($\mu_j = 500$ EPG and $k_j = 0.1$). We will re-examine the same subjects after intervention and both individual and pooled stool samples will be examined using the Kato-Katz thick smear ($f_j = 0.0417$ g). In this example, α was set at 0.05. When individual samples are examined a minimum of 14 individuals ($= N_{j_{ind}}$) will need to be screened to assess

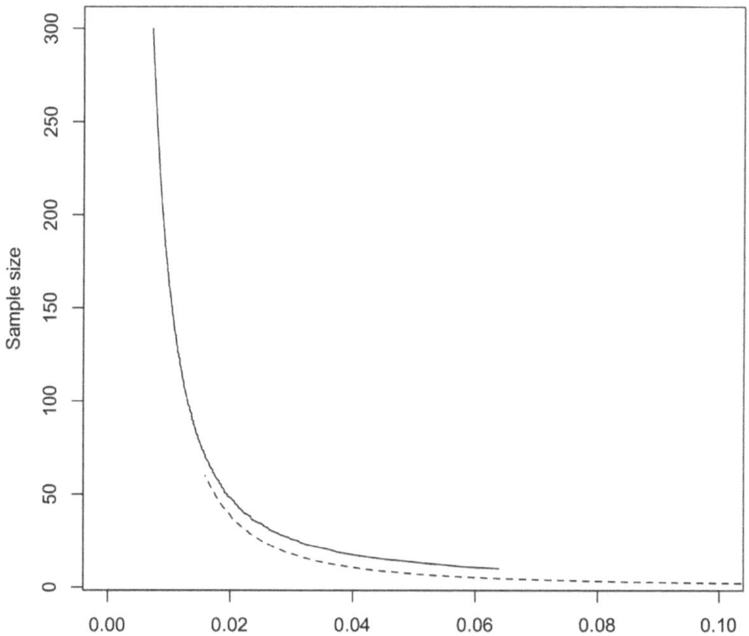

Figure 12 The number of individuals ($= N_{j_{ind}}$, straight line) and pools of 5 ($N_{j_{ind}}$, dashed line) as a function of the width of the 95% confidence interval ($\alpha = 0.05$) for the assessment of efficacy of an intervention e_j (= 0.98) based on faecal egg count reduction (FECR) in population j ($\mu_j = 500$ EPG and $k_j = 0.1$). Both individual samples ($FECR_{j_{ind}}$) and pooled samples ($FECR_{j_{pool}}$) were examined with a Kato-Katz thick smear ($f_j = 0.0417$ g). The same individuals were screened both prior to and after the intervention.

FECR with a precision of at least 0.05, when pools of 5 are examined 40 individuals will need to be screened ($= N_{j_{pool}} \cdot m_j = 8 \cdot 5$). In addition to the number of subjects that is required to guarantee a reliable assessment of e_j of an intervention, it is necessary also to ensure that e_j can be estimated at all. In the case of $\overline{Y_{\cdot j}} = 0$ or $\overline{U_{\cdot j}} = 0$, this will not be not possible. We will therefore calculate the sample size required that permits to assess e_j in a population j ($\overline{Y_{\cdot j}} > 0$ or $\overline{U_{\cdot j}} > 0$) with a probability of at least $1 - \varepsilon_{zero_j} \cdot \varepsilon_{zero_j}$ will be determined as described in Section 4.1.3, Assessment of absence of STH. For these sample sizes obtained to assess $FECR_j$ with a $W = 0.05$, the probability of ε_{zero_j} for examination of individual samples is less than 0.0005, for examination of pooled samples this was less than 10^{-7}. To obtain a comparable level of ε_{zero_j}, 28 individual subjects will need to be screened.

4.2.2 Assessment of FECR using an LQAS strategy

An LQAS strategy to assess FECR is most appropriate to verify whether the efficacy of drugs is still satisfactory. Based on two multicentre studies assessing the efficacy of ABZ (Vercruysse et al., 2011) and mebendazole (MBZ) (Levecke et al., 2014b), the WHO has defined the expected minimum for these drugs in all future drug efficacy studies. Drug efficacies below these levels should be viewed as indicative of potential drug resistance (WHO, 2013). For both drugs, the decision thresholds t are set at 0.95 for *A. lumbricoides* and 0.50 for *T. trichiura*. For hookworms, the threshold is set at 0.90 when ABZ is administered and 0.70 when MBZ is administered. For analogy with assigning a treatment to a population j when μ_j exceeds a threshold t, it is possible to write the probability $\varepsilon_{satisfact_j}$ of falsely concluding that a drug has satisfactory efficacy and the probability $\varepsilon_{reduced_j}$ of falsely concluding that the drug has reduced efficacy. For the examination of individual stool samples of the same individuals both prior to and after the intervention, these probabilities can be written as,

$$\begin{aligned} \varepsilon_{satisfact_{j_{ind}}} &\leq P\left(FECR_{j_{ind}} \geq t \middle| e_j = ll\right) \\ &\leq P\left(1 - FECR_{j_{ind}} \leq 1 - t \middle| e_j = ll\right) \\ &\leq F\left(1 - t; \gamma_{j_{ind}}, \theta_{j_{ind}}, N_{j_{ind}} \middle| e_j = ll\right) \end{aligned} \tag{23}$$

$$\begin{aligned} \varepsilon_{reduced_{j_{ind}}} &\leq P\left(FECR_{j_{ind}} < t \middle| e_j = ul\right) = P\left(1 - FECR_{j_{ind}} > 1 - t \middle| e_j = ul\right) \\ &\leq 1 - P\left(1 - FECR_{j_{ind}} \leq 1 - t \middle| e_j = ul\right) \\ &\leq 1 - F\left(1 - t; \gamma_{j_{ind}}, \theta_{j_{ind}}, N_{j_{ind}} \middle| e_j = ul\right) \end{aligned}$$

for the examination of individual stool samples of different subjects prior to and after the intervention as,

$$\begin{aligned}
\varepsilon_{satisfact_{j_{ind\Delta}}} &\leq P\left(FECR_{j_{ind\Delta}} \geq t \middle| e_j = ll\right) \\
&\leq P\left(1 - FECR_{j_{ind\Delta}} \leq 1 - t \middle| e_j = ll\right) \\
&\leq F\left(1 - t; \gamma_{j_{ind\Delta}}, \theta_{j_{ind\Delta}}, N_{j_{ind\Delta}} \middle| e_j = ll\right) \\
\varepsilon_{reduced_{j_{ind\Delta}}} &\leq P\left(FECR_{j_{ind\Delta}} < t \middle| e_j = ul\right) = P\left(1 - FECR_{j_{ind\Delta}} > 1 - t \middle| e_j = ul\right) \\
&\leq 1 - P\left(1 - FECR_{j_{ind\Delta}} \leq 1 - t \middle| e_j = ul\right) \\
&\leq 1 - F\left(1 - t; \gamma_{j_{ind\Delta}}, \theta_{j_{ind\Delta}}, N_{j_{ind\Delta}} \middle| e_j = ul\right)
\end{aligned} \tag{24}$$

for the examination of pooled stool samples of the same subjects prior to and after the intervention as,

$$\begin{aligned}
\varepsilon_{satisfact_{j_{pool}}} &\leq P\left(FECR_{j_{pool}} \geq t \middle| e_j = ll\right) \\
&\leq P\left(1 - FECR_{j_{pool}} \leq 1 - t \middle| e_j = ll\right) \\
&\leq F\left(1 - t; \gamma_{j_{pool}}, \theta_{j_{pool}}, N_{j_{pool}} \middle| e_j = ll\right) \\
\varepsilon_{reduced_{j_{pool}}} &\leq P\left(FECR_{j_{pool}} < t \middle| e_j = ul\right) = P\left(1 - FECR_{j_{pool}} > 1 - t \middle| e_j = ul\right) \\
&\leq 1 - P\left(1 - FECR_{j_{pool}} \leq 1 - t \middle| e_j = ul\right) \\
&1 - F\left(1 - t; \gamma_{j_{pool}}, \theta_{j_{pool}}, N_{j_{pool}} \middle| e_j = ul\right)
\end{aligned} \tag{25}$$

and for the examination of pooled stool samples of different subjects prior to and after the intervention as,

$$\begin{aligned}
\varepsilon_{satisfact_{j_{pool\Delta}}} &\leq P\left(FECR_{j_{pool\Delta}} \geq t \middle| e = ll\right) \\
&\leq P\left(1 - FECR_{j_{pool\Delta}} \leq 1 - t \middle| e_j = ll\right) \\
&\leq F\left(1 - t; \gamma_{j_{pool\Delta}}, \theta_{j_{pool\Delta}}, N_{j_{pool\Delta}} \middle| e_j = ll\right) \\
\varepsilon_{reduced_{j_{pool\Delta}}} &\leq P\left(FECR_{j_{pool\Delta}} < 1 - t \middle| e_j = ul\right) = P\left(1 - FECR_{j_{pool\Delta}} > 1 - t \middle| e_j = ul\right) \\
&\leq 1 - P\left(1 - FECR_{j_{pool\Delta}} \leq 1 - t \middle| e_j = ul\right) \\
&\leq 1 - F\left(1 - t; \gamma_{j_{pool\Delta}}, \theta_{j_{pool\Delta}}, N_{j_{pool\Delta}} \middle| e_j = ul\right)
\end{aligned} \tag{26}$$

Subsequently, the required sample size N_j can be determined for which $\varepsilon_{satisfact_j}$ at $e_j = ll$ and $\varepsilon_{reduced_j}$ at $e_j = ul$ does not exceed a predefined value. In the following hypothetical example we will determine the sample size N_j required to correctly classify the efficacy of ABZ against *A. lumbricoides* as unsatisfactory when e_j is up to 0.90 (= *ll*), and to correctly classify the efficacy as satisfactory if *e* is at least 0.98 (*ul*) with a probability of 0.90, implying that both $\varepsilon_{satisfact_j}$ and $\varepsilon_{reduced_j}$ equal 0.1. We will conduct the survey in a population *j* ($\mu_j = 500$ EPG and $k_j = 0.1$) and screen both individual and pooled samples ($m_j = 5$) using a single Kato-Katz thick smear ($f_j = 0.0417$ g). The same individuals are screened both prior to and after the intervention. Figure 13 illustrates the sample size as a function of $\varepsilon_{satisfact_j}$ and $\varepsilon_{reduced_j}$. When samples are examined individually at least 19 individuals are required ($N_{j_{ind}}$ for $\varepsilon_{satisfact_{j_{ind}}} \leq 0.1$ equals 19, $N_{j_{ind}}$ for $\varepsilon_{reduced_{j_{ind}}} \leq 0.1$ equals 18). When samples are pooled into pools of 5, 60 individuals are required ($N_{j_{pool}}$ for $\varepsilon_{satisfact_{j_{pool}}} \leq 0.1$ equals 11, $N_{j_{pool}}$ for $\varepsilon_{reduced_{j_{pool}}} \leq 0.1$ equals 12). For these sample sizes ε_{zero_j} for examination of individual samples is less than $2.5 \cdot 10^{-5}$, for examination of pooled samples this was less than $4.5 \cdot 10^{-13}$. To obtain a comparable level of ε_{zero_j}, 46 individual subjects will need to be screened.

4.3 Estimation of true prevalence in absence of a gold standard

Accurate assessment of true prevalence (*TP*) depends largely on the ability of the diagnostic technique to correctly identify the presence (sensitivity) and absence (specificity) of eggs in stool. The eggs excreted by STHs have a unique and distinct morphology, enabling a specific diagnosis (no false positives). However, diagnostic methods often lack sensitivity (false negatives), consequently the apparent prevalence (*AP*) of subjects excreting eggs underestimates in most cases *TP*. There is a broad literature available on how to estimate the *TP* in absence of a gold standard, of which the majority focuses on latent class models. In these models unknown (latent) categorical (class) variables are estimated based either on observed data (maximum likelihood approach) or on a combination of observed data and *a priori* scientific knowledge (Bayesian approach). It is not the scope of the current study to discuss the different estimation procedures in detail, for this we refer the reader to Hui and Zou (1998), Enøe et al. (2000), Diggle (2011) and Speybroeck et al. (2013). Instead, we would like to indicate that both approaches are fed with binary inputs only (positive or

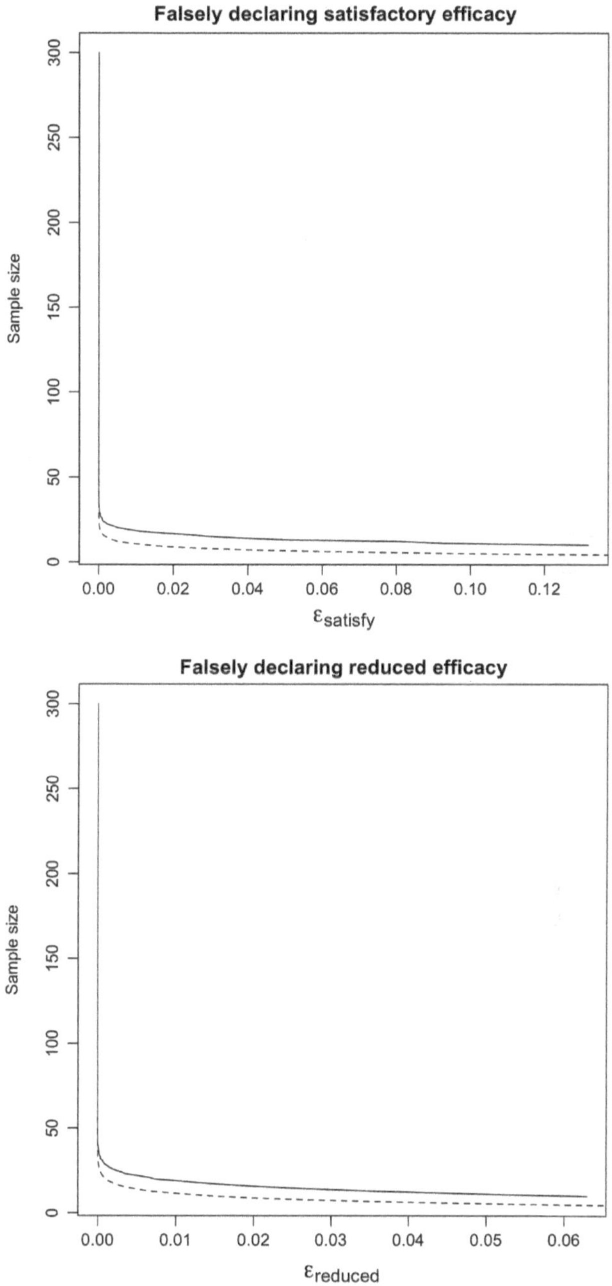

Figure 13 The number of individuals ($= N_{j_{ind}}$, straight line) and pools of 5 ($N_{j_{pool}}$, dashed line) as a function of the probability of falsely classifying drugs $\varepsilon_{satisfact_j}$ as satisfactory when $e_j = 0.90$ (top graph) and reduced $\varepsilon_{reduced_j}$ when $e_j = 0.98$ (bottom graph) with a probability not higher than 0.1 in a population j ($\mu_j = 500$ EPG and $k_j = 0.1$). Both individual samples and pooled samples were examined with a Kato-Katz thick smear ($f_j = 0.0417$ g). The same individuals are screened both prior and after the intervention.

negative test result), which are typically summarized in cross tables when at least two diagnostic methods are applied. In doing so for STHs, they totally ignore the underlying FECs which, as illustrated by Figure 4, largely explain the variation in sensitivity of diagnostic techniques between STH species, individuals and populations (see also Levecke et al., 2011b). In addition, from a programmatic point of view it is also important to periodically re-evaluate the prevalence of both any STH infection and the prevalence of low, moderate and high intensity infections after consecutive rounds of MDA. Prevalence of any STHs is currently the recommended parameter to scale down the frequency of MDA (WHO, 2011). Morbidity caused by STHs is generally more pronounced for moderate and high intensity infections, and hence a shift towards infections of low intensity is an additional parameter to evaluate the impact of MDA on public health. We will discuss how to estimate the *TP* of individual STH species, any STH and the three levels of infection intensity, taking into account the variation in sensitivity between individuals due to varying FECs.

4.3.1 Estimation of true prevalence of individual STH species

In our mathematical framework, the true prevalence TP_j of a particular STH species in a population j with mean population FEC μ_j and aggregation parameter k_j equals $1-P(X_{ij}=0)=1-\frac{\Gamma(x+k_j)}{\Gamma(k_j)\cdot x!}\cdot\left(\frac{k_j}{k_j+\mu_j}\right)^{k_j}\cdot\left(1-\frac{k_j}{k_j+\mu_j}\right)^{x}=1-\left(\frac{k_j}{k_j+\mu_j}\right)^{k_j}$.

When individual stool samples are examined, we can substitute μ_j by $\overline{Y_{\cdot j}}$; subsequently k_j can be substituted by $\overline{Y_{\cdot j}}/(Var[Y_{ij}]/\overline{Y_{\cdot j}}-1/f_j-1)$ (derived from Eqn (2)) resulting in the following equation to estimate TP_j,

$$E\left[TP_{j_{ind}}\right]=1-\left(\frac{\overline{Y_{\cdot j}}\Big/\left(Var\left[Y_{ij}\right]\Big/\overline{Y_{\cdot j}}-1\Big/f_j-1\right)}{\overline{Y_{\cdot j}}\Big/\left(Var\left[Y_{ij}\right]\Big/\overline{Y_{\cdot j}}-1\Big/f_j-1\right)+\overline{Y_{\cdot j}}}\right)^{\overline{Y_{\cdot j}}/\left(Var\left[Y_{ij}\right]/\overline{Y_{\cdot j}}-1/f_j-1\right)} \tag{27}$$

By analogy with this equation, the expected $TP_{j_{pool}}$ based on observed FECs U_{lj} obtained by screening f_j g of pooled stool of m_j individual samples can be estimated as,

$$E\left[TP_{j_{pool}}\right] =$$

$$1 - \left(\frac{\overline{U_{\cdot j}}\Big/\left(m_j\cdot\left(Var\left[U_{lj}\right]\Big/\overline{U_{\cdot j}} - 1\Big/f_j\right) - 1\right)}{\overline{U_{\cdot j}}\Big/\left(m_j\cdot\left(Var\left[U_{lj}\right]\Big/\overline{U_{\cdot j}} - 1\Big/f_j\right) - 1\right) + \overline{U_{\cdot j}}}\right)^{\overline{U_{\cdot j}}\Big/\left(m_j\cdot\left(Var\left[U_{lj}\right]\Big/\overline{U_{\cdot j}}-1\Big/f_j\right)-1\right)} \tag{28}$$

In the following theoretical example, we sampled 250 subjects from an infinite population j to assess the prevalence of *A. lumbricoides*. In this population, μ_j and k_j for *A. lumbricoides* equal 500 EPG and 0.1, respectively. Both individual and pooled samples ($m_j = 10$) were screened using a single Kato-Katz thick smear ($f_j = 0.0417$ g). The histogram of the true FECs (X_{ij}), the observed individual (Y_{ij}) and pooled FECs (U_{lj}), and their respective AP, mean and variance are provided in Figure 14. The TP_j in this population j equals 57.3% ($= 1 - (0.1/(0.1 + 500))^{0.1}$). By substituting expected values and variances in Eqns (27) and (28) by the observed means and variances, we obtain a $TP_{j_{ind}}$ (and corresponding 95%CI) of 64.1% (33.2; 94.0) and a $TP_{j_{pool}}$ 63.2% (31.1; 98.7) based on examination of individual and pooled stool samples, respectively. The 95%CI were obtained by a bootstrap analysis which took into account the correlation between mean and variance separately for Y_{ij} and U_{lj}.

4.3.2 Estimation of true prevalence of any STH

We can determine the TP of any STH by combining the above described Eqns (27) and (28) with those described by de Silva and Hall (2010) that allowed to estimate the prevalence of any STH based on the prevalence estimates of the individual STH species. We adapted their formulae below. In this formula $TP_{j_{sth}}$ represents the TP of any STH in a population j, whereas TP_{j_a}, TP_{j_t} and TP_{j_h} represent the TP of *A. lumbricoides*, *T. trichiura* and hookworm, respectively. The constant 1.06 is a correction factor suggested by de Silva and Hall after validation of their formulae with field data.

$$\begin{aligned} E\left[TP_{j_{sth}}\right] = \Big(E\left[TP_{j_a}\right] + E\left[TP_{j_t}\right] + E\left[TP_{j_h}\right] - \Big(E\left[TP_{j_a}\right]\cdot E\left[TP_{j_t}\right] + E\left[TP_{j_a}\right]\cdot E\left[TP_{j_h}\right] \\ + E\left[TP_{j_t}\right]\cdot E\left[TP_{j_h}\right] + E\left[TP_{j_a}\right]\cdot E\left[TP_{j_t}\right]\cdot E\left[TP_{j_h}\right]\Big)\Big)\Big/1.06 \end{aligned} \tag{29}$$

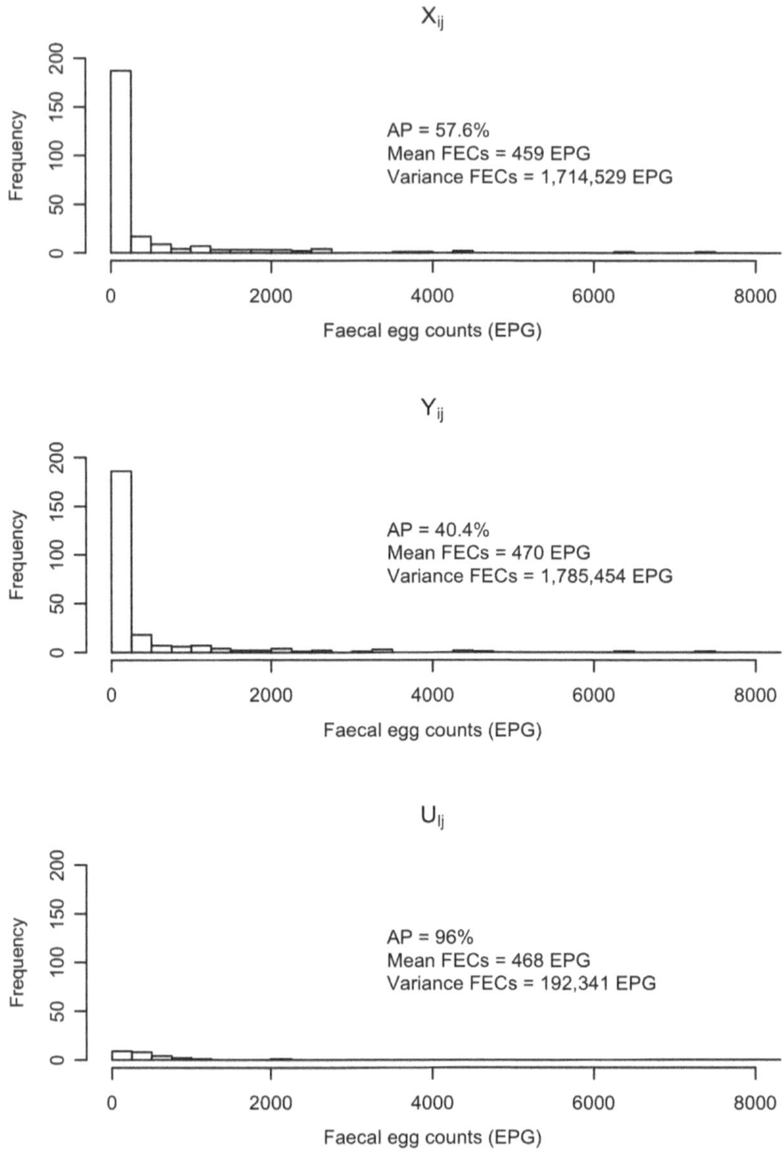

Figure 14 The histograms of the true *Ascaris lumbricoides* faecal egg counts (FECs; expressed in eggs per g of stool (EPG)) (X_{ij}), the observed individual (Y_{ij}) and pooled FECs (U_{lj}) from a random sample of 250 subjects from an infinitive population *j* ($\mu_j = 500$ EPG and $k_j = 0.1$). Both individual and pooled samples ($m_j = 10$) were screened using a single Kato-Katz thick smear ($f_j = 0.0417\ g$). The mean and variance of the different FECs, and their corresponding apparent prevalence (*AP*) are provided.

If only two STH species are present in a population j, for example *A. lumbricoides* and *T. trichiura*, we estimate the *TP* of any of those two STH infections $TP_{j_{at}}$ as,

$$E\left[TP_{j_{at}}\right] = \left(E\left[TP_{j_a}\right] + E\left[TP_{j_t}\right] - E\left[TP_{j_a}\right] \cdot E\left[TP_{j_t}\right]\right)/1.06 \tag{30}$$

In analogy, the *TP* of any other combination of two STH species infections can be written as,

$$E\left[TP_{j_{ah}}\right] = \left(E\left[TP_{j_a}\right] + E\left[TP_{j_h}\right] - E\left[TP_{j_a}\right] \cdot E\left[TP_{j_h}\right]\right)/1.06 \tag{31}$$

for estimating the *TP* of any *A. lumbricoides* and hookworm infection, and as,

$$E\left[TP_{j_{th}}\right] = \left(E\left[TP_{j_t}\right] + E\left[TP_{j_h}\right] - E\left[TP_{j_t}\right] \cdot E\left[TP_{j_h}\right]\right)/1.06 \tag{32}$$

for estimating the *TP* of any *T. trichiura* and hookworm infection.

4.3.3 Estimation of true prevalence of low, moderate and high intensity infections

The intensity of STH infections can be classified into low, moderate and high based on the individual FECs (in EPG). The FEC thresholds for each of the STH species are proposed by the WHO (1998) and are summarized in Table 4.

Because these thresholds are STH species-specific (due to the differences in fecundity and morbidity between these species) we only will work out the framework for *A. lumbricoides*. Subsequently, a framework can be developed for the two remaining STHs by replacing the *A. lumbricoides* thresholds by those for *T. trichiura* and hookworm, respectively.

For *A. lumbricoides*, the *TP* of low intensity infections in a population j with mean population FEC μ_j and aggregation parameter k_j can be written as,

$$\begin{aligned} E\left[TP_{j_{low_a}}\right] &= P\left(1 \geq X_{ij} \leq 4999\right) \\ &= P\left(X_{ij} \leq 4999\right) - P\left(X_{ij} = 0\right) \\ &= F\left(4999; \mu_j, k_j\right) - F\left(0; \mu_j, k_j\right) \end{aligned} \tag{33}$$

Table 4 The faecal egg count thresholds (expressed in number of eggs in 1 g of stool) proposed by WHO (1998) to classify the intensity of *Ascaris lumbricoides*, *Trichuris trichiura* and hookworm infections into low, moderate and high

	A. lumbricoides	***T. trichiura***	**Hookworm**
Low	1–4999	1–999	1–1999
Moderate	5000–49,999	1000–9999	2000–3999
High	≥50,000	≥10,000	≥4000

The *TP* of moderate intensity *A. lumbricoides* infections in this population j can be written as,

$$E\left[TP_{j_{moda}}\right] = P\left(5000 \geq X_{ij} \leq 49,999\right) = P\left(X_{ij} \leq 49,999\right) - P\left(X_{ij} \leq 4999\right) = F\left(49,999; \mu_j, k_j\right) - F\left(4999; \mu_j, k_j\right) \quad (34)$$

The *TP* of high intensity *A. lumbricoides* infections in this population can be written as,

$$E\left[TP_{j_{higha}}\right] = P\left(X_{ij} \geq 50,000\right) = 1 - P\left(X_{ij} \leq 49,999\right) = 1 - F\left(49,999; \mu_j, k_j\right) \quad (35)$$

In each of these formulae, F represents the cumulative distribution function (lower tail) of the negative binomial distribution with mean μ_j and aggregation parameter k_j. When individual stool samples are screened we replace μ_j by $\overline{Y_{\cdot j}}$ and k_j by $\overline{Y_{\cdot j}}/(Var[Y_{ij}]/\overline{Y_{\cdot j}} - 1/f_j - 1)$. When pooled samples are examined, we replace μ_j by $\overline{U_{\cdot j}}$ and k_j by $\overline{U_{\cdot j}}/(m_j \cdot (Var[U_{lj}]/\overline{U_{\cdot j}} - 1/f_j) - 1)$.

In the hypothetical example described in Section 4.3.1, estimation of the *TP* of individual STH species, the *TP* of low, moderate and high *A. lumbricoides* intensity infections equalled 54.9%, 2.4% and 0.0001%, respectively. By substituting expected values and variances in Eqns (33)–(35) by the observed means and variances, we obtain a *TP* (95%CI) of 62.2% (31.3; 93.6) (low intensity), 1.9% (0.4; 2.7) (moderate intensity) and 0.0001% (0; 0.0001) (high intensity) based on examination of individual stool samples, and a *TP* of 61.3% (28.4; 98.7) (low intensity), 1.9% (0.0001; 2.9) (moderate intensity) and 0.0001% (0; 0.0001) (high intensity) for the examination of pooled stool samples. The 95%CI were obtained by a bootstrap analysis which took into account the correlation between mean and variance separately for Y_{ij} and U_{lj}.

4.4 Estimating μ_j and k_j

For the different applications, it is essential to define the expected negative binomial distribution of the FECs by providing the expected values for μ_j and k_j for the population j in which the survey will be conducted. When detailed data are available, these parameters can be easily estimated based on the individual FEC data. However, when this level of information is not at hand, an alternative approach to estimate both parameters is required.

As illustrated in Figure 2, μ_j and k_j are correlated, and hence when prior knowledge is at hand for one variable, it is possible to estimate the other one. Because prior information on mean population FEC μ_j is generally at hand, we will estimate k_j as a function of μ_j.

To this end, we applied a linear model on the epidemiological survey conducted in Kenya to estimate the k_j in a school j based on its corresponding μ_j. We extracted the results obtained by duplicate Kato-Katz thick smear ($f_j = 2 \cdot 0.0417$ g). We used $\overline{Y_{\cdot j}}$ and $\overline{Y_{\cdot j}}/(Var[Y_{ij}]/\overline{Y_{\cdot j}} - 12 - 1)$ as a proxy of μ_j and k_j, respectively. Schools with μ_j equal to zero or $k_j < 0$ were omitted from the analysis. Figure 15 illustrates the predicted k_j based on μ_j for *A. lumbricoides*, *T. trichiura* and hookworm, respectively. The equations to estimate k_j as a function of μ_j are provided in Table 5. It is important to note that these equations obtained form a restrict range of μ_j, and hence any extrapolation from these equations beyond this range should be implemented with care.

5. CONCLUSIONS

In the present study, we described a general mathematical framework for egg counts in stool. Subsequently, we discussed how to estimate the sample size for assessing the population mean FEC and the impact of an intervention (measured as FECR) for any scenario of STH epidemiology (mean population FEC and aggregation of FEC between individuals) and diagnostic strategy (amount of stool examined ~ sensitivity of the diagnostic technique, and examination of individual/pooled samples). We illustrated how to calculate a sample size for (1) assessing population mean FEC with a predefined level of precision, (2) determining whether population mean FEC exceeds a predefined level of infection intensity, (3) correctly classifying a population free of STH with a predefined probability, (4) assessing impact of an intervention with a predefined level of precision and (5) examining whether the efficacy of a drug remains satisfactory, as well as estimating the *TP* of (6) each individual STH species separately, (7) any STH species and (8) low, moderate and high intensity infections in the absence of a gold standard.

Based on a prior knowledge on the local epidemiology, health decision-makers can now compare different diagnostic strategies and their corresponding technical and financial resources required, and hence optimize the use of funds allocated for monitoring MDA programmes to control STH. In the present study we focussed on STH infections, but this framework can be generalized to any other helminth infection, both in public or animal health, that is diagnosed by detection and quantification of eggs

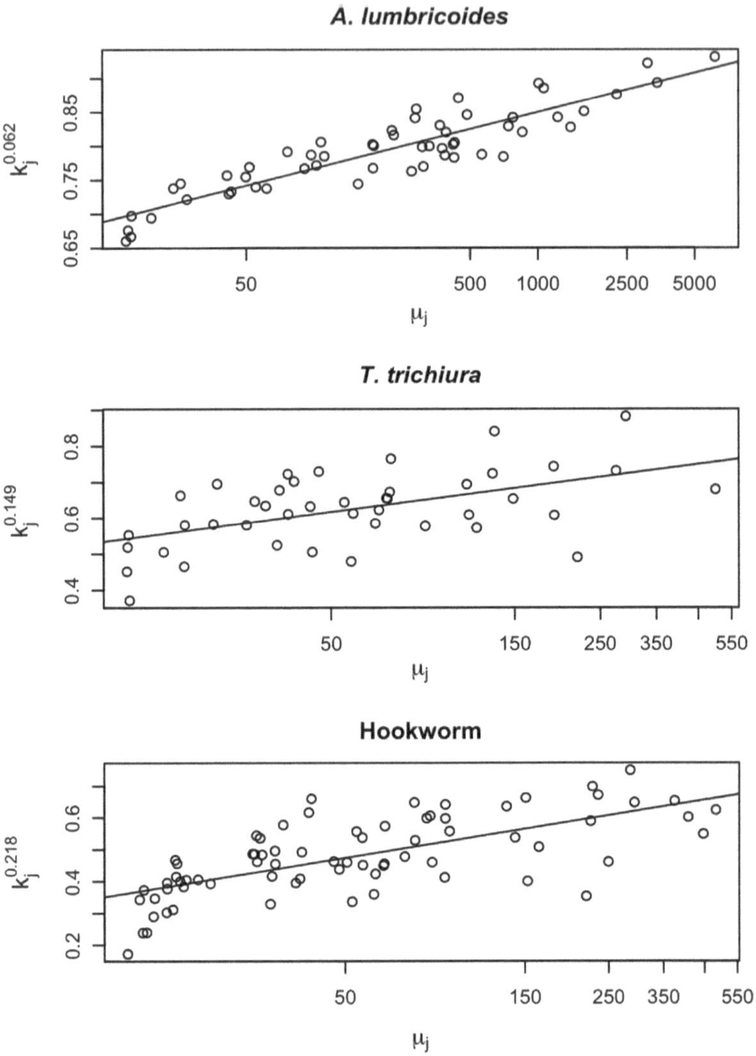

Figure 15 The scatter plots of aggregation parameter k_j as a function of mean school faecal egg counts (FECs; expressed in eggs per gram of stool (EPG)) μ_j, expressed in EPG of stool for *Ascaris lumbricoides*, *Trichuris trichiura* and hookworm. The dataset used was collected during an epidemiological survey in Kenya (Brooker et al., 2012).

in stool. For example, in public health the framework can be applied for *Schistosoma mansoni*, the causative agent of schistosomiasis (WHO, 2011). In animal health, we can apply the framework to gastrointestinal nematode infections, which up to date account for important losses in production of livestock (Charlier et al., 2014).

Table 5 Equations to estimate the aggregation parameter k_j based on the mean population faecal egg count (FEC; expressed in eggs per gram of stool (EPG)) μ_j, and the range of μ_j in which these equations were obtained for *Ascaris lumbricoides*, *Trichuris trichiura* and hookworm infections. The dataset used was collected during an epidemiological survey in Kenya (Brooker et al., 2012)

	***A. lumbricoides* n = 56**	***T. trichiura* n = 43**	**Hookworms n = 73**
k_j	$\sqrt[0.062]{0.060 + 0.036 \cdot \log(\mu_j)}$	$\sqrt[0.149]{0.381 + 0.060 \cdot \log(\mu_j)}$	$\sqrt[0.218]{0.150 + 0.083 \cdot \log(\mu_j)}$
min μ_j	14	14	13
max μ_j	6173	499	482

Although this study allows healthcare decision-makers to adapt their survey design according to both local helminth epidemiology and resources, there are some limitations that need to be acknowledged. First, it is important to note that this mathematical framework only holds true under the assumptions made, and hence that any sample size and *TP* obtained through the framework should be interpreted with caution. In the current framework we assumed that (1) individual FECs within a population follow a negative binomial distribution, (2) the number of eggs observed by microscopy in *f* g of stool follows a Poisson distribution, (3) the true FEC in a pooled stool sample equals the mean of the true FECs of the individual stool samples pooled and that (4) the impact of an intervention is equal for all subjects. As discussed in Section 2, sources of variability in egg counts, these assumptions explain a large proportion of the variation, but definitely not all of the variation observed in the field. For example, a zero-inflated negative binomial may be a more appropriate fit in scenarios where zero FECs are more frequent, and additional sources of variation will need to be parameterized to grasp all the variation observed in egg excretion (e.g. day-to-day variation), the egg counting process (e.g. variation in FEC between diagnostic techniques and laboratory technicians, and the impact of an intervention (see Figure 7). On the other hand, we should be aware that expanding the model by parameterizing the remaining variation, if possible at all, would compromise the accessibility of the framework for the end users: more parameters will need to be defined and prior information on these additional may not always be at hand. Second, if one wants to estimate the mean FEC in school children within a district by first sampling a number of schools followed by sampling a number of subjects per school (two-stage cluster sampling), one cannot use the framework to calculate the total number of children that needs to be screened over these schools. This is because the current model does not account for clustering of STH infections between populations, and hence ignores the additional variation in FEC between clusters (*in casu* schools). Given the additional complexity of clustered STH infections, we will work out an extended model in a follow-up study. Finally, it is obvious that the framework in its current form will not be attractive to a wide spectrum of possible end users. To bridge the gap between this study and the end users we plan to develop an online tool that provides sample size and *TP* estimates without the need of prior knowledge on the mathematical framework or any statistical software.

APPENDICES

A. The expected value and variance of faecal egg counts (FECs) based on individual stool samples

A.1 FECs prior to an intervention

A.1.1 The expected value

$$
\begin{aligned}
E[Y_{ij}] &= E\left[\frac{Z_{ij}}{f_j}\right] = \frac{E[Z_{ij}]}{f_j} \\
&= \frac{E[E[Z_{ij}|\lambda_{ij}, X_{ij}]]}{f_j} \\
&= \frac{E[f_j \cdot X_{ij}]}{f_j} \\
&= \frac{f_j \cdot E[X_{ij}]}{f_j} \\
&= \mu_j
\end{aligned}
$$

A.1.2 The variance

$$
\begin{aligned}
Var[Y_{ij}] &= Var\left[\frac{Z_{ij}}{f_j}\right] = \frac{Var[Z_{ij}]}{f_j^2} \\
&= \frac{1}{f_j^2} \cdot (E[Var[Z_{ij}|\lambda_{ij}, X_{ij}]] + Var[E[Z_{ij}|\lambda_{ij}, X_{ij}]]) \\
&= \frac{1}{f_j^2} \cdot \left(E[f \cdot X_{ij}] + Var[f_j \cdot X_{ij}]\right) \\
&= \frac{1}{f_j^2} \cdot \left(f_j \cdot E[X_{ij}] + f_j^2 \cdot Var[X_{ij}]\right) \\
&= \frac{1}{f_j^2} \cdot \left(f_j \cdot \mu_j + f_j^2 \cdot \left(\mu_j + \mu_j^2 / k_j\right)\right) \\
&= \frac{f_j^2 \cdot \mu_j}{f_j^2} \cdot \left(\frac{1}{f_j} + 1 + \frac{\mu_j}{k_j}\right) \\
&= \mu_j \cdot \left(\frac{1}{f_j} + 1 + \frac{\mu_j}{k_j}\right)
\end{aligned}
$$

A.2 FECs after an Intervention

A.2.1 The expected value

$$
\begin{aligned}
E\left[Y'_{ij}\right] &= E\left[\frac{Z'_{ij}}{f_j}\right] = \frac{E\left[Z'_{ij}\right]}{f_j} \\
&= \frac{E\left[E\left[Z'_{ij}\middle|\lambda'_{ij}, X'_{ij}\right]\right]}{f_j} \\
&= \frac{E\left[f_j \cdot X'_{ij}\right]}{f_j} \\
&= \frac{f_j \cdot E\left[X'_{ij}\right]}{f_j} \\
&= E\left[\left(1 - e_j\right) \cdot X_{ij}\right] \\
&= \left(1 - e_j\right) \cdot E\left[X_{ij}\right] \\
&= \left(1 - e_j\right) \cdot \mu_j
\end{aligned}
$$

A.2.2 The variance

$$
\begin{aligned}
Var\left[Y'_{ij}\right] &= Var\left[\frac{Z'_{ij}}{f_j}\right] = \frac{Var\left[Z'_{ij}\right]}{f_j^2} \\
&= \frac{1}{f_j^2} \cdot \left(E\left[Var\left[Z'_{ij}\middle|\lambda'_{ij}, X'_{ij}\right]\right] + Var\left[E\left[Z'_{ij}\middle|\lambda'_{ij}, X'_{ij}\right]\right]\right) \\
&= \frac{1}{f_j^2} \cdot \left(E\left[f_j \cdot X'_{ij}\right] + Var\left[f_j \cdot X'_{ij}\right]\right) \\
&= \frac{1}{f_j^2} \cdot \left(f_j \cdot E\left[X'_{ij}\right] + f_j^2 \cdot Var\left[X'_{ij}\right]\right) \\
&= \frac{1}{f_j^2} \cdot \left(f_j \cdot E\left[\left(1 - e_j\right) \cdot X_{ij}\right] + f_j^2 \cdot Var\left[\left(1 - e_j\right) \cdot X_{ij}\right]\right) \\
&= \frac{1}{f_j^2} \cdot \left(f_j \cdot \left(1 - e_j\right) \cdot E\left[X_{ij}\right] + f_j^2 \cdot \left(1 - e_j\right)^2 \cdot Var\left[X_{ij}\right]\right) \\
&= \frac{1}{f_j^2} \cdot \left(f_j \cdot \left(1 - e_j\right) \cdot \mu_j + f_j^2 \cdot \left(1 - e_j\right)^2 \cdot \left(\mu_j + \mu_j^2 / k_j\right)\right) \\
&= \frac{f_j^2 \cdot \left(1 - e_j\right) \cdot \mu_j}{f_j^2} \cdot \left(\frac{1}{f_j} + \left(1 - e_j\right) \cdot \left(1 + \mu_j / k_j\right)\right) \\
&= \left(1 - e_j\right) \cdot \mu_j \cdot \left(\frac{1}{f_j} + \left(1 - e_j\right) \cdot \left(1 + \mu_j / k_j\right)\right)
\end{aligned}
$$

B. The expected value and variance of FECs based on pools of m_j individual stool samples

B.1 FECs prior to an intervention

B.1.1 The expected value

$$
\begin{aligned}
E[U_{lj}] &= E\left[\frac{Z_{lj}}{f_j}\right] = \frac{E[Z_{lj}]}{f_j} \\
&= \frac{E[E[Z_{lj}|\lambda_{lj}, V_{lj}]]}{f_j} \\
&= \frac{E[f_j \cdot Vlj]}{f_j} \\
&= \frac{f_j \cdot E[Vlj]}{f_j} \\
&= E\left[\frac{\sum_{i=1}^{m_j} X_{ij}}{m_j}\right] \\
&= \frac{m_j}{m_j} \cdot E[X_{ij}] \\
&= \mu_j
\end{aligned}
$$

B.1.2 The variance

$$
\begin{aligned}
Var[U_{lj}] &= Var\left[\frac{Z_{lj}}{f_j}\right] = \frac{Var[Z_{lj}]}{f_j^2} \\
&= \frac{1}{f_j^2}\cdot\left(E\left[Var[Z_{lj}|\lambda_{lj}, V_{lj}]\right] + Var\left[E[Z_{lj}|\lambda_{lj}, V_{lj}]\right]\right) \\
&= \frac{1}{f_j^2}\cdot\left(E[f_j\cdot V_{lj}] + Var[f_j\cdot V_{lj}]\right) \\
&= \frac{1}{f_j^2}\cdot\left(f_j\cdot E[V_{lj}] + f_j^2\cdot Var[V_{lj}]\right) \\
&= \frac{1}{f_j^2}\cdot\left(f_j\cdot E\left[\frac{\sum_{i=1}^{m_j} X_{ij}}{m_j}\right] + f_j^2\cdot Var\left[\frac{\sum_{i=1}^{m_j} X_{ij}}{m_j}\right]\right) \\
&= \frac{1}{f_j^2}\cdot\left(\frac{f_j\cdot m_j}{m_j}\cdot E[X_{ij}] + \frac{f_j^2\cdot m_j}{m_j^2}\cdot Var[X_{ij}]\right) \\
&= \frac{1}{f_j^2}\cdot\left(f_j\cdot\mu_j + \frac{f_j^2}{m_j}\cdot\left(\mu_j + \mu_j^2/k_j\right)\right) \\
&= \frac{f_j^2\cdot\mu_j}{f_j^2}\cdot\left(\frac{1}{f_j} + \frac{1+\mu_j/k_j}{m_j}\right) \\
&= \mu_j\cdot\left(\frac{1}{f_j} + \frac{1+\mu_j/k_j}{m_j}\right)
\end{aligned}
$$

B.2 FECs after an intervention

B.2.1 The expected value

$$
\begin{aligned}
E\left[U'_{lj}\right] &= E\left[\frac{Z'_{lj}}{f_j}\right] = \frac{E\left[Z'_{lj}\right]}{f_j} \\
&= \frac{E\left[E\left[Z'_{lj}\middle|\lambda'_{lj}, V'_{lj}\right]\right]}{f_j} \\
&= \frac{E\left[f_j \cdot V'_{lj}\right]}{f_j} \\
&= \frac{f_j \cdot E\left[V'_{lj}\right]}{f_j} \\
&= E\left[\frac{\sum_{i=1}^{m_j} X'_{lj}}{m_j}\right] \\
&= \frac{m_j}{m_j} \cdot E\left[X'_{lj}\right] \\
&= E\left[\left(1 - e_j\right) \cdot X_{ij}\right] \\
&= \left(1 - e_j\right) \cdot E\left[X_{ij}\right] \\
&= \left(1 - e_j\right) \cdot \mu_j
\end{aligned}
$$

B.2.2 The variance

$$Var\left[U'_{lj}\right] = Var\left[\frac{Z'_{lj}}{f_j}\right] = \frac{Var\left[Z'_{lj}\right]}{f_j^2}$$

$$= \frac{1}{f_j^2}\cdot\left(E\left[Var\left[Z'_{lj}\middle|\lambda'_{lj}, V'_{lj}\right]\right] + Var\left[E\left[Z'_{lj}\middle|\lambda'_{lj}, V'_{lj}\right]\right]\right)$$

$$= \frac{1}{f_j^2}\cdot\left(E\left[f_j\cdot V'_{lj}\right] + Var\left[f_j\cdot V'_{lj}\right]\right)$$

$$= \frac{1}{f_j^2}\cdot\left(f_j\cdot E\left[V'_{lj}\right] + f_j^2\cdot Var\left[V'_{lj}\right]\right)$$

$$= \frac{1}{f_j^2}\cdot\left(f_j\cdot E\left[\frac{\sum_{i=1}^{m_j} X'_{ij}}{m_j}\right] + f_j^2\cdot Var\left[\frac{\sum_{i=1}^{m_j} X'_{ij}}{m_j}\right]\right)$$

$$= \frac{1}{f_j^2}\cdot\left(\frac{f_j\cdot m_j}{m_j}\cdot E\left[X'_{ij}\right] + \frac{f_j^2\cdot m_j}{m_j^2}\cdot Var\left[X'_{ij}\right]\right)$$

$$= \frac{1}{f_j^2}\cdot\left(f_j\cdot E\left[(1-e_j)\cdot X_{ij}\right] + \frac{f_j^2}{m_j}\cdot Var\left[(1-e_j)\cdot X_{ij}\right]\right)$$

$$= \frac{1}{f_j^2}\cdot\left(f_j\cdot(1-e_j)\cdot E\left[X_{ij}\right] + \frac{f_j^2\cdot(1-e_j)^2}{m_j}\cdot Var\left[X_{ij}\right]\right)$$

$$= \frac{1}{f_j^2}\cdot\left(f_j\cdot(1-e_j)\cdot\mu_j + \frac{f_j^2\cdot(1-e_j)^2}{m_j}\cdot\left(\mu_j + \mu_j^2/k_j\right)\right)$$

$$= \frac{f_j^2\cdot(1-e)\cdot\mu_j}{f_j^2}\cdot\left(\frac{1}{f_j} + \frac{1-e_j}{m_j}\cdot\left(1+\mu_j/k_j\right)\right)$$

$$= (1-e_j)\cdot\mu_j\cdot\left(\frac{1}{f_j} + (1-e)\cdot\frac{1+\mu_j/k_j}{m_j}\right)$$

ACKNOWLEDGMENTS

Bruno Levecke is a Postdoctoral fellow of FWO (www.fwo.be). The work of Johannes Charlier received funding from the E.U. FP7 GLOWORM project (Grant Agreement No. 288975CP-TP-KBBE.2011.1.3-04).

REFERENCES

Addiss, D.G., Rheingans, R., Twum-Danso, N.A., Richards, F.O., 2003. A framework for decision-making for mass distribution of Mectizan in areas endemic of *Loa loa*. Filaria J. 2 (Suppl. 1), S9.

Albonico, M., Rinaldi, L., Sciascia, S., Morgoglione, M.E., Piemonte, M., et al., 2013. Comparison of three copromicroscopic methods to assess albendazole efficacy against soil-transmitted helminth infections in school-aged children on Pemba Island. Trans. R. Soc. Trop. Med. Hyg. 107, 493–501.

Anderson, R., Hollingsworth, T.D., Truscott, J., Brooker, S.J., 2012. Optimisation of mass chemotherapy to control soil-transmitted helminth infection. Lancet 379, 289–290.

Barda, B., Zepherine, H., Rinaldi, L., Cringoli, G., Burioni, R., et al., 2013. Mini-FLOTAC and Kato-Katz: helminth eggs watching on the shore of Lake Victoria. Parasit. Vectors 6, 220.

Bergquist, R., Johansen, M.V., Utzinger, J., 2009. Diagnostic dilemmas in helminthology: what tools to use and when? Trends Parasitol. 25, 151–156.

Bethony, J., Brooker, S.J., Albonico, M., Geiger, S.M., Loukas, A., et al., 2006. Soil-transmitted helminth infections: ascariasis, trichuriasis, and hookworm. Lancet 367, 1521–1532.

Bogoch, I.I., Raso, G., NGoran, E.K., Marti, H.P., Utzinger, J., 2006. Differences in microscopic diagnosis of helminths and intestinal protozoa among diagnostic centres. Eur. J. Clin. Microbiol. Infect. Dis. 25, 344–347.

Booth, M., Vounatsou, P., NGoran, E.K., Tanner, M., Utzinger, J., 2003. The influence of sampling effort and the performance of the Kato-Katz technique in diagnosing *Schistosoma mansoni* and hookworm co-infections in rural Côte d Ivoire. Parasitology 127, 525–531.

Brooker, S.J., Hotez, P.J., Bundy, D.A.P., 2010. The global atlas of helminth infection: mapping the way forward in neglected tropical disease control. PLoS Negl. Trop. Dis. 4, e779.

Brooker, S.J., Kabatereine, N.B., Myatt, M., Stothard, R.J., Fenwick, A., 2005. Rapid assessment of *Schistosoma mansoni*: the validity, applicability and cost-effectiveness of the lot quality assurance sampling method in Uganda. Trop. Med. Int. Health 10, 647–658.

Brooker, S.J., Pullan, R.L., Gitonga, C.W., Ashton, R.A., Kolaczinski, J.H., et al., 2012. *Plasmodium*—helminth coinfection and its sources of heterogeneity across East Africa. J. Infect. Dis. 205, 841–852.

Casella, G., Berger, B.L., 2002. Statistical Inference, second ed. Duxbury Press. Pacific Grove: Duxbury Thomson Learning.

Charlier, J., van der Voort, M., Kenyon, F., Skuce, P., Vercruysse, J., 2014. Chasing helminths and their economic impact on farmed ruminants. Trends Parasitol. 30, 361–367.

Cringoli, G., Rinaldi, L., Albonico, M., Bergquist, R., Utzinger, J., 2013. Geospatial (s)tools: integration of advanced epidemiological sampling and novel diagnostics. Geospat. Health 7, 399–404.

Cringoli, G., Rinaldi, L., Maurelli, M.P., Utzinger, J., 2010. FLOTAC: new multivalent techniques for qualitative and quantitative copromicroscopic diagnosis of parasites in animals and humans. Nat. Protoc. 5, 503–515.

de Silva, N., Hall, A., 2010. Using the prevalence of individual species of intestinal nematode worms to estimate the combined prevalence of any species. PLoS Negl. Trop. Dis. 4, e655.

Diggle, P.J., 2011. Estimating prevalence using an imperfect test. Epidemiol. Res. Int. 608719.

Dobson, R.J., Sangster, N.C., Besier, R.B., Woodgate, R.G., 2009. Geometric means provide a biased efficacy result when conducting a faecal egg count reduction test (FECRT). Vet. Parasitol. 161, 162–167.

Enøe, C., Georgiadis, M.P., Johnson, W.O., 2000. Estimation of sensitivity and specificity of diagnostic tests and disease prevalence when the true disease state is unknown. Prev. Vet. Med. 45, 6181.
Gregory, R.D., Woolhouse, M.E., 1993. Quantification of parasite aggregation: a simulation study. Acta Trop. 54, 131–139.
Hui, S.L., Zhou, X.H., 1998. Evaluation of diagnostic tests without gold standards. Stat. Methods Med. Res. 7, 354–370.
Hürlimann, E., Schur, N., Boutsika, K., Stensgaard, A.-S., Himpsl, Laserna de, et al., 2011. Toward an open-access global database for mapping, control, and surveillance of neglected tropical diseases. PLoS Negl. Trop. Dis. 5, e1404.
Humphries, D., Mosites, E., Otchere, J., Twum, W.A., Woo, L., et al., 2011. Epidemiology of hookworm infection in Kintampo North Municipality, Ghana: patterns of malaria coinfection, anemia, and albendazole treatment failure. Am. J. Trop. Med. Hyg. 84, 792–800.
Katz, N., Chaves, A., Pellegrino, J., 1972. A simple device for quantitative stool thick-smear technique in schistosomiasis mansoni. Rev. Inst. Med. Trop. São Paulo 14, 397–400.
Knopp, S., Glinz, D., Rinaldi, L., Mohammed, K.A., NGoran, E.K., et al., 2009. FLOTAC: a promising technique for detecting helminth eggs in human feces. Trans. R. Soc. Trop. Med. Hyg. 103, 1190–1194.
Knopp, S., Mgeni, A.F., Khamis, I.S., Steinmann, P., Stothard, J.R., et al., 2008. Diagnosis of soil-transmitted helminths in the era of preventive chemotherapy: effect of multiple stool sampling and use of different diagnostic techniques. PLoS Negl. Trop. Dis. 2, e331.
Krauth, S.J., Coulibaly, J.T., Knopp, S., Traoré, M., NGoran, E.K., et al., 2012. An in-depth analysis of a piece of shit: distribution of *Schistosoma mansoni* and hookworm eggs in human stool. PLoS Negl. Trop. Dis. 6, e1969.
Levecke, B., Behnke, J.M., Ajjampur, S.S., Albonico, M., Ame, S.M., et al., 2011b. A comparison of the sensitivity and fecal egg counts of the McMaster egg counting and Kato-Katz thick smear methods for soil-transmitted helminths. PLoS Negl. Trop. Dis. 5, e1201.
Levecke, B., Brooker, S.J., Knopp, S., Steinmann, P., Stothard, R.J., et al., 2014a. Effect of sampling and diagnostic effort on the assessment of schistosomiasis and soil-transmitted helminthiasis and drug efficacy: a meta-analysis of six drug efficacy trials and one epidemiological survey. Parasitology 141, 1826–1840.
Levecke, B., Speybroeck, N., Dobson, R., Vercruysse, J., Charlier, J., 2011a. Novel insights in the fecal egg count reduction test for monitoring drug efficacy against soil-transmitted helminths in large-scale treatment programs. PLoS Negl. Trop. Dis. 5, e1427.
Levecke, B., Montresor, A., Albonico, M., Ame, S.M., Behnke, J.M., et al., 2014b. Anthelminthic efficacy of single-oral dose 500 mg mebendazole in school children in six countries where soil-transmitted helminths are endemic: a meta-analysis. PLoS Negl. Trop. Dis. 8, e3204.
Ministry of Agriculture, Fisheries and Food, 1986. Manual of Veterinary Parasitological Laboratory Techniques (Reference Book; 418), third ed. Her Majesty's Stationery Office, London.
Montresor, A., 2011. Cure rate is not a valid indicator for assessing drug efficacy and impact of preventive chemotherapy interventions against schistosomiasis and soil-transmitted helminthiasis. Trans. R. Soc. Trop. Med. Hyg. 105, 361–363.
Montresor, A., Engels, D., Chitsulo, L., Gabrielli, A., Albonico, M., et al., 2011. The appropriate indicator should be used to assess treatment failure in STH infections. Am. J. Trop. Med. Hyg. 85, 579–580.
Morgan, E.R., Cavill, L., Curry, G.E., Wood, R.M., Mitchell, E.S., 2005. Effects of aggregation and sample size on composite faecal egg counts in sheep. Vet. Parasitol. 131, 79–87.

Murray, C.J.L., Vos, T., Lozano, R., Naghavi, M., Flaxman, A.D., et al., 2012. Disability-adjusted life years (DALYs) for 291 diseases and injuries in 21 regions, 19902010: a systematic analysis for the Global Burden of Disease Study 2010. Lancet 9859, 2179–2223.
Neglected Tropical Disease Partner Website, 2013. Uniting to Combat Neglected Tropical Diseases. Ending the Neglect and Reaching 2020 Goals. Available at: http://www.unitingtocombatntds.org (accessed 10.04.14.).
Olives, C., Valadez, J.J., Brooker, S.J., Pagano, M., 2012. Multiple category-lot quality assurance sampling: a new classification system with application to schistosomiasis control. PLoS Negl. Trop. Dis. 6, e1806.
Pullan, R.L., Brooker, S.J., 2012. The global limits and population at risk of soil-transmitted helminth infections in 2010. Parasit. Vectors 5, 81.
Pullan, R.L., Smith, J.L., Jasrasaria, R., Brooker, S.J., 2014. Global numbers of infection and disease burden of soil-transmitted helminth infections in 2010. Parasit. Vectors 7, 37.
Sinniah, B., 1982. Daily egg production of *Ascaris lumbricoides*: the distribution of eggs in the feces and the variability of egg counts. Parasitology 84, 167–175.
Speybroeck, N., Devleesschauwer, B., Joseph, L., Berkvens, D., 2013. Misclassification errors in prevalence estimation: Bayesian handling with care. Int. J. Public Health 58, 791–795.
Torgerson, P.R., Paul, M., Lewis, F.I., 2012. The contribution of simple random sampling to observed variations in faecal egg counts. Vet. Parasitol. 188, 397–401.
Torgerson, P.R., Schnyder, M., Hertzberg, H., 2005. Detection of anthelmintic resistance: a comparison of mathematical techniques. Vet. Parasitol. 128, 291–298.
Vercruysse, J., Behnke, J.M., Albonico, M., Ame, S.M., Angebault, C., et al., 2011. Assessment of anthelminthic efficacy of albendazole in school children in seven countries where soil-transmitted helminths are endemic. PloS Negl. Trop. Dis. 5, e948.
World Health Organization, 1991. Basic Laboratory Methods in Medical Parasitology. World Health Organization, Geneva, Switzerland.
World Health Organization, 1998. Guidelines for the Evaluation of Soil-transmitted Helminthiasis and Schistosomiasis at Community Level. A Guide for Control Programme Managers. World Health Organization, Geneva, Switzerland.
World Health Organization, 2011. Helminth Control in School-age Children: A Guide for Managers of Control Programmes, Second ed. World Health Organization, Geneva, Switzerland.
World Health Organization, 2012a. Accelerating Work to Overcome the Global Impact of Neglected Tropical Diseases: A Roadmap for Implementation. World Health Organization, Geneva, Switzerland.
World Health Organization, 2012b. Soil-transmitted helminthiases: number of children treated in 2012. Week. Epidemiol. Rec. 89, 133–140.
World Health Organization, 2012c. Soil-transmitted Helminthiases: Eliminating Soil-transmitted Helminthiases as a Public Health Problem in Children: Progress Report 2001–2010 and Strategic Plan 2011–2020. World Health Organization, Geneva, Switzerland.
World Health Organization, 2013. Assessing the Efficacy of Anthelminthic Drugs against Schistosomiasis and Soil-Transmitted Helminthiasis. World Health Organization, Geneva, Switzerland.
Ye, X.P., Donnelly, C.A., Fu, Y.L., Wu, Z.X., 1997. The non-randomness of the distribution of *Trichuris trichiura* and *Ascaris lumbricoides* in feces and the effect of stirring fecal specimens. Trop. Med. Int. Health 2, 261–264.

CHAPTER SIX

Modelling Lymphatic Filariasis Transmission and Control: Modelling Frameworks, Lessons Learned and Future Directions

Wilma A. Stolk*[,1], Chris Stone[§], Sake J. de Vlas*

*Department of Public Health, Erasmus MC, University Medical Center Rotterdam, The Netherlands
[§]Department of Epidemiology and Public Health, Swiss Tropical and Public Health Institute, Basel, Switzerland
[1]Corresponding author: E-mail: w.stolk@erasmusmc.nl

Contents

Advances in Parasitology, Volume 87
ISSN 0065-308X
http://dx.doi.org/10.1016/bs.apar.2014.12.005

Abstract

Mathematical modelling provides a useful tool for policy making and planning in lymphatic filariasis control programmes, by providing trend forecasts based on sound scientific knowledge and principles. This is now especially true, in view of the ambitious target to eliminate lymphatic filariasis as a public health problem globally by the year 2020 and the short remaining timeline to achieve this. To meet this target, elimination programmes need to be accelerated, requiring further optimization of strategies and tailoring to local circumstances. Insights from epidemiological transmission models provide a useful basis. Two general models of lymphatic filariasis transmission and control are nowadays in use to support decision-making, namely a population-based deterministic model (EPIFIL) and an individual-based stochastic model (LYMFASIM). Model predictions confirm that lymphatic filariasis transmission can be interrupted by annual mass drug administration (MDA), but this may need to be continued much longer than the initially suggested 4–6 years in areas with high transmission intensity or poor treatment coverage. However, the models have not been validated against longitudinal data describing the impact of MDA programmes. Some critical issues remain to be incorporated in one or both of the models to make predictions on elimination more realistic, including the possible occurrence of systematic noncompliance, the risk of emerging parasite resistance to anthelmintic drugs, and spatial heterogeneities. Rapid advances are needed to maximize the utility of models in decision-making for the ongoing ambitious lymphatic filariasis elimination programmes.

1. INTRODUCTION

Lymphatic filariasis is a mosquito-borne parasitic disease that is caused by a filarial nematode (*Wuchereria bancrofti*, *Brugia malayi* or *Brugia timori*) and is, within the group of neglected tropical diseases, an important cause of chronic morbidity. The disease occurs in 73 countries throughout the tropics and subtropics of Asia, Africa, the Western Pacific and in some parts of the Caribbean and South America (WHO, 2013b). The World Health Organization (WHO) aims to eliminate lymphatic filariasis as a public health problem globally by 2020 (WHO, 1997) and established the Global Programme to Eliminate Lymphatic Filariasis in 2000 (WHO, 2013b). The first objective of this programme is to interrupt transmission by annual mass drug administration (MDA) with antifilarial drugs (diethylcarbamazine (DEC) or ivermectin (IVM) plus albendazole (ABZ)). From here onwards, we refer to this as elimination, in line with published definitions from the International Task Force for Disease Eradication (Centers for Disease Control, 1993). The second objective is to provide already affected people with access to a morbidity management care package to relieve suffering and prevent progression to more advanced stages of chronic disease. The latter is not further discussed in this paper.

Progress towards elimination through MDA varies between countries (WHO, 2013a,b). Thirteen countries have already achieved the targets and are now under post-MDA surveillance, while 17 countries—mostly in Africa—still have to start the implementation of mass treatment. The 43 other countries are in various stages of the programme with varying levels of success: many countries are reporting incomplete geographic coverage or low programme coverage in treated populations (WHO, 2013a). The late start of some countries and low coverage hinder the achievement of elimination by 2020. To meet the 2020 goals, it is urgent to roll out the elimination programmes to all areas in need and improve the achieved programmatic coverage. Acceleration strategies will be needed to meet the 2020 goals (Stolk et al., 2013).

A fundamental question is whether the planned interventions will be sufficient to achieve the ambitious goal of elimination. Guidelines for determining whether mass treatment can be stopped were developed and tested for their assumptions, accuracy and practicality (Chu et al., 2013; WHO, 2011). However, the validation of the guidelines for longer-term post-MDA surveillance needs further empirical evidence, since uncertainty remains about the microfilarial (mf) or worm-antigen prevalence threshold levels, below which transmission cannot be sustained even in the absence of any further treatments. Also, it is practically impossible to survey every location in at-risk areas, leaving the possibility that pockets of continued transmission remain within a treated area.

Mathematical models for the transmission and control of infectious diseases present potentially valuable tools for the planning and evaluation of interventions (Garnett et al., 2011; Basáñez et al., 2012). Transmission dynamics models have also been developed for lymphatic filariasis and applied to address practical questions relating to the control and elimination of lymphatic filariasis, as reviewed elsewhere (Stolk et al., 2006; Michael et al., 2004). Since these earlier reviews, models have been applied to address a number of practical questions that gained relevance with the progress of the Global Programme to Eliminate Lymphatic Filariasis. Also, advanced modern techniques have been introduced in modelling studies, e.g. for the estimation of model parameters and investigation of uncertainty boundaries (Poole and Raftery, 2000; Coelho et al., 2011). For non-modellers it is often difficult to judge the quality of different modelling exercises, to fully understand the relevance and implications of using different modelling approaches or variations in model structure and parameterisations, and hence to judge the reliability of the modelling results and interpret the findings appropriately.

In view of the short remaining time to the 2020 target year for elimination, it is now timely and opportune to review the recent body of work critically with a focus on lessons learned regarding the impact of interventions and robustness of conclusions. Specific objectives of this paper are to describe the main features of available transmission dynamics modelling frameworks that have been used to simulate and predict the effectiveness of MDA interventions; to summarize lessons learned regarding transmission dynamics, the effectiveness of interventions and feasibility of elimination; and to discuss remaining challenges and priorities for future modelling work.

2. BRIEF HISTORY OF LYMPHATIC FILARIASIS MODELLING

Models have been in use for lymphatic filariasis for about half a century (WHO, 1990; Basáñez et al., 2012). The earliest models focused on explaining age-prevalence patterns of infection, by studying the rates of acquisition and loss of infection in an equilibrium situation (Hayashi, 1962; Hairston and Jachowski, 1968). In a next phase, models were developed to enhance quantitative understanding of the transmission from vector to humans, and the extremely high exposure to infective bites that is needed to establish patent infection (Hairston and Meillon, 1968; Gubler and Bhattacharya, 1974). These models only covered some aspects of the epidemiology of lymphatic filariasis, and did not model the human acquisition of new worms as a function of infection prevalence or infection intensity.

Anderson and May (1985) wrote a key publication on modelling of helminthic infection, and the general framework that was introduced still forms the basis of many currently applied models. The first full transmission dynamics model for lymphatic filariasis was published by Rochet (1990), aiming to describe the observed changes in infection levels in mosquitoes and humans during a 5-year vector control programme in Pondicherry, India. Several years later, a deterministic, population-based modelling framework was developed, tracking changes in the average parasite load at population level over age and time. This model was introduced under the name EPIFIL (Norman et al., 2000). An earlier publication presented a simplified version of the EPIFIL model, describing changes in infection status over age in case of constant force of infection (Chan et al., 1998). Plaisier et al. (1998) first published a probabilistic or stochastic individual-based modelling framework, named LYMFASIM, which tracks the life histories of individuals in a dynamic population and simulates infection transmission between these individuals while taking account of inter-individual differences in some key characteristics. LYMFASIM and EPIFIL have been tested against the same data from the vector control programme in Pondicherry, but were later also applied to other regions simulating vector control or MDA (Michael et al., 2001; Gambhir et al., 2010; Singh et al., 2013; Stolk et al., 2013). These two models have thereafter been used in a number of publications, modelling the impact of various interventions in various settings, as we review in more detail below. Several extensions and modifications of the EPIFIL model have been published, including the incorporation of population genetics to investigate the evolution of drug resistance in filarial worms (Schwab et al., 2006), and co-infection with malaria to investigate interactions between the two diseases and the effects of vector control in co-endemic regions (Slater et al., 2013; Stone et al., 2014).

A search on Google revealed some additional model publications (not indexed in Medline), describing how treatment affects transition of humans through different lymphatic filariasis disease stages (susceptible, exposed, diseased) or describing LF transmission as part of a malaria–filariasis co-infection model (Bhunu and Mushayabasa, 2012; Gweryina and Luga, 2014). The latter models contain less detail regarding the lymphatic filariasis transmission cycle and remain purely theoretical in the sense that model outcomes were not compared with empirical data.

Table 1 summarizes key characteristics of the transmission dynamics models described above. In the remainder of this chapter, we compare the two most used multipurpose modelling frameworks with respect to their structure, ability to reproduce observed data and predictions on the impact of interventions. The model presented by Bhunu and Mushayabasa (op. cit.) is not included in this comparison, because the model describes the lymphatic filariasis transmission cycle in less detail than the other models (e.g. not considering infection intensity) and was not tested against data. The Rochet model does not allow simulation of MDA and has not been applied in other studies; this model is therefore also not included in further comparisons.

Table 1 Characteristics of the four available modelling frameworks for lymphatic filariasis transmission and control

Name modelling framework	Rochet	LYMFASIM	EPIFIL	Bhunu
Publication				
Initial publication of modelling framework	Rochet (1990)	Plaisier et al. (1998)	Norman et al. (2000)	Bhunu and Mushayabasa (2012)
Intended use	Single use	Multiple applications	Multiple applications	Single use
Later publications based on same modelling framework (with or without modifications)	NA	Stolk et al. (2003), Stolk et al. (2005), Stolk et al. (2006), Stolk et al. (2008), Stolk et al. (2013)	Chan et al. (1998)[a], Michael (2000), Michael et al. (2001), Michael et al. (2004), (Gambhir and Michael (2008)[a,b], Gambhir et al. (2010), Singh et al. (2013)[b], Schwab et al. (2006)[b,c], Schwab et al. (2007)[b,c], Slater et al. (2013)[d], Stone et al. (2014)[e]	Gwerina and Luga (2014)[f]
Basic model characteristics				
Number and type of spatial locations modelled or implied	One place	One place	One place	One place
Population-based or individual-based	Population	Individual	Population	Population
Way of representing infection in hosts	Density (prevalence as a function of density)	Density and prevalence	Density (prevalence as a function of density)	Prevalence

Role of chance	Deterministic	Stochastic	Deterministic	Deterministic
Interventions that are being considered	Vector control	Vector control, mass treatment, selective treatment	Vector control, mass treatment	Selective treatment
Features included in the model				
Human population demographics	NA	Yes (birth and death rate dynamically modelled; age composition)	Yes (birth and death rate, age composition)	Yes (birth and death rate)
Heterogeneities in the human population	NA	Age, sex, life expectancy, level of exposure to mosquitoes, compliance to MDA, ability to develop immune response, efficacy of treatment	Age	Disease status
Mosquito population density	Dynamically modelled in relation to environmental factors	Fixed input	Fixed input	
Exposure to mosquitoes	Homogeneous	Heterogeneous (dependent on age, sex, personal attractiveness to mosquitoes)	Heterogeneous (dependent on age)	Homogeneous
Uptake of infection by mosquitoes	Constant (one per bite if bite is taken on infected host)	Varying with infection intensity on hosts	Varying with infection intensity on hosts	Constant

(Continued)

Table 1 Characteristics of the four available modelling frameworks for lymphatic filariasis transmission and control—cont'd

Name modelling framework	Rochet	LYMFASIM	EPIFIL	Bhunu
Infection in mosquitoes	Prevalence	Density (average L3 load per mosquito)	Density (average L3 load per mosquito)	Prevalence
Excess mortality of infected flies	Yes	No	No	No
Parasite acquisition in humans	Proportional to number of infective bites	Proportional to mean number of L3 larvae in mosquitoes; optional dependency on an human individual level of acquired immunity	Proportional to mean number of L3 larvae in mosquitoes; optional dependency on the average level of acquired immunity	Proportional to number of infective bites
Infection in human	Density (larval stages, adult worms)	Density (immature or mature worms, mf)	Density (mature worms, Mf[g])	Prevalence
Acquired immunity in humans	No	Yes	Yes	No
Other	NA[h]	Mf[g] count sampling for direct comparability of model predictions to data	Morbidity	Morbidity

[a]Simplified version of EPIFIL with constant force of infection (i.e. not dependent on intensity of infection), aimed to study steady state situation
[b]EPIFIL extended with a component that describes the mating probability of worms
[c]EPIFIL extended with components that describes mechanisms of selection of drug-resistant parasites
[d]Simplified version of EPIFIL that ignores age variation in exposure to mosquitoes, but population stratified for malaria co-infection status
[e]Simplified (ODE) version of EPIFIL that ignores immunity and age variation in exposure to mosquitoes, and uses a susceptible-exposed-infective framework for malaria and lymphatic filariasis infection in mosquitoes
[f]Extended version of the Bhunu model that takes account of malaria co-infection
[g]Mf as microfilarial
[h]NA as not applicable

3. COMPARISON OF MODELLING FRAMEWORKS: STRUCTURE AND PARAMETER VALUES

3.1 Deterministic, population-based modelling framework (EPIFIL)

3.1.1 Modelling approach

Norman et al. (2000) were the first to publish a deterministic transmission dynamics model for lymphatic filariasis, named EPIFIL. The model is formulated as a system of partial differential equations, which it has in common with earlier models of filarial worm infection (Dietz, 1982; Anderson and May, 1985). It is a population-based or compartmental model, rather than an individual-based one, and does not include stochasticity. In return, it is not very computationally expensive and remains somewhat tractable. The model was originally implemented in FORTRAN (Norman et al., 2000) while later versions used MATLAB (Gambhir et al., 2010; Singh et al., 2013) and uses a finite difference method to obtain solutions at discrete time points. The model simulates the average parasite density in a human population of unspecified size, but with a specified age composition; it does not distinguish individuals. This model allows the simulation of changes in infection status over time as well as age, making it suitable for simulations of various control measures. The inclusion of age structure in models of filariasis is helpful in two main ways: many data sources will provide information on age-specific infection rates, and heterogeneity in exposure to mosquito bites may, for a large part, be determined by age-specific differences in exposure, e.g. resulting from differences in body mass (Dye and Hasibeder, 1986).

3.1.2 Structure of the modelling framework

A diagrammatic overview of the model is provided (Figure 1).

The model (Gambhir et al., 2000) is described by a system of three partial differential equations and one ordinary differential equation:

$$\frac{\partial W}{\partial t} + \frac{\partial W}{\partial a} = \lambda \frac{V}{H} \psi_1 \psi_2 s_2 h(a) L^* e^{-\beta I} - \mu W$$

$$\frac{\partial M}{\partial t} + \frac{\partial M}{\partial a} = \alpha \phi(W, k) W - \gamma M$$

$$\frac{\partial I}{\partial t} + \frac{\partial I}{\partial a} = W - \delta I$$

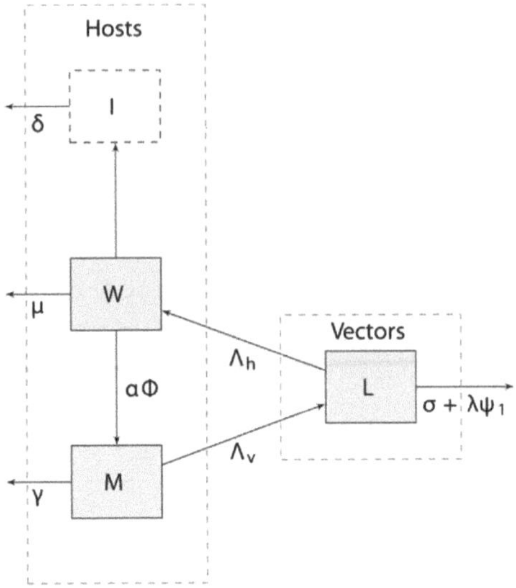

Figure 1 A diagrammatic overview of the state variables and rate parameters used in EPIFIL, where *W* represents the mean filarial worm burden in the human population, *M* the mean microfilarial intensity and *L* the mean intensity of infective stage (L3) filarial larvae in a mosquito species. *I* represents the mean level of immunity in the human population to filarial worm establishment.

$$\frac{dL}{dt} = \lambda \kappa g \int_a \pi(a)(1-f(M)da - (\sigma + \lambda\psi_1)L$$

The mean worm burden in the human population, *W*, is given by a birth–death process, where the deaths are simply due to the mortality rate, μ, with estimates for the mean life span of *W. bancrofti* ranging from 4 to 12 years. The birth process represents inoculation of infective, L3 larvae through mosquito bites and is governed by factors related to the biting rate on humans – the biting frequency, λ, and the vector to human ratio, *V:H* and a term related to age-dependent exposure to mosquito bites, *h(a)*, which is used to incorporate the common assumption that exposure increases in children with age until a stable maximum level that further remains unchanged, reflecting changes in body surface. The other parameters in the birth term relate to inefficiencies in transmission and establishment of adult worms – the probability that an infective larva breaks through the labium of the mosquito's proboscis and escapes the mosquito during a blood meal, ψ_1; the probability that a released larva successfully finds and enters the puncture site, ψ_2; the probability that larvae that entered the human successfully establish and develop into adult worms in the absence of an immune response, s_2; and an immune-mediated decrease in establishment, which depends on the strength of acquired immunity, β.

The average mf density in blood of humans, M, depends on the mf mortality rate, γ, and the fecundity of adult female worms, α. A factor of ½ related to the sex ratio of worms is not explicitly included here, but presumably incorporated in the parameter estimate. Later models using this modelling framework included a probability of a given female worm having mated, $\varphi(W,k)$ (Gambhir et al., 2010; Gambhir and Michael, 2008), based on functions initially derived for models of schistosome infection (May, 1977). Under the assumption that the parasites are polygamous, which is interpreted as at least one male worm being present within a human being a sufficient condition for all females within that host to be mated, the following mating function was used:

$$\phi(W,k) = 1 - \left(1 + \frac{W}{2k}\right)^{-k-1}$$

which results from combining the probability that all worms within a host are female with the moment-generating function of a negative binomial distribution (May, 1977), where k is the overdispersion parameter. Whether this mating system truly is applicable to filarial worms is somewhat debatable, and this could potentially affect the location/value of the breakpoint (May, 1977).

The use of the negative binomial here and in the function describing the uptake of microfilariae by mosquitoes (see below) therefore hinges on the assumption that the distribution of adult worms and microfilariae among humans can be appropriately described with this function, which is typically thought to be the case. It also assumes that the use of a single clumping or aggregation factor, k, among all age groups is realistic. In EPIFIL, this clumping factor is itself a function:

$$k(M) = k_0 + k_1 M$$

which allows for a greater degree of overdispersion at lower mf intensities (in younger age groups or posttreatment).

Although from a biological perspective mean worm or mf intensity levels are more appropriate to work with, data are often reported in the form of age-specific levels of prevalence, and criteria for cessation of control activities in relation to elimination programmes are likewise based on community prevalence. Translating from the mean burden to prevalence is easily done by defining prevalence as the proportion of the population harbouring one or more microfilariae:

$$\mathrm{p}(M) = 1 - \mathrm{P}(0|M, k)$$

where n is the microfilarial count

$$P(n|\mathrm{M}, k) = \left(1 - \left(\frac{M}{M+k}\right)\right)^{k} \frac{\Gamma(k+n)}{\Gamma k} \frac{(M/(M+k))^{n}}{n!}$$

so that the (age-specific) prevalence is given by the following expression (May, 1977):

$$p(M) = 1 - \left(1 + \frac{M}{k}\right)^{-k}$$

The third partial differential equation relates to an immune state in the human population, I, which is a function of cumulative exposure to adult worms and may be lost at a decay rate, δ. Different functions for the effect of immunity on the establishment of adult worms were used in more recent models, dividing the effect into an immunosuppression response and an immunity to establishment (Gambhir et al., 2010; Singh et al., 2013). Possible effects of immunity on adult worm mortality or fecundity have not yet been included.

The ordinary differential equation in the model describes the mean burden of third instar (infective) filarial larvae, L, in mosquitoes. The argument is made that due to the discrepancy in life spans of the filarial life stages (several years for adult worms, 10 months for microfilariae, and ca. 1–3 weeks for most mosquitoes) one can assume that the stage occurring within the vector will reach equilibrium within a monthly time step. The equilibrium level of third instar larvae, L^*, is given by:

$$L^* = \frac{\lambda \kappa g \int_a \pi(a)(1 - f(M))da}{\sigma + \lambda \psi_1}$$

which is the value used in the birth process of adult worms. Third instars are thus lost due to the natural (but not parasite-induced) mosquito mortality, σ, and by exiting the mosquito during a blood meal at rate, $\lambda^* \psi_1$. Infection is acquired as a function of the biting rate, λ, and the probability of ingesting microfilariae when biting an infective human, g. The rest of the terms relate to the yield of third instars from ingested microfilariae. For an individual mosquito, this relationship is given as:

$$L = \kappa\left(1 - e^{-rM/\kappa}\right)$$

where r and κ are constants that reflect the increase in L3 uptake as microfilarial density increases from zero, and the saturation level (i.e., maximal L3 load), respectively (Norman et al., 2000).

This yield function is based on entomological studies (for *Culex quinquefasciatus*, see Subramanian et al., 1998) that associate the mf density of specific humans to the number of third instar larvae in mosquitoes fed on those humans 12 days later. The model does not explicitly account for the losses that would occur during the extrinsic incubation period due to mosquito mortality; this may be captured implicitly within ψ_1 if the mosquito excess mortality is not dependent on the parasite density or within the yield function described below.

The yield function is combined with the moment-generating function of the negative binomial distribution to obtain a population-level average, where

$$f(M)_c = \left(1 + \frac{M}{k(M)}\left(1 - e^{-r/\kappa}\right)\right)^{-k(M)}$$

See Gambhir and Michael (2008) for the derivation. In the case of transmission by *Anopheles* spp., the yield function is written to capture the notion of facilitation, as opposed to limitation for *Culex* and *Aedes* spp. These concepts putatively result from differences between mosquito genera in the cibarial and pharyngeal armatures, midgut barriers and immune responses to the parasites within the haemocoel. In the case of facilitation, each microfilaria passing the midgut has been suggested to make the midgut more permeable to subsequent microfilariae, so that as the density of microfilariae ingested increases, the third instar larvae yield increases. In limitation, the opposite is thought to be true, so that as microfilariae pass the midgut, subsequent passages become less likely (Pichon, 2002). A third form, proportionality, is thought to be present in *Mansonia* spp., common vectors of Brugian filariasis.

For anopheline species, the function that has been used (Gambhir and Michael, 2008) to capture the uptake of third instar larvae by an individual mosquito is:

$$L = \kappa\left(1 - e^{-rM/\kappa}\right)^2$$

and for the population:

$$f(M)_{\mathrm{A}} = \left(\frac{2}{\left(1 + \frac{M}{k(M)}\left(1 - e^{-r/\kappa}\right)\right)^{k(M)}} - \frac{1}{\left(1 + \frac{M}{k(M)}\left(1 - e^{-2r/\kappa}\right)\right)^{k(M)}} \right)$$

These population-averaged yields are based on age-specific mf densities, and have to be taken over the age (a) distribution of humans, which can be described by a function, such as the following that was used for Papua New Guinea (Singh et al., 2013):

$$\pi(a) = 0.0361 e^{0.0304a}$$

An additional assumption is that the human age distribution is stable, which could be reasonable over the time periods one would typically wish to project the model.

3.1.3 Basic reproduction number

From these equations we can derive the expression of the basic reproduction number, R_0, using a next-generation matrix approach (Diekmann et al., 2010; Heesterbeek and Roberts, 1995). For the *Culex* spp. version of the model R_0 is:

$$R_0 = \frac{e^{-\frac{r}{\kappa}}\left(e^{-\frac{r}{\kappa}} - 1\right)\kappa\, g\bar{h}\frac{V}{H}\alpha\lambda^2 s_2\psi_1\psi_2}{\gamma\mu(\sigma + \lambda\psi_1)}$$

where $\bar{h}$ represents the average of the age-dependent exposure to mosquito bites, $h(a)$, over all ages. Note that this expression differs somewhat from that given in one of the more recent EPIFIL-based publications (Gambhir et al., 2010). Adding a worm mating function leads to an R_0 much smaller than 1, which is to be expected because this function introduces an Allee effect, or positive density dependence at low population densities (Courchamp et al., 1999; Gambhir et al., 2015). The above function thus has to be interpreted as the number of female worms produced over the lifetime of a single female worm in the absence of density-dependent processes (Heesterbeek and Roberts, 1995).

3.1.4 Simulation of interventions

Norman et al. (2000) describe how control strategies are simulated. Vector control is simulated by assuming a certain reduction in biting rate for a specified period of time. Mass treatment is simulated by defining moments of treatment and the achieved coverage level. Treatment is assumed to cause an immediate reduction in the mf level, assuming a combination of three possible efficacy mechanisms: (1) immediate killing of a proportion of adult worms; (2) immediate killing of a proportion of microfilariae; (3) temporary interruption of mf production during a specified period, while after the end of this period the worm is immediately as productive as before treatment.

3.2 Stochastic, individual-based modelling framework (LYMFASIM)

3.2.1 Modelling approach

Plaisier et al. (1998) published the first description of LYMFASIM, a stochastic, individual-based modelling framework for the transmission and control of lymphatic filariasis. This model is based on the technique of stochastic microsimulation (Habbema et al., 1996). This simulates a dynamic human population, which consists of a discrete number of individuals. The population composition changes over time due to processes of birth and death. The models keep track of the life histories of each individual, focussing on the acquisition and loss of adult worms, mf density and participation in mass treatment. Each simulation starts with a predefined initial population with specified size and age distribution, and a short-term external force of infection. It has a long burn-in period (typically >100 years) to allow both the population composition and infection levels to establish at a dynamic equilibrium, determined by input parameters. The individual-based structure allows that heterogeneities within the human and parasite population are taken into account. The characteristics of individuals are determined by chance and stochastic processes partly determine an individual's infection load. As a result, the outcome of a single simulation run represents only one of many possible outcomes. Repeated runs will give slightly different results. The variability between runs reflects natural variation in real-world settings, conditional on the appropriateness of the model structure. Simulations always have to be repeated to estimate the mean outcome and gain an insight into variability.

The model was originally implemented in a C++-computer program and later re-implemented in JAVA. Input parameters are read from an XML-file, which can be modified by the user. The program code is being maintained by the Department of Public Health of Erasmus Medical Centre, Rotterdam, the Netherlands. The program is available on request from the corresponding author.

3.2.2 Structure of the modelling framework

The model simulates a dynamic population, consisting of a discrete number of individuals. This is typically thought of as representing the population of a single community. Both the size and age—sex composition can change over time due to births and deaths of individuals, which are modelled as stochastic events in the life course of individuals. The expected number of newborns per time step depends on the number of females per age group and age-specific fertility rates. Random numbers define the realized number of newborns entering the population per time step. The age of death varies between individuals and is defined by drawing a random value from a life table. Immigration and emigration are not considered in the LYMFASIM model.

The simulated individuals are characterized by a set of fixed characteristics, relating to personal risk factors for infection and behaviours relevant to control, such as gender, attractiveness to mosquitoes, ability to develop acquired immunity or tendency to comply with mass treatment. These characteristics are determined by randomly drawing a value from prespecified probability distributions. Fertility rates and exposure to mosquito bites further vary with age. Because of their personal characteristics, individuals may be predisposed to heavy or light infections. The infection status of individuals is the outcome of interest of the model. LYMFASIM simulates the transmission of parasites from person to person and tracks changes in the number of worms per individual over time in 1-month time steps.

A schematic representation of the variables and processes involved in transmission is provided in Figure 2, where M_i is the number of adult worms per individual, m_i gives the number of microfilariae scaled to the 20 μL blood volume normally used for diagnosis and $\overline{L3}$ gives the average number of L3 larvae present in biting mosquitoes. The processes involved in transmission are here described in more detail, starting with the monthly transmission potential to which an individual is exposed (mtp_i) that specifies the number of L3 larvae that are released to a person per month. The subscript i denotes that the mtp value is individual-specific, as a consequence of variation in exposure to mosquito bites. On average, only a small proportion (called the success ratio, sr) of the released larvae will survive to develop further into adult worms; a chance process defines how many L3 larvae survive per month. The force of infection acting on individual i, defined as the number of inoculated L3 larvae that develops successfully into an adult worm per month, is calculated as follows in the absence of acquired anti-L3 immunity:

$$foi_i(t) = mtp_i(t) * sr$$

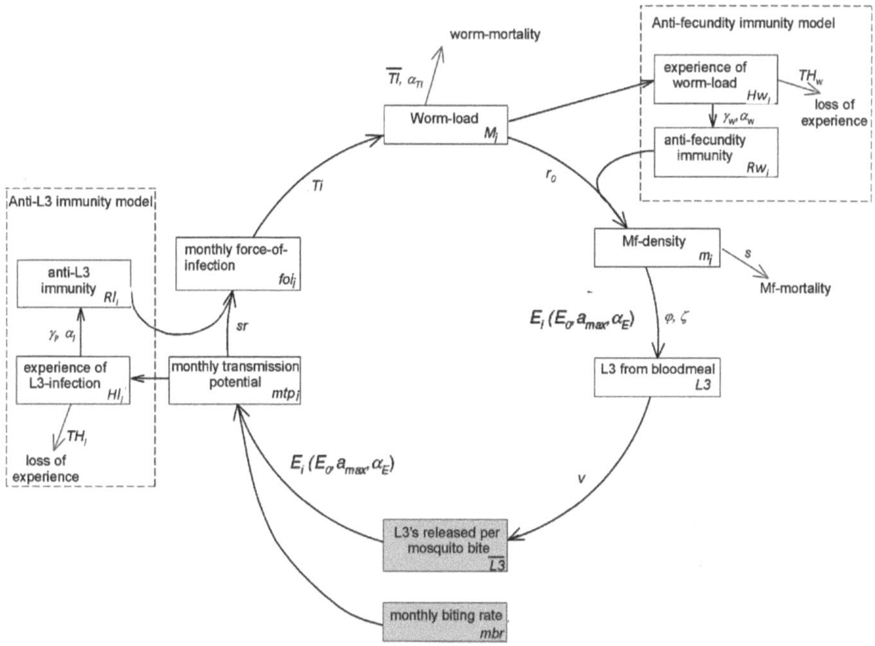

Figure 2 Schematic representation of LYMFASIM. Immune regulation is optional in LYMFASIM. The shaded boxes are entomological variables; these variables do not vary between individuals. The unshaded boxes are human variables; the index *i* indicates that the value of a parameter or variable varies between human individuals. Along the arrows, the symbols of relevant parameters are shown.

The life course of surviving worms is simulated at individual worm level. Worms are immature during a period *Ti* and their average life span is *Tl* (including the duration of the immature period). The life span varies between worms according to a Weibull distribution with shape parameter α_{Tl}. The model assumes that all adult females are inseminated and produce microfilariae, if at least one male worm is present in the human body. Parameter r_0 gives the mf production per female worm in the absence of fecundity-reducing acquired immunity, and is expressed as the number of microfilariae per month and per 20 μL of peripheral blood. Parameter *s* specifies the monthly survival of microfilariae and $1/(1-s)$ gives the mean mf life span (in months). The number of microfilariae are not simulated at the individual level; the model merely calculates the average mf density in the blood per individual (expressed in microfilariae per 20 mL night finger prick blood).

LYMFASIM has the option to include acquired immunity, which either reduces the probability of L3 larvae to develop into adult worms (anti-L3 immunity) or reduces the mf output by female adult worms (anti-fecundity immunity). Both mechanisms result in a lower mf count in the elderly compared to young adults. This type of immunity was included in the model variants for the *C. quinquefasciatus—W. bancrofti* vector parasite complex in India, but not in the African model variant (see Section 4 for further explanation). Anti-L3 immunity modifies the success ratio, *sr*. In the presence of acquired immunity the force of infection acting on an individual is described by the following set of equations:

$$foi_i(t) = mtp_i(t) * sr * (1 - Rl_i(t))$$

$$Rl_i(t) = 1 - \exp\left[-\gamma_l * \rho_{l,i} * Hl_i(t)\right]$$

$$Hl_i(t) = mtp_i(t) + \beta_l * Hl_i(t-1)$$

This implies the following. Rl_i varies between 0 (no immunity) and 1 (complete resistance), depending on the experience of L3 infection *Hl*, and an individual's ability to develop an effective immune response ($\rho_{l,i}$). The latter is a fixed characteristic of humans generated from a continuous probability distribution function with mean 1 (gamma distribution in published applications of this model). The exponential distribution can be replaced by other functions if wanted. Immunity wanes in the absence of boosting with factor βl.

Anti-fecundity immunity works to reduce the mf productivity by female adult worms, so that r_0 is replaced by $r_i(t)$:

$$r_i(t) = r_0 * (1 - Rw_i(t))$$

$$Rw_i(t) = 1 - \exp\left[-\gamma_w * \rho_{w,i} * Hw_i(t)\right]$$

$$Hw_i(t) = M_i(t) + \beta_w * Hw_i(t-1)$$

These immune mechanisms differ from the one implemented in EPIFIL with respect to the trigger of immunity or the processes modified by it; parameters therefore also have a somewhat different interpretation. The number of microfilariae that is taken up by a mosquito when biting on individual *i* depends on the mf density in the human blood of that individual. The association between the mf density in the blood of person *i* and the number of L3 developing in mosquitoes after a blood meal on this person ($L3_i$), is described by some mathematical function, and does not involve stochasticity. Different functions can be assumed to adjust this curve to the mosquito species involved in transmission. The average number of L3 larvae taken up by mosquitoes is given by the weighted average of the uptake from all individuals, the weights reflecting the relative exposure of each individual to mosquito bites (E_i, see below). The average L3 load in mosquitoes is then given by:

$$\overline{L3}(t) = \sum_{i=1}^{N(t)} (E_i(t) * v * L3_i(t)) \Big/ \sum_{i=1}^{N(t)} E_i(t)$$

where $N(t)$ is the total population size, $E_i(t)$ is the relative exposure of person i and the factor v (≤ 1) is included to account for the survival of mosquitoes in field settings and the probability that an L3 larva is released during feeding, which is assumed to be independent of the infection load in mosquitoes.

The relative exposure of an individual (E_i) indicates how many mosquito bites a person gets per month, and is given by:

$$E_i = Ea_i(a, s) * Ei_i$$

$$Ea_i(a, s) = Ea_i(a, s) * Ei_i$$

It is expressed as a fraction of the average number of mosquito bites received per adult male per month (monthly biting rate, *mbr*). An individual's exposure varies from the average in relation to age (and optionally) sex described by a function, $Ea_i(a,s)$, and an additional individual exposure index (Ei_i) describing an individual's attractiveness for mosquitoes. The model ignores sex differences in exposure and adopts the common assumption that exposure increases with body surface during growth in childhood and stabilizes in adults (Duerr et al., 2005; Smith et al., 2006). This is approximated by assuming a linear increase in relative exposure from E_0 at birth to the adult (maximum) level that is achieved at age a_{max}. The random, not age-related variability in exposure is described by a gamma distribution with shape parameter α_E and mean 1. An individual's relative exposure does not only determine his/her contribution to the mean L3 load of mosquitoes, but also the number of L3 larvae received: the monthly transmission potential (mtp_i) is calculated as the mean number of L3 larvae released per mosquito bite (L3, see above) multiplied by the *mbr* and the individual's relative exposure (Ei). The *mbr* is assumed to be constant over time, ignoring seasonal variation and other time trends.

3.2.3 Simulation of interventions

To simulate the impact of MDA, the user must specify the exact moments of treatment (year, month), the drug or administration regimen applied with its efficacy, the fraction of people treated per round (coverage) and the compliance pattern. The three main effects of treatment are: (1) a fraction of adult worms is killed; (2) a fraction of female adult worms is permanently sterilized (i.e. they stop producing microfilariae) and (3) a fraction of microfilariae is killed. The fraction of parasites affected can be constant or can vary according to a chosen probability distribution function. In addition, for each of the three mechanisms, the user may specify a fraction of treated patients with no or full effect of treatment (i.e. the fraction of parasites affected is respectively 0 or 1). All stochastic variables related to the effects of treatment are by default assumed to be independent and to be generated for each person at each treatment. As an alternative, the treatment efficacy can be attributed as a fixed characteristic to an individual, who in that case always responds in the same way to treatment. Temporal reductions in the mf production can also be simulated (Plaisier et al., 1998), but these are not included here.

The compliance pattern describes the tendency of persons to participate in repeated treatment rounds. In case of random compliance, all individuals have the same probability to be treated (equal to the fraction covered). In case of systematic compliance, each person in the population is characterized by a time-invariable compliance factor (a random number between 0 and 1), which results in a treatment probability of either 1 (for compliance factor =< coverage) or 0 (for compliance factor > coverage). Consequently, if coverage is constant over time, some individuals will always be treated while the remaining persons are never treated. In the case of a semi-systematic compliance pattern, the compliance factor indicates a person's tendency to participate. Random numbers define whether an individual is actually treated or not. The latter pattern is presumably most realistic (Plaisier et al., 2000).

LYMFASIM also allows the simulation of selective treatment. In that case, treatment is only provided to those persons who were mf positive in the most recent survey (which may take place in the same month as treatment, see below). Coverage and compliance play no role. Vector control can be modelled as a percentage reduction of the monthly biting rate during a specified period. The number of such periods, their duration and the reduction in monthly biting rate can be chosen.

3.2.4 Model output

At chosen times epidemiological surveys can be simulated to determine the mf counts for all individuals in the population. The results can be summarized by different population-level indicators of infection: the mf prevalence, the geometric or arithmetic mean number of microfilariae per blood smear or the frequency distribution of mf counts in the population. Survey results can be tabulated by age group and gender. The mf counts are subject to random variation and may follow a Poisson, geometric or a negative binomial distribution with clumping factor k and the simulated mf density as mean expected outcome. In previous LYMFASIM applications, mf counts in thick blood smears were assumed to follow a negative binomial distribution and Subramanian et al. (2004) estimated the k-parameter of the distribution for mf counts in 20-μL thick blood smears. If a large volume of blood is used for the thick smear, the value of k increases proportionally with blood volume. The probability distribution may take a different shape for other methods to measure mf density in blood, which can be quantified on the basis of data with repeated measurements. Accurate representation of this variability is particularly relevant if the model is fitted to data and inferences are made about processes and parameter values. Any remaining uncertainty affects the accuracy of inferences and thus of predictions made by resulting models.

4. VALIDATION AND LESSONS LEARNED REGARDING TRANSMISSION DYNAMICS

In this section, we discuss evidence for the validity of EPIFIL and LYMFASIM model variants, as obtained by comparing model outcomes to empirical data. Evidence comes from calibration studies, in which parameter values were estimated by fitting model outcomes to data. The main focus in published calibration studies has been to obtain a better understanding of the processes involved in transmission and estimate the value of unknown parameters.

Initially both LYMFASIM and EPIFIL were developed using the data from a detailed study on the impact of vector control, performed in Pondicherry, India (Chan et al., 1998; Norman et al., 2000; Subramanian et al., 2004). These studies aimed to make inferences about the role of acquired immunity in shaping age-infection patterns as was suggested in the literature (Anderson and May, 1985; Woolhouse, 1992) and to estimate unknown parameters. Norman et al. calibrated a single EPIFIL model variant based on cross-sectional pre-control data with age-specific mf prevalence and mf counts estimates, and provided point estimates of the fitted variables, after which the validity of model-predicted trends in infection indicators during and after a 5-year control period was tested against the observed data (Chan et al., 1998; Norman et al., 2000). In view of uncertainty regarding the role of acquired immunity, Subramanian et al. (2004) tested the goodness of fit of several competing model variants (a model without acquired immunity, and two models with different immune mechanisms). These model variants were directly fitted to longitudinal data on mf prevalence and intensity, cross-tabulating the observed mf densities of 5,071 individuals (categorized into 0, 1–5, or 6+ microfilariae/smear) that were observed just before and at the end of the 5-year vector control programme, and confidence intervals were deduced for a number of key model parameters. The data from Pondicherry show an increase in prevalence and intensity of infection in children to a peak level in young adults, followed by declining levels in the elderly. In both EPIFIL and LYMFASIM model the level of exposure to mosquitoes was assumed to increase with age until a maximum in teenage or young adults, after which it remains constant. Given that assumption, both models concluded that the declining infection prevalence and intensity in the elderly results from some form of acquired immunity. Alternative explanations for a declining prevalence and intensity of infection in older age groups (such as reduced exposure to mosquitoes) were not considered.

The LYMFASIM variants for Pondicherry (fitted against longitudinal data) gave accurate predictions of the expected trends in infection prevalence and intensity over time during and until 6 years after a vector control programme (Subramanian et al., 2004). The EPIFIL model (fitted against cross-sectional pre-control data only) appeared to be a bit too optimistic about the long-term impact of control, predicting lower mf prevalence and lower mf densities than those observed after the control programme (Norman et al., 2000).

Later calibration studies aimed at developing EPIFIL and LYMFASIM models for the African context, examined whether the role of acquired immunity could be confirmed. The models were now applied to data from multiple locations, although this only concerned cross-sectional pre-control data, reflecting the absence of detailed longitudinal studies with long enough follow-up. A simplified version of EPIFIL with constant force of infection was used and several competing model variants (with and without acquired immunity) were fitted to cross-sectional pre-control data from three communities in Kenya and Tanzania with varying endemicity levels (Michael et al., 2001). Two villages with low and intermediate endemicity levels did not show the typical decline in mf prevalence and intensity that would be expected in case of strong acquired immunity, although this pattern was found in a village with higher endemicity level. It was therefore concluded that, if at all, acquired immunity only plays an important role in areas with intense transmission and high pre-control endemicity levels in this region. Stolk et al., however, argued that published age-specific prevalence and intensity data do not consistently show the patterns that are typically expected in high endemic areas, and contested the role of acquired immunity in their Africa model (Stolk et al., 2008, 2004). They proposed the distinction between 'global' structures and parameters (values independent of the region and community under study, e.g. factors related to parasite life course and reproduction within the human host), 'regional' parameters (values that may vary between regions, e.g. human population composition and vector species-dependent factors) and 'local' parameters (e.g. parameters describing exposure level and heterogeneity in exposure). They showed that the model explained cross-sectional pre-control data from different communities reasonably well if only local parameters were varied.

More recently a Bayesian approach has been used (Gambhir et al., 2010; Singh et al., 2013). Gambhir et al. extended the EPIFIL model with density-dependent mechanisms that were not part of earlier EPIFIL model variants, namely the worm mating probability (which declines when worm density falls) and a mechanism of immunosuppression. The Bayesian approach allows estimation of the likelihood of possible values for multiple parameters, hence identifying multiple sets of parameter values that all could explain the data. With this method, they reanalyzed the data from Pondicherry and the three communities in Kenya/Tanzania, together with data from five communities from the Dreikikir area in Papua New Guinea. Each community was analyzed separately, testing many different sets of parameter values and identifying for each parameter the probability distribution and mean of acceptable values. They showed that parameters of density-dependent functions relating to immunity, parasite establishment, as well as parasite aggregation, varied significantly, contributing to locally varying filarial elimination thresholds. The authors did not aim at identifying a single model that could be fitted to all communities after adjustment of local parameters. For the data set in Papua New Guinea, sequential fits were performed and succeeded in obtaining a better match between the credible interval of simulations and the confidence interval of the data over a 5-year period (Singh et al., 2013).

For an overview and description of the parameters, and typical values used in different variants of EPIFIL and LYMFASIM we refer to the original publications indicated above. We conclude that uncertainty remains about the role of acquired immunity and the role for other density-dependent mechanisms in the regulation of parasite densities and specific parameter values. This uncertainty is carried over to predictions of the impact of interventions and efforts required to achieve elimination, as we discuss in the following section.

5. LESSONS LEARNED REGARDING INTERVENTION IMPACT

5.1 Vector control

The EPIFIL and LYMFASIM models were initially developed and validated using epidemiological and entomological data from a long-term vector control programme in Pondicherry, India. Some of the first publications with these models showed the expected impact of a 5-year vector control programme (Michael et al., 2004; Norman et al., 2000; Subramanian et al., 2004). Under the assumption that vector control results in 80—90% reduction in biting rates, as was achieved in Pondicherry, the models predict a gradual decline in the mf intensity and mf prevalence as was observed. The EPIFIL predictions were somewhat too optimistic, while the LYMFASIM model fitted the data well. If control measures are not sustained and biting rates go back to their pre-control level, the mf prevalence also increases again to its pre-control level (Subramanian et al., 2004). The prevalence may further decline or stabilize if the reduction in biting rates would be sustained after cessation of the progamme (Michael et al., 2004; Norman et al., 2000). How uncertainty in model structure and parameter values affects the predicted long-term impact of vector control was not described in these papers.

With a modified version of EPIFIL, Stone et al. (2014) recently estimated the required duration of vector control interventions (long-lasting insecticidal nets, with or without larval source management) for reducing lymphatic filariasis mf prevalence to below 1%, in areas where anophelines transmit both malaria and lymphatic filariasis. Far lower coverage levels were required to reduce R_0 to below 1 for lymphatic filariasis than for malaria, but these had to be maintained for much longer periods (in the simulations, up to 3–4 decades) in order to drain the infectious reservoir. These predictions remain to be validated. An open question is to what extent this predicted duration depended on the assumption of an exponential adult worm mortality function. Allowing for worm senescence (as in LYMFASIM, where a Weibull function is assumed) may lead to predictions closer to the typical life span of filarial worms.

Other studies concentrated on estimating the threshold biting rate, i.e. the minimum biting rate that is required by the parasite population for effective reproduction and population persistence. If biting rates are brought and sustained below that threshold, the parasite population will eventually go extinct. The value of the threshold depends on characteristics of the parasite, vector, host and their interactions. Several theoretical studies illustrated that the threshold biting rate depends on the extent to which transmission intensity is regulated by density-dependent processes (which could result in a considerably higher efficiency of transmission/reproduction when biting rate and mean infection level are low) and on the aggregation of parasites within the human population (Duerr et al., 2005). Unfortunately, knowledge on the operation of density-dependent mechanisms regulating the parasite density in human hosts is incomplete, due to the difficulties in measuring the adult worm burden.

Michael et al. (2007, 2006a) analyzed data from the literature on the association between annual biting rate and overall mf prevalence using a general model; they estimated the threshold biting rate at 833 bites per person per year (p.p.p.y) for areas where *Culex* and *Aedes* mosquito species are responsible for transmission and at 2,375 bites p.p.p.y where *Anopheles* spp. act as main vectors. Using an extended version of the EPIFIL-modelling framework, Gambhir and Michael (2008) arrived at an even lower estimate for *Culex*-transmitted infection (threshold biting rate = 108 bites p.p.p.y), but a somewhat higher value for *Anopheles*-transmitted infection (threshold biting rate = 3,252 bites p.p.p.y). More detailed analysis showed that the value of the threshold biting rate does not only depend on vector species but is also influenced by other local factors; further, prediction intervals for the threshold biting rate remain considerable, due to parameter uncertainty (Gambhir et al., 2010; Singh et al., 2013). Nevertheless, from this body of work we conclude that – especially for areas with *Culex*-transmitted infection, like India – it may be difficult and costly to bring biting rates via vector control measures below the very low threshold biting rate and sustain this reduction long enough to achieve elimination. Vector control alone therefore does not present a practical strategy for the elimination of lymphatic filariasis.

5.2 Mass drug administration

Extensive modelling work has been done to support decision-making in ongoing elimination programmes, which aim to interrupt transmission by annual MDA. Policy makers and programme managers are confronted with the following practical questions. Is it possible to eliminate lymphatic filariasis by mass treatment alone, and — if so — how long must MDA be continued to achieve this goal? Are programmes performing well and on track to achieve their goal and what strategy adjustments could be used to accelerate the achievement of elimination or improve programme performance? When can MDA safely be stopped without an unacceptably high risk of recrudescence? In this paragraph, we discuss lessons learned on each of these issues and remaining uncertainties.

The EPIFIL and LYMFASIM models have been applied to assess the potential impact of MDA and feasibility of elimination. Both models predict the strong decline in mf prevalence that occurs after initiation of repeated MDA if population coverage is sufficiently high. The models also show clearly that premature programme cessation leads to reinfection of the population and a gradual increase of mf prevalence back to its pre-control level. See Figure 3 for an example from LYMFASIM, which shows predicted trends in mf prevalence during and after a hypothetical MDA programme for the Pondicherry (India) setting, assuming that 65% of the population is treated per round and that each treatment immediately kills (or permanently sterilizes) one-third of adult worms and eliminates 90% of microfilariae in all treated individuals. Under these assumptions, four treatment rounds reduce the mf prevalence from about 8.5% to 2.5%, but after cessation of mass treatment the mf prevalence gradually increases again back to its pre-control levels. After eight rounds of mass treatment, the mf prevalence is reduced to about 0.5% and this appears to be sufficient to cause the eventual elimination of the parasite population.

In Table 2, we present EPIFIL- and LYMFASIM-based estimates of the number of annual MDA rounds that are required to achieve elimination. We focussed on predictions for currently recommended single dose treatment regimens (i.e. annual treatment with a combination of ivermectin (IVM) + albendazole (ABZ) for the African context, and diethylcarbamazine (DEC) + ABZ for other endemic regions). This compiled table clearly shows that the required duration of annual MDA for elimination increases strongly with the pre-control endemicity level. This is a consequence of the associated higher worm burdens (in terms of the average number of adult

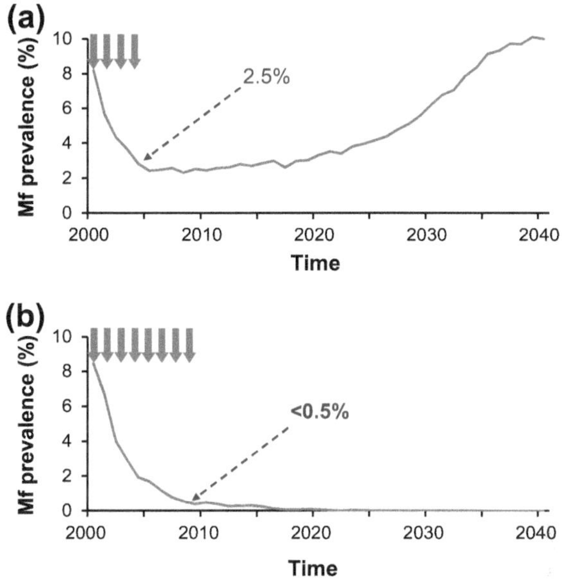

Figure 3 LYMFASIM predictions of the expected long-term impact of annual mass drug administration in Pondicherry with 4 (a) and 8 (b) treatment rounds. Each treatment is assumed to immediately kill one-third of all adult worms and 90% of microfilariae in all treated individuals, with 65% of the population being treated per round. Compliance with offered treatment was simulated as a partially systematic process. That is to say, it is neither completely random (where each person has the same chance to get treated in each round) nor completely systematic (where all individuals either take all or none of the treatments), but somewhere in between. Figures represent the outcomes of one single simulation run.

worms per host) and higher residual transmission levels between treatment rounds. Similarly, the required duration decreases with increasing treatment coverage or the use of more efficacious drugs (especially with respect to killing or permanent sterilization of adult worms): the higher coverage implies a faster decline in parasite burden and less residual transmission in between treatments. The explorative predictions from Stolk et al. (2013) suggest that the required duration of MDA will in general be higher in Africa than in India, which is probably largely caused by the different treatment regimens: the combination of IVM + ABZ, that is used in African countries, is assumed to be considerably less efficacious than the combination of DEC + ABZ. However, vector and host factors may also play a role.

The results of Table 2 should be interpreted with caution. The models have not yet been validated against empirical data from successful and unsuccessful MDA programmes. While the models probably capture the

Table 2 Model-based estimates of the number of annual mass drug treatment rounds required to achieve elimination, for different settings (region; vector/parasite complex; pre-control endemicity level) and characteristics of the mass drug administration (MDA) program (frequency of treatment, employed treatment regimen, assumed treatment efficacy, coverage and compliance patterns)

Model	**EPIFIL**		**LYMFASIM**				
Source	Michael et al. (2004)		Stolk et al. (2003)	Stolk et al. (2005)		Stolk et al. (2013)	Stolk et al. (2013)
Model variant							
Region	India		India				Africa
Vector/parasite complex	C. q./W. b.		C. q./W. b.				An./W. b.
Immunity	No		Yes				No
Vector control	No		No				No
Compliance with MDA	NA		Semi-systematic				
Treatment efficacy assumptions							
Treatment regimen	DEC + ABZ	IVM + ABZ	IVM	IVM + ABZ	DEC + ABZ	DEC + ABZ	IVM + ABZ
% of adult worms killed or permanently sterilized	55%	35%	0%	35%	65%	65%	35%
% of microfilariae killed	95%	99%	100%	100%	70%	70%	100%
Temporary reduction in fecundity: unproductive period	6 months	9 months	NA	NA	NA	NA	NA
Temporary reduction in fecundity: irreversible reduction in female worm fecundity	NA	NA	77%	NA	NA	NA	NA

Number of treatment rounds required for elimination, given the coverage achieved and pre-control endemcity level:

	Pre-control endemicity level																	
Coverage	2.5%	5%	10%	15%	2.5%	5%	10%	15%	8.50%	8.50%	8.50%	7.7% (5%)	11.5% (7.5%)	15% (10%)	12.5% (9%)	20% (14%)	27.5% (20%)
55%	-	-	-	-	-	-	-	-	12	-	-	3	5	9	7	11	>20
60%	7	9	10	12	-	-	-	-	10	-	-	-	-	-	-	-	-
65%	-	-	-	-	-	-	-	-	8	10	6	-	-	-	-	-	-
70%	6	7	8	9	-	-	-	-	7	-	-	2	3	5	5	7	13
80%	5	6	7	8	6	8	11	12	5	6	4	-	-	-	-	-	-
85%	-	-	-	-	-	-	-	-	4	-	-	2	2	3	4	6	9
90%	4	5	6	7	-	-	-	-	4	-	-	-	-	-	-	-	-

ABZ, albendazole; An., *Anopheles*; C. q., *Culex quinquefasciatus*; IVM, ivermectin; W. b., *Wuchereria bancrofti* qualitative associations between outcome and its determinants quite adequately, quantitative predictions are subject to remaining structural and parameter uncertainties in the model. The unknown role of acquired immunity (see above) presents a good example of structural uncertainty in the model. Multiple parameters in both models are difficult to quantify directly from empirical evidence, and their values may have been based on expert opinion or indirect estimation by fitting the model to data. Univariate sensitivity analysis or variable importance analysis shows that model predictions of the required duration of MDA are sensitive to assumed parameter values (Stolk et al., 2008, 2003, 2013). Probabilistic multivariate sensitivity analysis was not carried out. For policy support, it is crucial that models are validated against longitudinal data collected during MDA programmes and ideally also several years after cessation of MDA. Such data should ideally be collected under the controlled circumstances of a research project, to reduce uncertainties regarding their interpretation and maximize the information that can be obtained from those data. Remaining uncertainty after model validation should be captured, for example, using the Bayesian techniques from recent EPIFIL publications (Gambhir et al., 2010; Singh et al., 2013), and the influence of this uncertainty on predicted duration of MDA required for elimination should be studied.

Model-predicted trends in infection indicators during MDA can be used as a reference curve, against which achievements of actual programmes can be compared to assess whether they are performing well. This was discussed and illustrated by Michael et al. (2006b, 2007). Such a comparison might reveal that observed declines in infection indicators are smaller than predicted by models. This could indicate either that the programme was poorly implemented (e.g. resulting in low coverage) or that local circumstances are more unfavourable than anticipated with intense transmission or possibly even resistance of parasites to anthelmintic drugs. In both cases, the control strategy may need to be adapted to ensure that programme targets are timely achieved. If poor performance results from low coverage, action should be taken to improve those coverage levels. This might require ensuring that sufficient treatment doses are available, the implementation of health information and education campaigns and other sensitization activities, or changing treatment distribution strategies and routes (time period in which treatment takes place; house to house versus centralized distribution, distribution by volunteers or health workers, etc.). Local barriers to the uptake of treatment should be analyzed in order to define the best remedial action. In case of delayed start of MDA and for areas with very intense transmission, acceleration strategies may have to be considered. Simulations performed with LYMFASIM suggest that doubling the frequency of treatment (from yearly to 6-monthly) will effectively reduce the required duration by about 50%: the same number of treatment rounds is required, but is simply given over a shorter period (Stolk et al., 2013). Vector control may also be considered as an adjunct strategy to MDA. EPIFIL simulations suggest that joint implementation of MDA and vector control could reduce the required duration of interentions by about 20–25% (Michael et al., 2004).

At the start of lymphatic filariasis elimination programmes, policy makers and programme managers usually have a crude idea of how long MDA must be continued to achieve the goal of elimination. The actual required duration, however, depends on prevailing transmission conditions and programme performance. For example, the required duration may be longer if delays occur in the implementation of MDA or coverage levels are lower than anticipated. The decision to stop MDA must therefore be based on empirical evidence that the infection levels have declined to zero or almost zero. Theoretical work on helminth infections showed that infection levels (e.g. measured through mf prevalence or mean count in a blood sample) do not necessarily have to be brought down to zero to achieve elimination: if the infection level is brought below the transmission breakpoint, the mating probability of male and female worms becomes so low that parasites can no longer effectively reproduce so that the parasite population will locally go extinct without further control measures (May, 1977; Anderson and May, 1985; Duerr et al., 2005; Gambhir et al., 2015) Models are useful tools to establish the value of such breakpoints, which can serve as target endpoints for MDA programmes. Through a mathematical analytical approach of the transmission dynamics, the breakpoint was estimated at approximately 0.5% mf prevalence for bancroftian filariasis transmitted by *Culex* spp. vectors (Michael et al., 2006a). Breakpoints may be different in areas with other vector species.

The stochastic simulation model LYMFASIM was also used to investigate the existence and location of transmission breakpoints (unpublished work, (Stolk, 2007)). This work illustrates some important issues with the theoretical breakpoint concept. Firstly, as illustrated in Figure 3, there may be a long time gap between the last treatment round and the moment infection indicators are reaching zero. Low-level transmission may occur in the posttreatment period but is at such low rates that the parasite population eventually moves to extinction. Figure 4 shows that the mf prevalence observed 1 year after the last treatment round in an MDA programme is a good predictor of elimination, but not perfect. A lower mf prevalence

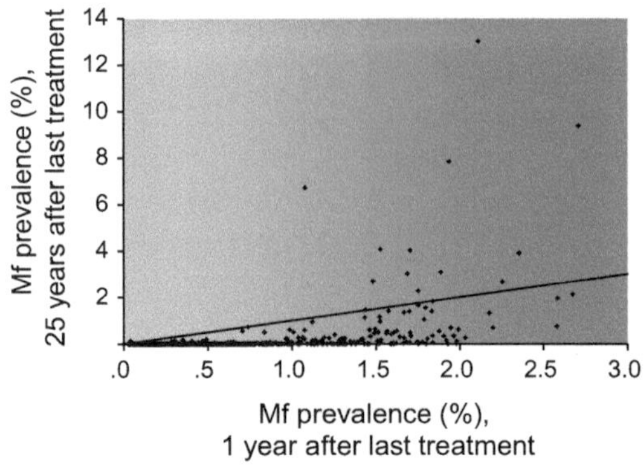

Figure 4 LYMFASIM predictions on the association between the mf prevalence as measured 1 and 25 years after the last treatment in an annual mass drug administration programme, based on the outcome of 650 simulations (each dot indicates the results of 1 simulation run). In all simulation runs, the transmission conditions were the same: the vector was *Culex quinquefasciatus*, monthly biting rate was 1000 bites per person per month, resulting in an mf prevalence before the start of control of 8.7%. Because of chance phenomena in the LYMFASIM model, there is some variation in the pre-control mf prevalence before the start of treatment (5th–95th percentile 5.3–11.7%). The number of treatment rounds simulated was varied from 4 to 10 (90–95 runs per scenario), always assuming that 80% of the population was treated and that each treatment results in immediate death of 65% of adult worms and 70% of mf in treated individuals. Because of the variation in the duration of mass drug administration and due to chance phenomena in the LYMFASIM model, the simulated mf prevalence 1 and 25 years after the last treatment round varies between runs (range: 0–2.7%). There is a clear association between the simulated mf prevalence at 1 and 25 years after the last treatment round. Elimination was said to occur if the prevalence was <0.001% 30 years after the last treatment round. The dashed line represents situations where the mf prevalence at 1 and 25 years following cessation are exactly the same. If the prevalence after 25 years is above this line, it is indicative of recrudescence. If the prevalence after 25 years is zero or very close to zero, elimination has already occurred or is expected to occur after longer follow-up. The background shading indicates where the probability of success is high (green (light grey in print versions)) or lower (red (dark grey in print versions)).

1 year after the last treatment round is associated with a higher probability of elimination: the eventual probability of elimination was 98%, 95%, 65%, 38% and 0% respectively, if the mf prevalence 1 year after the last treatment was below 0.5%, between 0.5% and 1.0%, between 1.0% and 1.5%, between 1.5% and 2.0% or higher than 2%. These simulations match closely the estimated breakpoint of 0.5% mf prevalence. Figure 5 presents a similar analysis, based on the model variant developed for rural African settings with *Anopheles*-transmitted infection, comparing settings with different transmission

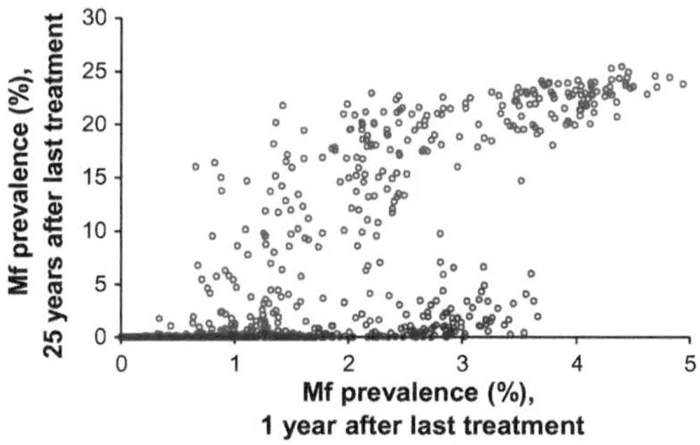

Figure 5 Similar to Figure 4, contrasting the outcomes for areas with different transmission conditions. All simulations are for rural African settings, where bancroftian filariasis is transmitted by *Anopheles* spp. mosquitoes. Red (light grey in print versions) points are for settings with a monthly biting rate equal to 1000 (pre-control mf prevalence 26% on average), blue (dark grey in print versions) points for settings with a monthly biting rate equal to 600 (pre-control mf prevalence 18% on average).

intensity. In settings with a monthly biting rate equal to 1,000, the mf prevalence must be brought down to levels below 1% in order to have a high probability of elimination. If the monthly biting rate is only 600, the model suggests that it may be sufficient to bring down mf prevalence to about 2%. This dependency on local transmission conditions makes it difficult to define to what level the mf prevalence must be reduced in order to achieve elimination. If a single target is wanted, it should reflect the situation in areas with the most intense transmission.

Model-based estimates of the breakpoint mf prevalence also depend strongly on whether the models accurately reflect the processes involved in transmission. Incorrect representation of density-dependent processes in the model could lead to over- or underestimation of the breakpoint. As described above for the estimation of threshold biting rates, Bayesian techniques were also used to estimate the prediction interval for the breakpoint in relation to parameter and structural uncertainty in the models and in relation to difference between settings (Gambhir et al., 2010; Singh et al., 2013).

6. CRITICAL ISSUES IN MODELLING ELIMINATION

Eliminating lymphatic filariasis is a very ambitious goal. The MDA strategy is based on a simple principle: by reducing infection levels below a certain breakpoint level, the worm density becomes too low to enable effective reproduction, and eventually the worm population will die out.

The models discussed predict that this is a feasible strategy and may be achieved in a relatively short time frame. But, will it really be that simple? Are the available models not too simplistic? Modelling requires that simplifying assumptions are made and for many applications of the models that may be good enough. But more complexity may be needed when it comes to simulating the prospects for elimination. In this section, we identify some issues that are critical for the success of MDA programmes and elimination prospects, discussing the extent to which these topics are covered by modelling studies and identifying future needs.

6.1 Coverage and compliance patterns

The Global Programme to Eliminate Lymphatic Filariasis primarily relies on MDA to interrupt transmission of the infection, although vector control measures may be taken to complement the approach. The effectiveness of MDA programmes is compromised if treatment does not reach all people who are eligible for treatment. Incomplete coverage is perhaps not that problematic if noncompliance occurs completely at random in the population, because untreated people may still be treated in a next round and moreover also still benefit indirectly from a reduction in transmission. However, it becomes problematic if entire communities are left untreated (i.e. incomplete geographic coverage), if noncompliance clusters geographically (drug coverage lower than elsewhere in some communities, subcommunities or households) and especially if noncompliance occurs systematically in the same individuals over multiple treatment rounds. Such problems are likely to occur due to problems of drug availability, poor performance of health workers and household perspective on the use of mass treatment (WHO, 1996; Plaisier et al., 2000; Babu and Mishra, 2008; Ssemanda et al., 2010). Spatial issues are further discussed below; here we restrict the discussion to the impact of coverage and compliance patterns.

The impact of poor coverage on programme effectiveness has been described in multiple studies and is also apparent from Table 2. It is important to realize that the EPIFIL model implicitly makes the common assumption that all people equally benefit from mass treatment, although it would be possible to stratify the population according to their compliance with mass treatment (Turner et al., 2013). The individual-based model LYMFASIM does take account of differences in compliance with treatment between individuals, and noncompliance can be assumed to occur either completely at random (where each person has the same chance to get treated in each round), or completely systematic (where all individuals either take all or none of the treatments) or in a semi-systematic pattern somewhere in between those extremes. The latter was shown to be most realistic (Plaisier et al., 2000). The importance of correct representation of this mechanism in simulation models has been illustrated by Plaisier et al. (2000). They used a simple mathematical model to examine the reduction in mf density achieved by five rounds of MDA, comparing scenarios with different coverage levels for each of the three noncompliance patterns described above. The model-predicted reduction in mean mf density after five rounds of mass treatment was lowest for the model with systematic noncompliance, higher in case of semi-systematic noncompliance and highest if noncompliance occurred at random. This work highlights the importance of preventing systematic noncompliance in real-life MDA programmes. Further, it is important that modelling frameworks take account of such compliance patterns, if they are applied to estimate the number of treatment rounds and coverage that are needed to achieve elimination in specific settings.

6.2 Resistance to anthelmintic drugs

Due to its reliance on prolonged yearly mass distributions of anthelmintic drugs (DEC or IVM in combination with ABZ), the concomitant selection for drug-resistant phenotypes and the implications thereof on the elimination programme have to be taken seriously (McCarthy, 2005). Experiences from veterinary medicine show that such resistance can spread rapidly and widely, despite initial beliefs that the long generation times of helminths would delay or prevent this. High prevalences of resistance are now common in certain farmed ruminants, and include instances of multiple drug resistance (Geerts and Gryseels, 2000). The experience has been that resistance may go unnoticed until the opportunity for resistance management has gone, because its spread often follows a sigmoidal progression: a long latent phase followed by an exponential increase (Churcher et al., 2010; Geerts and Gryseels, 2000). A number of reports suggest that tolerance or resistance of *W. bancrofti* to ABZ or DEC may exist. For instance, treatment failure in patients could not be explained by inadequate drug doses or differences in serum levels, suggesting parasite tolerance to DEC (Eberhard et al., 1991). A stronger suggestion that resistance may be selected for by ongoing MDA programmes comes from the finding that a point mutation of the β-tubulin gene associated in a number of nematodes with resistance to benzimidazoles (including ABZ) was already present in *W. bancrofti* at fairly high frequencies prior to drug distribution and increased dramatically over two annual rounds. In Burkina Faso the frequency of worms homozygous for this allele increased from 15% to 86% during that period (Schwab et al., 2005). The mechanisms of resistance in human filarial nematodes are however poorly understood, owing in large part to our poor knowledge of the mode of action of IVM and DEC (Prichard, 2005).

A number of modelling studies have been performed that looked into the probability of resistance emerging during MDA programmes. A standard population genetics model, not taking into account the population dynamics of filarial worms, was used to gain qualitative insights into the spread of IVM and ABZ resistance. The results suggested that while inevitable, the spread of resistance would likely be slow. Resistance to IVM went to fixation at a steady rate, due to its assumed dominant nature, while ABZ resistance increased later, but more suddenly, as a result of its presumed recessive nature (Michael et al., 2004). A one-locus, two-allele genetic structure was included in a version of EPIFIL in order to investigate the effects of combination treatments on the spread of ABZ resistance. Key insights were that selection pressures on ABZ resistance were far weaker when using IVM and ABZ than when using DEC and ABZ. An increase in resistant gene frequencies was particularly pronounced when nonrandom mating was assumed and was also highly dependent on the coverage (proportion of the population treated) (Schwab et al., 2006). The implications for elimination or lymphatic filariasis control may however not be dramatic unless resistance to either IVM or DEC develops within the same population. Schwab et al. note that in their simulations, despite the increase in ABZ-resistant genotypes, the trend in microfilaraemia intensity is only minimally affected. The probability of multiple drug resistance developing has not been modelled. Going forward, mathematical models may have to be used more extensively to inform antlemintic resistance management strategies in the framework of the lymphatic filariasis elimination programme.

6.3 Spatial heterogeneities in the spread of infection

Spatial aspects of the spread of infection have as yet been ignored in the available modelling frameworks for lymphatic filariasis. An obvious and well-recognized risk to elimination is the possibility of reinstigation of infection by immigration of infected humans or mosquitoes, and models can be used to study the implications of influx of infection from other areas on elimination prospects and required duration of MDA. Less obvious is the effect of spatial heterogeneities and nonhomogeneous mixing within the area under study. Because of the relatively short flight range of mosquitoes, the assumption that each mosquito can bite on any individual in the population (i.e. well-mixed distribution of mosquitoes over the human population) may not hold in real life. In real life, there may be subpopulations of mosquitoes covering geographic subcommunities or household clusters. The transmission conditions may differ between clusters, and transmission 'hot spots' may exist with high vector density due to the proximity of breeding sites, the vicinity of other hosts or household factors. This was demonstrated for malaria (Bousema et al., 2010) and could be similarly important for lymphatic filariasis, since the same mosquito species can be involved in transmission. To clear infection from such hot spots, longer duration or intensification of interventions may be required to eliminate infection from clusters with higher transmission intensity. The same is true if there are geographic clusters with systematically lower uptake of interventions than other clusters, which has been described for marginalized populations.

Programmes may be unaware of the existence or location of hot spots, basing their planning and decisions on the community average situation. If transmission hot spots and areas with low uptake of interventions are not targeted with extra interventions and are not covered by monitoring and transmission assessment surveys, there is a risk that interventions are stopped prematurely and that residual infection remains in subgroups of a population that is assumed to be sufficiently treated. From these clusters, infection can easily be reintroduced into other settings. This needs to be acknowledged and properly addressed with targeted interventions. Further, it remains important to target surveillance activities on presumed hot spots of infection transmission and high-risk subpopulations.

The available models for lymphatic filariasis transmission need to be extended to include spatial heterogeneities with respect to local transmission conditions or the operational effectiveness of interventions, in order to investigate how such spatial heterogeneities are affecting elimination prospects and what extra measures are needed to prevent this. Such models are already available for some other vector-borne diseases (Dommar et al., 2014; Favier et al., 2005; Prosper et al., 2012). In view of the ambitious elimination target, it is now timely to include this in models for lymphatic filariasis.

7. CONCLUSION

Mathematical modelling provides a useful tool for policy making and planning in lymphatic filariasis control programmes, by providing trend forecasts based on sound scientific knowledge and principles. Two general models for lymphatic filariasis transmission and control have been developed and were applied in multiple studies to investigate the long-term impact of control strategies and assess elimination prospects. Model predictions suggest that lymphatic filariasis transmission can indeed be interrupted by annual MDA alone, but do not support the suggestion that this can be achieved everywhere by 4–6 yearly rounds of MDA. The number of treatment rounds may be considerably larger in areas with high transmission intensity or poor treatment coverage.

Considering the short remaining time to the 2020 target year for elimination, models should be used to their maximum potential to optimize interventions and help to solve the many practical questions that arise in operational programs. Closer linkages with policy makers and programme managers are useful to ensure that the right questions are being addressed in a timely manner. Two important practical questions still need to be addressed. Firstly, models need to be applied more systematically to define when interventions can be stopped. Different attempts were made to determine the breakpoint prevalence for specific settings, but models have also highlighted the dependency of this breakpoint on setting characteristics and there is no final conclusion on the endpoint that should be used in practice or even which diagnostic tool to use. Modelling can help to evaluate the adequacy of current algorithms for determining whether interventions can be stopped (Chu et al., 2013). Secondly, models should be used to study more systematically how elimination can be accelerated most cost-effectively, e.g. through implementation of complementary vector control (Michael et al., 2004), by increasing the frequency of treatment from yearly to 6-monthly (Stolk et al., 2013), or perhaps interventions targeted at high-risk groups and infection hot spots.

Considerable uncertainty remains in the models. To a large extent this relates to difficulties in measuring an individual's level of exposure to mosquitoes and the worm burden carried by this person. It is therefore unknown to what extent transmission processes are being regulated by density-dependent mechanisms. Given the many uncertainties it is critical that the models are more extensively validated against longitudinal empirical data. It is important to realize that no single data set can validate all different model components and data from different sources must be combined. For example, particular effort is still needed to test whether the models adequately mimic the processes determining between-setting differences in endemicity and intervention impact. Still, such external validation efforts will not take away all uncertainty. Therefore, it is equally important to better define remaining structural and parameter uncertainties, and study how these uncertainties affect model predictions on elimination prospects. The Bayesian technique that was introduced in this field by Gambhir et al. (2010) presents a very powerful tool in this respect and we recommend that this is further advanced and used more widely.

Modellers — like policy makers and programme managers — need to acknowledge the many complexities that play a role in transmission of lymphatic filariasis and could potentially limit the effectiveness of interventions locally. While parsimony is a laudable goal in modelling in view of tractability, public health decision-making requires that all critical processes are adequately represented in the model. The model-based assessment of elimination prospects likely presents one of the greatest challenges for infectious disease modellers, because complexities like systematic noncompliance, emerging resistance to anti-helminthic drugs and spatial heterogeneities can no longer be ignored.

In conclusion, rapid advances are needed in mathematical modelling of lymphatic filariasis: advanced modelling techniques need to be applied to incorporate critical complexities and make model predictions more realistic, and model validation and uncertainty analysis needs to be carried out according to the highest quality standards. In this respect we welcome the initiation of the Neglected Tropical Diseases Modelling Consortium (NTD Modelling Consortium, 2014), which will foster the collaboration between modelling groups and enable more rapid exchange of ideas, insights and expertise. This will help to advance the modelling work more rapidly and therefore to maximize the utility of models for decision support in the ongoing, ambitious attempt to eliminate lymphatic filariasis.

REFERENCES

Anderson, R.M., May, R.M., 1985. Helminth infections of humans: mathematical models, population dynamics, and control. Adv. Parasitol. 24, 1–101.

Babu, B.V., Mishra, S., 2008. Mass drug administration under the programme to eliminate lymphatic filariasis in Orissa, India: a mixed-methods study to identify factors associated with compliance and non-compliance. Trans. R. Soc. Trop. Med. Hyg. 102 (12), 1207–1213.

Basáñez, M.G., McCarthy, J.S., French, M.D., Yang, G.J., Walker, M., Gambhir, M., Prichard, R.K., Churcher, T.S., 2012. A research agenda for helminth diseases of humans: modelling for control and elimination. PLoS Negl. Trop. Dis. 6, e1548.

Bhunu, C.P., Mushayabasa, S., 2012. Transmission dynamics of lymphatic filariasis: a mathematical approach. ISRN Biomath. 2012, 1–9.

Bousema, T., Drakeley, C., Gesase, S., Hashim, R., Magesa, S., Mosha, F., Otieno, S., Carneiro, I., Cox, J., Msuya, E., Kleinschmidt, I., Maxwell, C., Greenwood, B.,

Riley, E., Sauerwein, R., Chandramohan, D., Gosling, R., 2010. Identification of hot spots of malaria transmission for targeted malaria control. J. Infect. Dis. 201, 1764–1774.
Centers For Disease Control, 1993. Recommendations of the international task force for disease eradication. MMWR Morb. Mortal. Wkly. Rep. 42, 1–38.
Chan, M.S., Srividya, A., Norman, R.A., Pani, S.P., Ramaiah, K.D., Vanamail, P., Michael, E., Das, P.K., Bundy, D.A., 1998. Epifil: a dynamic model of infection and disease in lymphatic filariasis. Am. J. Trop. Med. Hyg. 59, 606–614.
Chu, B.K., Deming, M., Biritwum, N.K., Bougma, W.R., Dorkenoo, A.M., El-Setouhy, M., Fischer, P.U., Gass, K., Gonzalez De Pena, M., Mercado-Hernandez, L., Kyelem, D., Lammie, P.J., Flueckiger, R.M., Mwingira, U.J., Noordin, R., Offei Owusu, I., Ottesen, E.A., Pavluck, A., Pilotte, N., Rao, R.U., Samarasekera, D., Schmaedick, M.A., Settinayake, S., Simonsen, P.E., Supali, T., Taleo, F., Torres, M., Weil, G.J., Won, K.Y., 2013. Transmission assessment surveys (TAS) to define endpoints for lymphatic filariasis mass drug administration: a multicenter evaluation. PLoS Negl. Trop. Dis. 7, e2584.
Churcher, T.S., Kaplan, R.M., Ardelli, B.F., Schwenkenbecher, J.M., Basáñez, M.G., Lammie, P.J., 2010. Mass treatment of parasitic disease: implications for the spread of anthelmintic resistance. In: Weber, J.T. (Ed.), Antimicrobial Resistance – Beyond the Breakpoint. Issues in Infectious Diseases Series Vol. 6: 120–137. Basel (Switzerland), Karger.
Coelho, F.C., Codeco, C.T., Gomes, M.G., 2011. A Bayesian framework for parameter estimation in dynamical models. PLoS One 6, e19616.
Courchamp, F., Clutton-Brock, T., Grenfell, B., 1999. Inverse density dependence and the Allee effect. Trends Ecol. Evol. 14, 405–410.
Diekmann, O., Heesterbeek, J.A., Roberts, M.G., 2010. The construction of next-generation matrices for compartmental epidemic models. J. R. Soc. Interface 7, 873–885.
Dietz, K., 1982. The population dynamics of onchocerciasis. In: Anderson, R. (Ed.), Population Dynamics of Infectious Diseases: Theory and Applications. Chapman & Hall, London.
Dommar, C.J., Lowe, R., Robinson, M., Rodo, X., 2014. An agent-based model driven by tropical rainfall to understand the spatio-temporal heterogeneity of a chikungunya outbreak. Acta Trop. 129, 61–73.
Duerr, H.P., Dietz, K., Eichner, M., 2005. Determinants of the eradicability of filarial infections: a conceptual approach. Trends Parasitol. 21, 88–96.
Dye, C., Hasibeder, G., 1986. Population dynamics of mosquito-borne disease: effects of flies which bite some people more frequently than others. Trans. R. Soc. Trop. Med. Hyg. 80, 69–77.
Eberhard, M.L., Lammie, P.J., Dickinson, C.M., Roberts, J.M., 1991. Evidence of nonsusceptibility to diethylcarbamazine in *Wuchereria bancrofti*. J. Infect. Dis. 163, 1157–1160.
Favier, C., Schmit, D., Muller-Graf, C.D., Cazelles, B., Degallier, N., Mondet, B., Dubois, M.A., 2005. Influence of spatial heterogeneity on an emerging infectious disease: the case of dengue epidemics. Proc. Biol. Sci. 272, 1171–1177.
Gambhir, M., Bockarie, M., Tisch, D., Kazura, J., Remais, J., Spear, R., Michael, E., 2010. Geographic and ecologic heterogeneity in elimination thresholds for the major vector-borne helminthic disease, lymphatic filariasis. BMC Biol. 8, 22.
Gambhir, M., Michael, E., 2008. Complex ecological dynamics and eradicability of the vector borne macroparasitic disease, lymphatic filariasis. PLoS One 3, e2874.
Gambhir, M., Singh, B.K., Michael, E., 2015. The Allee effect and elimination of Neglected Tropical Diseases: a mathematical modelling study. Adv. Parasitol. 87, 1–32.
Garnett, G.P., Cousens, S., Hallett, T.B., Steketee, R., Walker, N., 2011. Mathematical models in the evaluation of health programmes. Lancet 378, 515–525.
Geerts, S., Gryseels, B., 2000. Drug resistance in human helminths: current situation and lessons from livestock. Clin. Microbiol. Rev. 13, 207–222.

Gubler, D.J., Bhattacharya, N.C., 1974. A quantitative approach to the study of Bancroftian filariasis. Am. J. Trop. Med. Hyg. 23, 1027–1036.
Gweryina, R.I., Luga, T., 2014. Modelling the potential impact of chemotherapy and treatment on the dynamics of malaria-filariasis coinfection. Int. J. Mod. Math. Sci. 10, 163–188.
Habbema, J.D.F., De Vlas, S.J., Plaisier, A.P., Van Oortmarssen, G.J., 1996. The microsimulation approach to epidemiologic modeling of helminthic infections, with special reference to schistosomiasis. Am. J. Trop. Med. Hyg. 55, 165–169.
Hairston, N.G., Jachowski, L.A., 1968. Analysis of the *Wuchereria bancrofti* population in the people of American Samoa. Bull. World Health Organ. 38, 29–59.
Hairston, N.G., Meillon, B.D., 1968. On the inefficiency of transmission of *Wuchereria bancrofti* from mosquito to human host. Bull. World Health Organ. 38, 935–941.
Hayashi, S., 1962. A mathematical analysis on the epidemiology of bancroftian and malayan filariasis in Japan. Jpn. J. Exp. Med. 32, 13–43.
Heesterbeek, J.A., Roberts, M.G., 1995. Threshold quantities for helminth infections. J. Math. Biol. 33, 415–434.
May, R.M., 1977. Togetherness among schistosomes: its effects on the dynamics of the infection. Math. Biosci. 35, 301–343.
McCarthy, J., 2005. Is anthelmintic resistance a threat to the program to eliminate lymphatic filariasis? Am. J. Trop. Med. Hyg. 73, 232–233.
Michael, E., 2000. The population dynamics and epidemiology of lymphatic filariasis. In: Nutman, T.B. (Ed.), Lymphatic Filariasis. Imperial College Press, London, pp. 41–84.
Michael, E., Malecela-Lazaro, M.N., Kabali, C., Snow, L.C., Kazura, J.W., 2006a. Mathematical models and lymphatic filariasis control: endpoints and optimal interventions. Trends Parasitol. 22, 226–233.
Michael, E., Malecela-Lazaro, M.N., Kazura, J.W., 2007. Epidemiological modelling for monitoring and evaluation of lymphatic filariasis control. Adv. Parasitol. 65, 191–237. Elsevier Ltd.
Michael, E., Malecela-Lazaro, M.N., Maegga, B.T., Fischer, P., Kazura, J.W., 2006b. Mathematical models and lymphatic filariasis control: monitoring and evaluating interventions. Trends Parasitol. 22, 529–535.
Michael, E., Malecela-Lazaro, M.N., Simonsen, P.E., Pedersen, E.M., Barker, G., Kumar, A., Kazura, J.W., 2004. Mathematical modelling and the control of lymphatic filariasis. Lancet Infect. Dis. 4, 223–234.
Michael, E., Simonsen, P.E., Malecela, M., Jaoko, W.G., Pedersen, E.M., Mukoko, D., Rwegoshora, R.T., Meyrowitsch, D.W., 2001. Transmission intensity and the immuno-epidemiology of bancroftian filariasis in East Africa. Parasite Immunol. 23, 373–388.
Norman, R.A., Chan, M.S., Srividya, A., Pani, S.P., Ramaiah, K.D., Vanamail, P., Michael, E., Das, P.K., Bundy, D.A., 2000. EPIFIL: the development of an age-structured model for describing the transmission dynamics and control of lymphatic filariasis. Epidemiol. Infect. 124, 529–541.
Ntd Modelling Consortium, 2014. Available. http://www.ntdmodelling.org/ (accessed 17.08.14.).
Pichon, G., 2002. Limitation and facilitation in the vectors and other aspects of the dynamics of filarial transmission: the need for vector control against *Anopheles*-transmitted filariasis. Ann. Trop. Med. Parasitol. 96, S143–S152.
Plaisier, A.P., Stolk, W.A., Van Oortmarssen, G.J., Habbema, J.D., 2000. Effectiveness of annual ivermectin treatment for *Wuchereria bancrofti* infection. Parasitol. Today 16, 298–302.
Plaisier, A.P., Subramanian, S., Das, P.K., Souza, W., Lapa, T., Furtado, A.F., Van Der Ploeg, C.P.B., Habbema, J.D.F., Van Oortmarssen, G.J., 1998. The LYMFASIM

simulation program for modeling lymphatic filariasis and its control. Methods Inf. Med. 37, 97–108.
Poole, D., Raftery, A.E., 2000. Inference for deterministic simulation models: the Bayesian melding approach. J. Am. Stat. Assoc. 95, 12.
Prichard, R.K., 2005. Is anthelmintic resistance a concern for heartworm control? what can we learn from the human filariasis control programs? Vet. Parasitol. 133, 243–253.
Prosper, O., Ruktanonchai, N., Martcheva, M., 2012. Assessing the role of spatial heterogeneity and human movement in malaria dynamics and control. J. Theor. Biol. 303, 1–14.
Rochet, M.J., 1990. A simple deterministic model for bancroftian filariasis transmission dynamics. Trop. Med. Parasitol. 41, 225–233.
Schwab, A.E., Boakye, D.A., Kyelem, D., Prichard, R.K., 2005. Detection of benzimidazole resistance-associated mutations in the filarial nematode *Wuchereria bancrofti* and evidence for selection by albendazole and ivermectin combination treatment. Am. J. Trop. Med. Hyg. 73, 234–238.
Schwab, A.E., Churcher, T.S., Schwab, A.J., Basáñez, M.G., Prichard, R.K., 2006. Population genetics of concurrent selection with albendazole and ivermectin or diethylcarbamazine on the possible spread of albendazole resistance in *Wuchereria bancrofti*. Parasitology 133, 589–601.
Schwab, A.E., Churcher, T.S., Schwab, A.J., Basáñez, M.G., Prichard, R.K., 2007. An analysis of the population genetics of potential multi-drug resistance in *Wuchereria bancrofti* due to combination chemotherapy. Parasitology 134, 1025–1040.
Singh, B.K., Bockarie, M.J., Gambhir, M., Siba, P.M., Tisch, D.J., Kazura, J., Michael, E., 2013. Sequential modelling of the effects of mass drug treatments on anopheline-mediated lymphatic filariasis infection in Papua New Guinea. PLoS One 8, e67004.
Slater, H.C., Gambhir, M., Parham, P.E., Michael, E., 2013. Modelling co-infection with malaria and lymphatic filariasis. PLoS Comput. Biol. 9, e1003096.
Smith, T., Maire, N., Dietz, K., Killeen, G.F., Vounatsou, P., Molineaux, L., Tanner, M., 2006. Relationship between the entomologic inoculation rate and the force of infection for *Plasmodium falciparum* malaria. Am. J. Trop. Med. Hyg. 75, 11–18.
Ssemanda, E.N., Munoz, B., Harding-Esch, E.M., Edwards, T., Mkocha, H., Bailey, R.L., Sillah, A., Stare, D., Mabey, D.C., West, S.K., 2010. Mass treatment with azithromycin for trachoma control: participation clusters in households. PLoS Negl. Trop. Dis. 4.
Stolk, W.A., 2007. When can mass drug administration for lymphatic filariasis be stopped? A modeling perspective [Conference presentation]. In: 56th Annual Meeting of the American Society of Tropical Medicine and Hygiene. Philadelphia, Pennsylvania, USA.
Stolk, W.A., De Vlas, S.J., Borsboom, G.J., Habbema, J.D., 2008. LYMFASIM, a simulation model for predicting the impact of lymphatic filariasis control: quantification for African villages. Parasitology 135, 1583–1598.
Stolk, W.A., De Vlas, S.J., Habbema, J.D., 2006. Advances and challenges in predicting the impact of lymphatic filariasis elimination programmes by mathematical modelling. Filaria J. 5, 5.
Stolk, W.A., De Vlas, S.J., Habbema, J.D.F., 2005. Anti-*Wolbachia* treatment for lymphatic filariasis. Lancet 365, 2067–2068.
Stolk, W.A., Ramaiah, K.D., Van Oortmarssen, G.J., Das, P.K., Habbema, J.D.F., De Vlas, S.J., 2004. Meta-analysis of age-prevalence patterns in lymphatic filariasis: no decline in microfilaraemia prevalence in older age groups as predicted by models with acquired immunity. Parasitology 129, 605–612.
Stolk, W.A., Subramanian, S., Oortmarssen, G.J., Das, P.K., Habbema, J.D.F., 2003. Prospects for elimination of bancroftian filariasis by mass drug treatment in Pondicherry, India: a simulation study. J. Infect. Dis. 188, 1371–1381.

Stolk, W.A., Ten Bosch, Q.A., De Vlas, S.J., Fischer, P.U., Weil, G.J., Goldman, A.S., 2013. Modeling the impact and costs of semiannual mass drug administration for accelerated elimination of lymphatic filariasis. PLoS Negl. Trop. Dis. 7, e1984.

Stone, C.M., Lindsay, S.W., Chitnis, N., 2014. How effective is integrated vector management against malaria and lymphatic filariasis where the diseases are transmitted by the same vector? PLoS Negl. Trop. Dis. 8, e4493.

Subramanian, S., Krishnamoorthy, K., Ramaiah, K.D., Habbema, J.D.F., Das, P.K., Plaisier, A.P., 1998. The relationship between microfilarial load in the human host and uptake and development of *Wuchereria bancrofti* microfilariae by *Culex quinquefasciatus*: a study under natural conditions. Parasitology 116, 243–255.

Subramanian, S., Stolk, W.A., Ramaiah, K.D., Plaisier, A.P., Krishnamoorthy, K., Van Oortmarssen, G.J., Amalraj, D., Habbema, J.D.F., Das, P.K., 2004. The dynamics of *Wuchereria bancrofti* infection: a model-based analysis of longitudinal data from Pondicherry, India. Parasitology 128, 467–482.

Turner, H.C., Churcher, T.S., Walker, M., Osei-Atweneboana, M.Y., Prichard, R.K., Basáñez, M.G., 2013. Uncertainty surrounding projections of the long-term impact of ivermectin treatment on human onchocerciasis. PLoS Negl. Trop. Dis. 7, e2169.

WHO, 1990. Informal Consultation on Epidemiological Modelling for Research and Control of Lymphatic Filariasis. World Health Organization, Geneva.

WHO, 1996. Community directed treatment with ivermectin: report of a multi-country study. In: UNDP/World Bank/WHO Special Programme for Research and Training in Tropical Diseases (TDR), in Collaboration with Onchocerciasis Control Programme in West Africa (OCP) and the African Programme for Onchocerciasis Control (APOC).

WHO, 1997. Elimination of Lymphatic Filariasis as a Public Health Problem—Resolution of the Executive Board of the WHO (WHA50.29). Fiftieth World Health Assembly, Geneva, Switzerland.

WHO, 2011. Global programme to eliminate lymphatic filariasis: monitoring and epidemiological assessment of mass drug administration. In: A Manual for National Elimination Programmes. Geneva, Switzerland. World Health Organization.

WHO, 2013a. Global programme to eliminate lymphatic filariasis: progress report for 2012. Wkly. Epidemiol. Rec. 88, 389–399.

WHO, 2013b. Sustaining the Drive to Overcome the Global Impact of Neglected Tropical Diseases. Second WHO Report on Neglected Tropical Diseases. World Health Organization, Geneva, Switzerland.

Woolhouse, M.E., 1992. A theoretical framework for the immunoepidemiology of helminth infection. Parasite Immunol. 14, 563–578.

CHAPTER SEVEN

Modelling the Effects of Mass Drug Administration on the Molecular Epidemiology of Schistosomes

Poppy H.L. Lamberton*, Thomas Crellen*,§, James A. Cotton§, Joanne P. Webster*,¶,1

*Department of Infectious Disease Epidemiology, School of Public Health, Faculty of Medicine, Imperial College London, St Mary's Campus, London, UK
§Wellcome Trust Sanger Institute, Wellcome Trust Genome Campus, Cambridge, UK
¶Department of Pathology and Pathogen Biology, Centre for Emerging, Endemic and Exotic Diseases (CEEED), Royal Veterinary College, University of London, London, UK
[1]Corresponding author: E-mail: jowebster@rvc.ac.uk

Contents

Abstract

As national governments scale up mass drug administration (MDA) programs aimed to combat neglected tropical diseases (NTDs), novel selection pressures on these parasites increase. To understand how parasite populations are affected by MDA and how to maximize the success of control programmes, it is imperative for epidemiological,

ISSN 0065-308X
http://dx.doi.org/10.1016/bs.apar.2014.12.006

molecular and mathematical modelling approaches to be combined. Modelling of parasite population genetic and genomic structure, particularly of the NTDs, has been limited through the availability of only a few molecular markers to date. The landscape of infectious disease research is being dramatically reshaped by next-generation sequencing technologies and our understanding of how repeated selective pressures are shaping parasite populations is radically altering. Genomics can provide high-resolution data on parasite population structure, and identify how loci may contribute to key phenotypes such as virulence and/or drug resistance. We discuss the incorporation of genetic and genomic data, focussing on the recently sequenced *Schistosoma* spp., into novel mathematical transmission models to inform our understanding of the impact of MDA and other control methods. We summarize what is known to date, the models that exist and how population genetics has given us an understanding of the effects of MDA on the parasites. We consider how genetic and genomic data have the potential to shape future research, highlighting key areas where data are lacking, and how future molecular epidemiology knowledge can aid understanding of transmission dynamics and the effects of MDA, ultimately informing public health policy makers of the best interventions for NTDs.

1. INTRODUCTION

Mass drug administration (MDA) is the recommended strategy of the World Health Organization (WHO) to control or, in certain cases, eliminate a subgroup of neglected tropical diseases (NTDs) based on the annual distribution of inexpensive and/or donated drugs (Lammie et al., 2006; Webster et al., 2014). Through MDA, intensive and prolonged selective pressures are, and will be, placed on these parasites, which may have implications for the long-term success of campaigns (Webster et al., 2008, 2014). Policies to optimize success become crucial as programmes in selected countries shift from morbidity control toward elimination as a public health burden (WHO, 2012).

Mathematical models have the potential to explore a number of epidemiological features of both interest and importance in parasite populations (Levin et al., 1997; Levin, 1992; Anderson and May, 1991). Mathematical models can provide profound logistical, financial, temporal and biological benefits through, for instance, testing potential disease control strategies prior to final design and implementation (Rivers et al., 2014) or improving current control methods (Turner et al., 2014; Luz et al., 2011). However, valuable models require accurate parameterizations and a reliable understanding of the disease dynamics under investigation and how these can, and do, respond to differential selection pressures. Unfortunately, to date, few strong predictive mathematical models exist for the majority of metazoan parasites under MDA pressure. This is linked to a wider neglect in the study of helminth biology and genetics (Prugnolle et al., 2005a) in comparison to protozoan and bacterial agents such as malaria and tuberculosis (Hotez and Pecoul, 2010). In the case of several of the NTDs, the extent to which mathematical models may be developed and influence public health policy is limited by this absence of biological data on parasite population structure and genetic diversity. Substantial progress is occurring particularly with the recent establishment of the NTD modelling consortium (NTD_Modelling_Consortium). However, what remains to be addressed are the key data that these models require, particularly on the complex transmission dynamics of these often multi-host diseases.

The landscape of infectious disease biology is being radically restructured through the advancement of population genetic techniques, the development of next-generation sequencing (NGS) and the abundance of data generated from whole genome sequencing (WGS) (Luikart et al., 2003; Oleksyk et al., 2010; Vitti et al., 2013). Through applying NGS technology to parasite population studies, well-designed research can reap the benefits of sequence data, such as locating regions of the genome under selection (Valentim et al., 2013), and understanding levels and directions of gene flow and infections within and between subpopulations (Kao et al., 2014). Such information will boost our capacity to model and predict the effects of MDA on disease transmission dynamics and selection and how best to maximize control success (Kim et al., 2014).

Within infectious disease epidemiology, bacteriology has been one of the disciplines to benefit earliest from the genomic revolution due to the low cost of sequencing smaller bacterial genomes. Recent work on drug-resistant *Mycobacterium tuberculosis*, for instance, has used large datasets to identify key targets of selection (Farhat et al., 2013; Zhang et al., 2013), to show how antibiotic resistance spreads at different spatial scales (Harris et al., 2010), and quantifying how often resistant clones spread between animal and human populations (Mather et al., 2013). Applying WGS to eukaryotic parasites is complicated by the increased size and complexity of genomes. Population genomics approaches in protozoans such as *Plasmodium falciparum* (Cheeseman et al., 2012; Takala-Harrison et al., 2013) and *Leishmania donovani* (Downing et al., 2011), have been driven by the need to identify the locus or loci responsible for drug resistance (Bright and Winzeler, 2013). Within parasitology, malaria research has paved the way in WGS and NGS (Volkman et al., 2012) since the *P. falciparum* genome was published over a decade ago (Gardner et al., 2002). For example, a major locus responsible for artemisinin resistance in *P. falciparum* was identified using WGS of both a laboratory-induced resistant line and field isolates (Ariey et al., 2014), and clear-cut examples of natural selection associated with resistance to artemisinin combination therapies in Asia have come from analysis of 91 *P. falciparum* genomes (Cheeseman et al., 2012).

Despite the recent advances in WGS and NGS, the challenges of genomic analyses of metazoan species are, however, still great. The *Schistosoma mansoni* genome of 380 megabases (Mb) (Protasio et al., 2012) is an order of magnitude larger than the 23 Mb genome of *P. falciparum* (Gardner et al., 2002) and one order again than the 4 Mb genome of *M. tuberculosis* (Cole et al., 1998). Under the 50 Helminth Genomes Initiative of the Wellcome Trust Sanger Institute (WTSI) and other projects elsewhere (Blaxter et al., 2012), further helminth draft genomes are becoming available. While sequencing output is increasing, large genomes are fragmented (Parkhill, 2013) and stitching these genomes together into complete scaffolds took years (Holroyd and Sanchez-Flores, 2012). Nevertheless, changes in sequencing and mapping technologies are gradually making this process easier and more rapid (Chin et al., 2013; Dong et al., 2013), and the *Schistosoma* genome is starting to be used to study anthelmintic resistance. For example, WGS facilitated the finding of the locus responsible for oxamniquine resistance in *S. mansoni* by allowing genome-wide single nucleotide polymorphisms (SNPs) in the progeny of a genetic cross to be mapped against a drug resistance phenotype (Valentim et al., 2013).

As genomic data becomes available and costs of WGS reduce, questions on how to maximize the information from such data become pertinent. Here we consider what is known today about the interdisciplinary approaches of molecular epidemiology and mathematical modelling. We discuss the current and newly developing techniques available in population genetics and genomics and how these could, and should, be incorporated into novel mathematical predictive models, primarily using *Schistosoma* spp. as a case study. Whilst many excellent mathematical models already exist on schistosome epidemiology, beyond the scope of this current article, here we focus on those incorporating genetics and genomics. We specifically address what information is urgently needed to parameterize mathematical models aimed to help us elucidate: (1) key questions relating to the basic biological and transmission potential of schistosome populations under differing MDA pressures; (2) transmission cut-off points under repeated MDA, informing when treatments can be stopped and/or scaled down without population infection recrudescing; (3) the potential gauge of if, and when, drug resistance will arise and how, where and how fast it may spread; (4) the role of non-human zoonotic reservoirs in maintaining transmission under differential selective pressures; and (5) the potential for novel introgressed hybrid schistosomes to emerge and establish. We conclude by highlighting priority areas for future cross-disciplinary research in this emerging area and how findings can be translated into policy.

2. SCHISTOSOMES AS A 'MODEL' FOR ANTHELMINTIC MDA

Schistosomes, the causative agent of schistosomiasis, are one of the major NTDs targeted by MDA. *Schistosoma* spp. are indirectly transmitted dioecious trematode macroparasites, with an asexual stage in an intermediate host snail and a sexual stage within a definitive mammalian host. Schistosomiasis infects >240 million people with >750 million at risk, of which over 90% are within sub-Saharan Africa (SSA) (Steinmann et al., 2006). Two forms of human schistosomiasis are recognized – urogenital and intestinal. Urogenital schistosomiasis, from infection by *Schistosoma haematobium*, is associated with haematuria, bladder damage and a risk of progression to kidney failure and bladder cancer. More than 150,000 people die annually from *S. haematobium*-related kidney failure. Intestinal schistosomiasis is predominantly caused by *S. mansoni* in SSA and *Schistosoma japonicum* in Asian foci, and is associated with bloody diarrhoea, hepatosplenomegaly and liver failure (Gryseels et al., 2006). Schistosomiasis is primarily treated and controlled through MDA with praziquantel (PZQ) to school-age children in endemic districts. The Schistosomiasis Control Initiative (SCI), based at Imperial College London, has been instrumental in providing over 100 million PZQ treatments across parts of SSA from 2003 to 2014 (Fenwick et al., 2009; Webster et al., 2014). Vast quantities of the drug are donated through private–public partnerships including by pharmaceutical companies such as Merck-KGaA who have pledged to increase their donations and provide 250 million PZQ tablets annually for SSA by 2016. As treatment efforts are stepped-up, stronger selection pressures will be imposed on populations of *Schistosoma* spp. (Webster et al., 2008, 2014).

The genomes of the three major schistosome species infecting humans have been published – *S. mansoni* (Berriman et al., 2009), *S. haematobium* (Young et al., 2012) and *S. japonicum* (Schistosoma japonicum Genome and

Functional Analysis, 2009), and have been made accessible through SchistoDB (Zerlotini et al., 2013). Unpublished draft genome data for a further eight *Schistosoma* spp. are also now available from the WTSI (http://www.sanger.ac.uk/research/initiatives/globalhealth/research/helminthgenomes/). While these data have already been put to use in studies on functional (Protasio et al., 2013; Valentim et al., 2013) and comparative genomics (Tsai et al., 2013), work on population genomics is in its relative infancy.

Microsatellite studies have instead been the mainstay of recent *Schistosoma* evolutionary analysis, including population genetic studies examining potential changes to microsatellite diversity of *S. mansoni* before and after PZQ MDA. A key 2005–2006 study in Tanzania (Norton et al., 2010), for instance, revealed a significant reduction in the allelic richness of *S. mansoni* following PZQ, both within individuals treated and most notably also at the population level, as revealed by a similar reduction in genetic diversity observed in the youngest cohort of previously untreated children. Data from Senegal collected in 2007–2008 in a similar study used microsatellites to analyse miracidia collected from 12 children pre- and post-treatment/s. Those authors, however, found no significant changes to allelic richness or expected heterozygosity in *S. mansoni* before and after two rounds of treatment with PZQ amongst their small sample group (Huyse et al., 2013). In addition, no reduction in genetic diversity following 4 years of MDA was observed in parasites from children in Western Kenya (Lelo et al., 2014). Likewise, a microsatellite study in Brazil (Blanton et al., 2011) which compared the similarity of *S. mansoni* populations that survived PZQ treatment with susceptible worms also found no significant difference according to the differentiation index (Jost, 2008). Such inconsistencies could be due to differences in sample sizes used or due to true biological differences. Future molecular and mathematical modelling analyses would help elucidate this, by, for instance, estimating the minimum number of miracidia per individual required and/or the number of individual host samples required to provide robust estimates of genetic diversity (French et al., 2012). It should also be acknowledged that the number of microsatellites needed to accurately infer an effect varies greatly with the nature of the question, the scale of the analyses, the specific parameters to be estimated as well as the sample size. Any apparent absence of differentiation or change at neutral markers is not, however, necessarily indicative of whole genome processes (Allendorf et al., 2010). A study on Atlantic cod (Pampoulie et al., 2006), for instance, showed that very little genetic differentiation (F_{st}) was observed over nine microsatellite loci, but that substantial variation was taking place at the *PanI* locus, which is known to be under natural selection. The level of detail that genomics can now provide, particularly in terms of parasite population structure, transmission dynamics, and genetic diversity and how these are affected by MDA, as well as the more specific effects of treatment on drug resistance and other genetic traits such as virulence, will greatly aid parameterization of transmission models thereby helping policy makers formulate informed decisions on how to maximize control of these NTDs while minimizing the risk of development and/or spread of drug resistance.

2.1 Micro-evolutionary processes

Some of the most genetically explicit models for multi-host parasites focus on basic population biology (Prugnolle et al., 2005a; Criscione et al., 2005). Analyses of genetic variation in parasites at different hierarchical levels enables elucidation of parameters such as gene flow, effective population sizes and breeding units, all information relevant for understanding the potential rate of spread of important traits such as drug resistance (Anderson and May, 1991). Schistosomes have, however, several unique aspects of their biology, distinguishing them from other trematodes. Prugnolle and colleagues used an infinite island model to explore the alternation of sexual and asexual reproduction in monoecious trematodes on the partitioning of genetic variance among and within definitive hosts (Prugnolle et al., 2005a). Variation in reproductive success of clones was found to be important in shaping the distribution of the genetic variability both within and among definitive hosts (Prugnolle et al., 2005a, 2005b) and F_{is} (a measure of inbreeding) increased with higher levels of self-fertilization (Prugnolle et al., 2005a), limiting the scope for gene flow within these populations. Schistosomes are believed to have no or low levels of inbreeding (Basch and Basch, 1984; Prugnolle et al., 2005c; Huyse et al., 2009), and hence it might be predicted that F_{is} values would be lower for these organisms than for monoecious trematodes, with future models fitted to field molecular data helping to elucidate this.

The same group then parameterized a model for dioecious trematodes and compared their theoretical findings with empirical *S. mansoni* data from Guadeloupe (Prugnolle et al., 2005b). These models examined differential life-history traits such as sex-biased dispersal or clonal reproductive success (Prugnolle et al., 2002). Such characterizing of host and parasite population genetic structure and estimating gene flow among populations is essential for understanding coevolutionary interactions between hosts and parasites (Prugnolle et al., 2005c). Their models, however, assume nonoverlapping generations. Adult schistosomes can live for, on average 6–15 and up to 40 years within a human host, and as hosts are repeatedly exposed over their lifetime, generations of the adult schistosome populations can and do overlap. It is unknown what the effects of overlapping generations are with regards to population structure among hosts and if overlapping generations might impact the genetic effects of drug treatment. Inclusion of overlapping generations in mathematical models would undoubtedly increase the mathematical complexity (Prugnolle et al., 2005a), but may be in many cases the most biologically realistic scenario. In addition it is known that schistosome males can competitively mate (Tchuem Tchuenté et al., 1995; Tchuem Tchuenté et al., 1996; Webster et al., 2007) and therefore a male which survived treatment could mate with either a newly infecting female, a surviving juvenile female or a mature female which was either resistant to the drug, or which was protected from the drug by a dying male. New males from subsequent generations could also competitively mate with females from earlier generations, leading to pairs from these overlapping generations. Such competition among males will lead to an increased variance in male reproductive success. This could plausibly increase with drug treatment, in comparison to natural death and generation turnover, as the new cohorts of males infecting the hosts posttreatment may be fitter than those which have been exposed to the drugs, but could mate with previously exposed, but surviving females. The inclusion of overlapping generations in mathematical model design could thus have profound implications on the predicted spread of drug resistance via sexual reproduction. Work by Xu et al. (2006) has indeed demonstrated the importance of incorporating mating structure into model design, where he showed the potential maintenance of drug-resistant strains of schistosomes where generations overlap in comparison to simpler models without mating structure. Furthermore, such mating structure models suggested that multiple strains of drug-resistant parasites are likely to be favoured as the treatment rate increases (Xu et al., 2006). Models produced to date have also suggested that the likelihood that these resistant strains will increase in frequency also depend on the interplay between their relative fitness, the costs of resistance and the degree of selection pressure by the drug treatments (Feng et al., 2001).

2.2 Macro-evolutionary processes

Molecular models for macro-epidemiological processes directly relevant to public health policy makers are scarce, despite the fact that population genetics and epidemiology both extend basic biological processes at the individual level to the population level, and clearly come together to model drug resistance (Levin et al., 1997). The difference in the spread of drug-resistant genes through a parasite population to that of drug-resistant parasites through a host population is that drug-resistant alleles replace sensitive alleles depending on their relative fitness, whilst drug-resistant parasite densities increase according to their absolute fitness (Paterson and Viney, 2000). This difference is of greater importance in complex parasite systems where high levels of genetic diversity are required to complete the life cycle, such as schistosomes (Rollinson et al., 2009), with stronger constraining selective pressures acting on the parasites at different life cycle stages than, for example, directly transmitted viral pathogens. Because the rate of increase of a parasite population (and therefore the density of the parasite population) depends on R_0, the epidemiology of anthelmintic resistance cannot be determined without using a model which incorporates the underlying genetics of resistance (Paterson and Viney, 2000). Smith and colleagues used estimates of the absolute fitness of different parasite genotypes within an epidemiological framework to model the effects of under dosing, treatment strategies and mating probabilities on anthelmintic resistance (Smith et al., 1999). Such a model demonstrates how population genetics can help build theoretical models of infectious disease to understand patterns of transmission in the field.

3. THE ROLE OF GENETICS AND GENOMICS IN MATHEMATICAL MODELS

3.1 Elucidate the basic biology and transmission potential of schistosomes

3.1.1 Population structure

In many countries the majority of MDA programmes, particularly for schistosomiasis and soil-transmitted helminths (STH), target school-aged children. The effects that such treatment strategies have on parasite transmission dynamics, as well as the potential selection pressures, depend on a range of factors. These include at least three directly relating to *refugia* (the proportion of the parasite population not exposed to the drug): (1) the proportion of school-aged children who attend school and therefore receive treatment, (2) the proportion of the population that are school aged and (3) the proportion of the total parasite population harboured by this targeted treatment group. Additional key factors include drug efficacy, life-history costs/trade-offs of resistance and parasite population diversity, structure and transmission dynamics, such as how different age groups are exposed to eggs or infective larvae produced by these school-aged children and *vice versa* (Anderson et al., 2013). These factors affect ongoing transmission and reinfection of treated, and untreated, individuals and influence the potential development, rate, and spread, of drug-resistant strains.

Models can aid understanding about potential benefits that school-based MDA may have on untreated groups. For example, if parasite population structure, elucidated through genetics and/or genomics, indicates that children and adults were found to be infected from a similar genetic pool of parasites (i.e. no genetic differentiation of parasites between children and adults), then treating children may potentially decrease infection in the untreated adult communities, recently termed the 'herd impact' of a treatment programme (Anderson et al., 2013). Conversely, if parasites in children only tend to circulate within their age groups, then infection intensities in adults will be unaffected by only treating the children.

Although measuring transmission rates in multi-host systems is difficult, data on genetic variation within an individual- and population-level hierarchy can also enable measurements of genetic structure and associated rates of gene flow (Weir and Cockerham, 1984). Genomic and gene flow data can likewise be used to build transmission trees, including those aimed to identify, for certain microparasites at least, the origins of any new infections (Kao et al., 2014). While these approaches may have lower resolution in metazoan parasites, due to their slower rates of molecular evolution, multilocus and genomic approaches will allow us to approximate intraspecific phylogenies and elucidate transmission between hosts, as has been demonstrated for *Ascaris* (Criscione et al., 2010). Such data become vitally important as control programmes move towards elimination, highlighting key individuals or subpopulations driving reinfections. Examples of where genetic data have already informed public health policy include a study on *Ascaris* in pigs and humans in Guatemala, where even though infections were sympatric, there appeared to be, using the molecular tools available at the time, little gene flow between the parasite populations indicating no transmission between the two host species (Anderson et al., 1993). Conversely, two other more recent studies have indicated cross transmission between the *Ascaris lumbricoides* and *Ascaris suum* species, with, furthermore, up to 4% and 7% of *Ascaris* appearing as hybrids, which raises a number of potential implications for long-term evolutionary dynamics (Criscione et al., 2007). Models have already been used in directly transmitted pathogens, including sexually transmitted diseases, where contact tracing data may not be complete, but where genome data can inform on infection networks (e.g. HIV transmission in a dental practice (Ou et al., 1992)) as well as theoretical pathogen models using evolutionary trees resulting from different evolutionary processes (Nee et al., 1994). In micro-parasites with rapid in-host evolution, sensitivity of phylogenetic-based networks may be reduced (Resik et al., 2007). Similar lack of contact tracing exists for indirectly transmitted pathogens such as STHs and schistosomes and microparasite models incorporating similar genomic data may be developed for more complex indirect transmission networks (Gupta et al., 1996), although these may also be highly dependent on host immunological factors (Anderson et al., 1989).

At present mathematical models of indirectly transmitted parasites often assume that exposure to eggs or larvae across all age groups is random and independent of the relative contribution of infective stages from each age group and maintained transmission. However, the spatial structure of concomitant parasite transmission between age groups is unlikely to be random and even less so with MDA programmes targeting specific groups of individuals. Therefore, models should incorporate such heterogeneous mixing (Chan et al., 1994). An additional complication is that transmission dynamics and population biology are likely to change if MDA reduces parasite transmission significantly (Klepac et al., 2013). Models for such organisms, with the exception of (Gurarie and King, 2005), also assume that hosts are exposed to a common source of infection. However, even on a very local scale, parasites have been shown, using population genetics, to have focal transmission (Criscione et al., 2010) independent of the among age-group population structures, negating some of the classic models for parasite transmission.

Above, we have discussed how population genetics can inform on specific treatment scenarios, and be used to predict gene flow between certain individual hosts, or host groups. On a larger and more basic scale, population genetics can inform on the very basis of what types of models should be used for certain diseases. At present some network contact models estimate transmission patterns purely from host behaviour. The use of population genetics and genomics enables elucidation of contact network structures, even at the most basic level. Questions such as whether transmission is random or nonrandom, and therefore which model is appropriate to fit the data, can be answered, as has already been initiated for *A. lumbricoides* in Nepal (Criscione et al., 2010). This is particularly important for multi-host parasites where several transmission mechanisms could result in a similar profile. To date, many of the theoretical challenges faced by epidemiologists and population geneticists were problems of scale, whilst the use of population genomics greatly diminishes these challenges.

3.1.2 Transmission rates

Changes in the transmission rate of macroparasites may be inferred from longitudinal changes in descriptive statistics (prevalence, infection intensity) or modelled through the force of infection (FOI: the rate at which human hosts acquire parasites). Uganda, for instance, has been treating individuals with PZQ since 2003 with significant reductions in prevalence, infection intensity and FOI after only three rounds of MDA in low-, moderate- and high-infection areas in both treated and untreated children (French et al., 2010). However, Basáñez and colleagues argue that presently there are very few models with the potential to inform on optimal methods at a clinical or epidemiological level to monitor such changes and they outline a number of areas for future model development. These include 'the design of treatment efficacy and effectiveness studies; phenotypic characterisation of responses to treatment; and design of sampling protocols for the study of parasite genetic structure under treatment, thereby facilitating prompt detection of anthelmintic resistance' (Basáñez et al., 2012). Examples of modelling work already carried out in this field include studies examining the effect of density-dependent forces known to act on parasites generally, including schistosomes (Medley and Anderson, 1985) and the effect of human and parasite sample sizes on measuring reductions in genetic diversity in *S. mansoni* infections post-treatment (French et al., 2012). Both of these factors are vitally important, particularly as control programmes progress and infection intensities reduce.

Genomic techniques can be employed to gain a better understanding of the nature of transmission in endemic settings, including those under differing MDA pressures (Volkman et al., 2012). Within a parasite population, the degree of outbreeding and recombination is linked to the proportion of people harbouring parasites of variable genotypes; this proportion is known to scale with the level of transmission (Anderson et al., 2000). A study on the Thai–Burmese border where transmission of *P. falciparum* has declined from 2000 to 2010 associated a loss of heterozygosity across 96 SNPs with the sampling year (Nkhoma et al., 2013) using logistic regression, therefore multiple genotype infections had significantly declined over time in line with falling transmission. A study in Senegal found a similar reduction in the heterozygosity of *P. falciparum* populations as a consequence of public health interventions (Daniels et al., 2013). Nkhoma and colleagues argue that the linear relationship between the carriage of multiple parasite genotypes within human hosts and transmission intensity means that population genetics can be used as a reliable and inexpensive method of tracking temporal changes in transmission intensity (Nkhoma et al., 2013). Despite very different life cycles, the effects seen in *P. falciparum* are likely to be replicated in *Schistosoma* spp. In both parasites, the total populations are subdivided within each definitive host (*Anopheles* in the case of *P. falciparum* and humans in the case of *Schistosoma* spp.), meaning that sexual reproduction for any individual parasite is possible only with a small fraction of organisms from the total population. This presents barriers to gene flow and outbreeding and so the reductions in heterozygosity are enhanced in both species. It may be expected, however, to occur more slowly in schistosomes, due to a longer generation time, the relatively higher intensities of infections in humans than in the mosquitoes and the different timing of sexual reproduction in the two life cycles.

3.1.3 Spatial heterogeneity

Modelling spatial and demographic heterogeneity with the aim of understanding fundamental processes underlying infection dynamics provides a framework for evaluating potential control strategies for infectious diseases (Paterson and Viney, 2000). Using well-documented and commonly modelled systems, such as measles, social and geographical structure in contact networks have been deduced with heterogeneity in transmission rates within families, between children at school and between communities (Keeling et al., 1997). Models for macroparasites may also highlight key social and geographical groups where transmission is high. However, in indirect life cycles, this is hard to do from host behaviour and contact monitoring. Parasite population genetics and genomics can ultimately inform on transmission structure. Heterogeneity in infection patterns are biological realities and must be incorporated into models (Paterson and Viney, 2000) improving their fit to empirical data. Kao et al. (2014) reviewed the use of WGS in contact tracing models to reveal points of control and predict the direction of the spread of diseases for microparasites (Kao et al., 2014). They discuss the complexities associated with inferring the epidemiological dynamics of multi-host pathogens, as is often the case for NTDs. Furthermore, these authors explore the difficulties of when microparasite mutation rates are low in comparison to generation times, but that such situations may be resolved using within-host genetic variation to infer properties of between-host transmission (Stack et al., 2013). While these methods may be less powerful in metazoan parasites, whole genome data will certainly reveal significant details of the spatial genetic structure for these organisms (Archie et al., 2009).

3.2 Evaluate transmission cut-off points and estimate when to stop MDA

Population genetic mathematical models can also inform on cut-off points for MDA through predicting the levels of infection where disease transmission should not recrudesce (Plaisier et al., 1997; Turner et al., 2013). This has been demonstrated for the NTD trachoma (*Chlamydia trachomatis*) in communities where MDA of antibiotic eye ointment are used (Lietman et al., 2011). This model included estimates of transmission parameters relating to reinfection from both within or outside the community. The values were robustly estimated from prevalence data at baseline and 24 month follow up. Nevertheless, future additional data on the level of gene flow over time and space and phylogenies of the infections appearing following treatment would enable control programmes to know the true influence of recrudescence and reinfection from within or outside the community. Such data would also inform on the distance that reinfection can occur from. This would aid policy makers in their decisions on when they can halt treatment in a central focus, depending on threshold infection levels in a range of surrounding foci. Population genomic data could also parameterize smaller spatial household models, identifying the extent of gene flow within and between infected subpopulations (Blake et al., 2009). Similar transmission potential parameters could be incorporated into other vector-borne/intermediate host disease elimination models, such as for *S. haematobium* on Zanzibar, for example, where they are aiming towards elimination (Rollinson et al., 2013). Of particular value would be the use of genetic and genomic data to understand if any detected cases remaining or reappearing were from local transmission or imported cases from the mainland. Such genomic data have already enabled reconstruction of transmission trees for directly transmitted non-NTDs, for example, understanding cross-species transmission in *Salmonella* (Mather et al., 2013).

Parasite diversity may also strongly affect transmission dynamics, reinfection rates posttreatment and associated thresholds for ending MDA. Strain diversity in trachoma, for instance, is known to affect reinfection levels, as heterogeneity exists in part to evade the human immune system. If many strains become eliminated, then the remaining ones may not be able to reach pretreatment levels due to independent factors such as strain-specific host immunity (Zhang et al., 2004). Incorporating diversity into models predicting MDA cut-off points, may thereby not only greatly enhance their accuracy, but also aid in the understanding of such interventions on the transmission dynamics of the infectious agent, as has recently demonstrated for dengue (Coudeville and Garnett, 2012).

3.3 Elucidate the potential evolution and spread of PZQ resistance

Early detection of anthelmintic resistance is vital for controlling the spread of such genotypes. When (Churcher and Basáñez, 2009) and how (French et al., 2012) best to examine human helminth parasite populations posttreatment, have recently been the subject of model-based studies. If PZQ resistance is recessive, drug-resistant alleles could spread through the population to relatively high levels before phenotypic manifestation (Churcher and Basáñez, 2009). There is also evidence for negative density-dependent fecundity in *S. mansoni* (Medley and Anderson, 1985) and models on helminth life cycles show that density-dependent fecundity, in comparison to that in parasite establishment or mortality (Churcher and Basáñez, 2008) may facilitate the spread of resistance as parasite population intensities decrease with treatment. One of the most important aspects of mathematical models is their direct use to public health policy makers. A recent model based on antibiotic resistance and drug use and how to communicate the extent of problems to policy makers (Laxminarayan and Klugman, 2011) could be adapted for other infections to maximize the speed of implanting changes in control strategies should they be required.

Some modelling work to date has taken into account the life cycle of schistosomiasis by using a time delay function and found that the presence of this delay, i.e. the average time between two adult generations, makes it more likely for resistant strains, for example, to invade and persist in a parasite population (Castillo-Chavez et al., 2008). Furthermore, population genetic analysis of the structure of *S. mansoni* and *S. haematobium* diversity across Africa using microsatellite markers, found that, on the basis of population structuring and high genetic diversity, should drug resistance evolve it would be slow to spread through schistosome populations, at least across large scales (Gower et al., 2013), as there were low levels of gene flow observed between samples from different countries (although high levels of gene flow between samples within countries). As with many parasite species, the complex nature of the schistosome life cycle violates assumptions of the Wright–Fisher model by: 1) having highly overlapping generations due to the long reproductive lifespan of adult schistosomes; and 2) non-random mating due to focal transmission and the inability to mate with schistosomes in other hosts. These effects make it difficult to make use of genetic variation at neutral sites to understand population size and structure (Balloux and Lehmann, 2012) and to predict how effective natural selection will be in increasing the frequency of positively selected alleles in the face of random drift or constraining selective pressures elsewhere in the complex life cycle. A population genetic model recently developed for malaria, which aids understanding of the emergence of resistance and its early spread (Kim et al., 2014), in this instance under combination drug therapy (something which is not currently available for schistosome treatment), will likely act as a key foundation for other indirectly transmitted parasite genetic models in the near future.

Recent mathematical models on large-scale PZQ administration have shown that the FOI of *S. mansoni* is reduced throughout communities; even in untreated PZQ naïve individuals, as control programmes progress (French et al., 2010). Such an FOI, however, is calculated from humans, through snails and back into humans, with little knowledge on the relative force between individual life cycle stages. In *S. japonicum*, an Asian species which has multiple definitive hosts, the driving FOI is the transmission from snail to mammal (Riley et al., 2008), with different mammalian hosts maintaining transmission in different geographical regions (Lu et al., 2010; Rudge et al., 2013). In an *S. mansoni* focus in Guadeloupe, similar molecular epidemiology studies have indicated that parasite migration is primarily driven by the rodent hosts (Prugnolle et al., 2005c). Little, however, is known about the relative forces and spatial heterogeneities in *S. mansoni* and *S. haematobium* between subpopulations of parasites in humans or on the effect of PZQ. Future work needs to expand current FOI models, to include drug-resistance parameters, and determine the driving forces of continued transmission despite repeated drug treatments. As schistosome control programmes progress, empirical and theoretical data warrant further research on PZQ efficacy, and how best to monitor and evaluate disease transmission, to maximize the lifespan of PZQ.

As some countries or regions in countries push towards elimination (Rollinson et al., 2013), biannual treatment with PZQ has been a suggested strategy to further reduce the FOI and halt transmission. One key area that mathematical models should evaluate is the effect of biannual MDA on the potential increased selection for resistance, versus the rate of reducing the FOI (French et al., 2010). Models should be fitted to known clearance data post-treatment to elucidate the relative contribution of non-clearance versus reinfection in maintaining high-infection intensities. These models should then incorporate the currently unknown parameters for PZQ resistance, starting with a simple single locus model and progressing to multilocus genotypes. This would inform policy makers on the potential risks of increasing treatment frequencies, versus the potential benefits of reducing FOI, for a range of scenarios for potential genotypic resistance to the drug.

In the absence of any current molecular markers for monitoring potential PZQ resistance, or even a full understanding of the molecular mechanisms of PZQ action (Chan et al., 2013), phenotypic tests have been trialled with some success in both the laboratory and the field (Liang et al., 2001; Lamberton et al., 2010). Accurate phenotypic measures of drug efficacy are a vital requirement for comparison of genomic sequences to locate potential genes and/or regions associated with drug resistance. Complexities arise with several chronic macroparasite infections as methods for phenotypically detecting resistance are often not properly standardized. In schistosomes, the WHO designated phenotype for drug tolerance is lowered 'cure rates' and/or 'egg reduction rates' as measured by Kato-Katz thick smears (Katz et al., 1972) or urine filtration, depending on the species, before and 14–21 days following treatment (WHO, 2013). Limitations in the sensitivity of such diagnostic tests (Lamberton et al., 2014) may restrict the inferences that can be drawn from any association testing between polymorphisms and reduced treatment efficacy. An important role for modelling will be to account for the limitations in diagnostic accuracy, which is increasingly performed through latent class analysis (Koukounari et al., 2013; Assefa et al., 2014). The most recent WHO manual for evaluating anthelmintic drug efficacy 'tentatively' places the egg reduction rate at approximately 90% for the three main *Schistosoma* species infecting humans when individuals are treated with 40 mg/kg of PZQ (WHO, 2013). On the basis of these criteria, treatment failures with PZQ, although rare, have been reported (Greenberg, 2013) with systematic non-clearers often observed in the field. Warning also comes from veterinary parasitology with anthelmintic resistance an inevitable consequence of mass anthelmintic treatment, with parasites in some regions being resistant to all major drug classes, leading to total anthelmintic failure (Kaplan and Vidyashankar, 2012; Webster et al., 2014), and threatening the profitability of whole sheep farming industries in Australia (Wolstenholme et al., 2004).

Care must, however, be taken in using genetic data from populations of larval, or egg, stages as a proxy for adult worm intensity or reproduction (Criscione and Blouin, 2005), or to measure transmission between hosts (Steinauer et al., 2010, 2013). Potential solutions to such issues have involved, for example, estimating the adult population sizes through kinship analysis to either partition miracidia into sibships or assign miracidia to parents of the parasites under investigation (Blouin, 2003; Jones and Ardren, 2003; Criscione and Blouin, 2005; Steinauer et al., 2013). This has also enabled the accurate incorporation of density-dependent factors into models by drawing comparisons between adult population sizes and egg counts at the individual level. Effective population size (N_e) is an important parameter in evolutionary biology because it quantifies genetic drift, and crucially with regard to MDA and potential drug resistance, the response to selection (Criscione and Blouin, 2005). N_e has a large influence on the overall level of genetic diversity in populations and selection for drug-resistance alleles might be more efficient in parasite populations with a large $N_{e.}$ The complex life cycles of many NTDs affect the type of model chosen to estimate N_e (Balloux and Lehmann, 2012) and make it more challenging to collect samples from appropriate life cycle stages to infer genetic estimates, and to collect the necessary demographic data, such as generation time, that may be needed to augment these genetic estimates. For example, in many NTDs, such as schistosomes, the eggs from sexual reproduction are passed into the external environment, so that offspring from different infrapopulations are mixed every generation. Criscione and Blouin use a model which subdivides breeders into infrapopulations, nested within a component population, to demonstrate basic demographic factors that control N_e in macroparasite species (Criscione and Blouin, 2005). They incorporate incomplete mixing, which increases reproductive success of some infrapopulations and discuss the effects of aggregation and crowding on per capita fecundity, both aspects vitally important in future models on the potential rate of spread of drug resistance. They also demonstrate a pronounced sex ratio effect on N_e due to separation of individuals among hosts. Such a model would be greatly enhanced by knowledge gained from kinship analysis through population genetics or genomics, briefly mentioned above (Blouin, 2003; Jones and Ardren, 2003), which would inform on reproductive success and density-dependent factors enabling the accurate estimation on the effective number of breeders in each infrapopulation, which could then be incorporated into models and would be even more important after

MDA when density-dependent pressures may be reduced on potentially resistant strains.

Apart from details of the population structure and reproductive biology of the parasite population, the speed and extent of spread of drug-resistance alleles will depend on the detailed genetic architecture of drug resistance. Factors include how many loci are involved, how loci interact in determining resistance phenotypes, whether and to what extent the loci are genetically linked and whether resistant alleles are dominant, recessive or something in-between, the relative fitness of resistant genotypes in the presence of drug treatment, the cost of resistance (reduced fitness in the absence of drug treatment) and the degree of selection pressure exerted by treatment on different populations. Basáñez et al. (2012) discuss models capturing some or all of these factors for some parasite species, including a number of schistosome-specific models, but nothing is known about loci underlying clinical failure of PZQ treatment or PZQ resistance, let alone any information to parameterize the other factors required for accurate, model-based prediction of the potential for drug resistance.

3.3.1 Using population genomics to understand the mode of PZQ resistance

Circumstantial evidence suggests that PZQ acts on voltage-gated Ca^{2+} channels (Doenhoff et al., 2008; Chan et al., 2013). While these insights are important, whole genome approaches enable an unbiased approach to understanding the genetics of drug resistance, avoiding potential biases from prior assumptions about mechanisms of action and/or resistance embodied in the choice of 'candidate genes' for more targeted genotyping approaches. Traditional 'forward' genetics approaches based on analysis of a genetic cross are possible in *Schistosoma* (Criscione et al., 2009) and have been successfully applied to understanding the genetics of drug resistance to oxamniquine in one isolate (Valentim et al., 2013). Reverse genetic approaches such as transgenesis and RNAi are also available, or under active development for *Schistosoma*, and while not yet sufficiently reliable for high throughput for 'discovery' of loci underlying resistance phenotypes, such techniques may be valuable in confirming the importance of candidate loci discovered by other approaches. More fundamentally, these kinds of laboratory genetics approaches allow detailed investigation of causal loci in individual resistant isolates, but unless diverse isolates can be looked at — and relatively few schistosomes isolates are now maintained in laboratories, and once introduced from the field rapid genetic bottlenecks occur

(Gower et al., 2007) – these loci may not be responsible for resistance observed in clinical conditions.

A complementary approach is to investigate genetic variation in natural populations. A conceptually simple approach is to look directly for genetic variation that is associated with a quantifiable phenotype, such as drug resistance (Borrmann et al., 2013; Takala-Harrison et al., 2013; Zhang et al., 2013). The key difficulties for using such approaches in schistosomes at present lie in the lack of a diagnostic test for reduced PZQ efficacy, in part due to the complexity of the parasite genome and life cycle and likely polygenic basis for such resistance. An alternative is to look for the impact of selection caused by drug treatment using genome-wide sequence data, an approach that implicitly assumes that chemotherapy is one of the most important selective forces on these genomes. Methods to detect recent selection are looking for the signals of a selective sweep (Smith and Haigh, 1974), in which a polymorphism under selection increases in frequency rapidly in a population, and the speed of this spread is sufficiently fast that an extended contiguous portion of a chromosome will spread through the population alongside the variant under selection, as there has not been enough time for this unit (haplotype) to be broken up through recombination. Different tests look for different patterns in the distribution of genetic variation across the genome and within populations generated by this process – these tests have been reviewed a number of times recently (Volkman et al., 2012; Oleksyk et al., 2010; Vitti et al., 2013). Examples of the patterns are enhanced linkage disequilibrium and depleted polymorphisms around a region, using tests such as the long-range haplotype test, which in malaria, for example, have detected drug-resistant loci for chloroquine (Wootton et al., 2002) and pyrimiethamine (Nair et al., 2003), or differences in allele frequencies in populations under selection and those without selection. More sophisticated statistical tests build on these approaches e.g. Cross-population Extended Haplotype Homozygosity (XP-EHH) (Sabeti et al., 2007; Grossman et al., 2010). These tests are often performed from in vitro cultures for drug-resistance markers, with linkage mapping using laboratory crosses to correlate segregation patterns in the progeny that are associated with the drug-resistant phenotype (Ferdig et al., 2004; Hayton and Su, 2008; Sanchez et al., 2011).

Knowledge on the effect of MDA on parasite genotypes, such as virulence and drug resistance, and their potential costs, are required to understand the risk of development and then spread. Understanding gene flow between subpopulations also informs on levels of *refugia* as touched upon previously, which is a key strategy for minimizing the risk of drug resistance developing. It is imperative that we understand the complex interactions between minimizing selective pressures, but treating the most heavily infected individuals, by maintaining an optimum level of *refugia* to maximize the short- and long-term control of these NTDs.

Models developed for the demographics of human and malaria movement (Pindolia et al., 2013) and the potential spread of drug-resistant malaria (Anderson, 2009; Lynch and Roper, 2011) indicate the type of predictive maps that can be produced for indirectly transmitted pathogens on regional (Pearce et al., 2009), national (Pindolia et al., 2013) and international (Lynch and Roper, 2011) scales. Given the often transitory migration lifestyles of many fishermen and their families in areas of high *Schistosoma* spp. endemicity, monitoring human movement, potentially through national migration statistics could be extrapolated to these metazoan parasite species.

3.4 Elucidate the potential role of non-human definitive hosts

Zoonotic reservoirs affect the transmission dynamics of a disease and can limit the extent to which a pathogen can be controlled or potentially eliminated (Taylor et al., 2001). Work on the zoonotic *S. japonicum* offers an example of how interdisciplinary research with population genetics and modelling can be done successfully. *S. japonicum* is unique among schistosome species in that it can infect over 40 mammalian hosts and with zoonotic transmission an important factor in its epidemiology (He et al., 2001). Despite a concerted public health effort over many decades, the Chinese government has been unable to eradicate *S. japonicum* (Wang et al., 2008). Population genetic analyses found close phylogenetic relationships between strains of *S. japonicum* in human and rodent hosts in hilly parts of Anhui Province, China using microsatellite markers, whilst schistosomes in the humans in the marshland areas were more closely related to those in bovines, which appear to drive reinfection in those areas (Rudge et al., 2009). Bovines were generally assumed to be the primary reservoir of human infection (Mcmanus et al., 2010), but population genetic analyses greatly increased our understanding of these complex interactions in zoonotic parasites. Knowledge of the genetic similarity and transmission potential between hosts was then used to develop a multi-host model of transmission for *S. japonicum* (Rudge et al., 2013). This showed that rodents were the only hosts with a basic reproductive number (R_0) > 1 in hilly regions of Anhui, China driving transmission and reinfection in treated humans. This explained the re-emergence of *S. japonicum* infection in some mountainous areas of China where the disease was thought to have been eliminated. In marshland areas, where bovines $R_0 > 1$, an earlier government organized cull of cattle may have aided transmission control, but similar extreme methods would not have helped in the hilly regions. Rudge and colleagues used a prevalence model framework which predicted that a reduction in rodent density by around 20% would lead to a 40–50% reduction in *S. japonicum* incidence in humans. This integration of techniques helps inform public health policy makers on how best to maximize control, through giving fixed targets which need to be reached to bring the overall R_0 below 1 and to halt the re-emergence of the disease. Without such population genetic analyses of the parasite strains circulating in these regions, the source of human reinfection after these mass treatment campaigns would have remained misunderstood, dramatically reducing the cost benefit of specific control methods in these regions. Conversely, controlling infections in these host species may potentially induce atypical *refugia* populations, such as wild animal reservoirs to become more epidemiologically important (Bockarie et al., 2013; Webster et al., 2014).

3.5 Elucidate the potential role of hybridization and introgression

Hybridization events can have important evolutionary outcomes on species and populations and may improve the fitness of resulting hybrids though the acquisition of adaptive traits. Schistosomatoidae are capable of trans-species mating within the definitive host. Schistosomes are known to hybridize frequently (Steinauer et al., 2010), and reports from Senegal have identified hybridization between *S. haematobium/S. mansoni* (Huyse et al., 2013), *S. haematobium/Schistosoma curassoni* and *S. haematobium/Schistosoma bovis* (Webster et al., 2013) and between *S. mansoni/Schistosoma rodhaini* in western Kenya (Steinauer et al., 2008). The implications of these events for MDA programs have yet to be established, but hybridization may have implications for disease morbidity and drug tolerance and therefore represent a public health concern.

Continued high transmission of *S. haematobium* despite often high MDA coverage have recently been molecularly identified as being due to, at least in part, hybrid infections with *S. bovis* (Huyse et al., 2009; Webster et al., 2013). Since these relatively new PZQ selective pressures have been imposed, any endemic equilibrium of coinfections and hybrids may now be changing. Whether or not these hybrids will maintain transmission and/or evolve into a potentially drug-resistant parasite species remains to be elucidated, however, mathematical models for *S. haematobium* and *Schistosoma mattheei* indicate that such evolution is highly unlikely (Kruger, 1990), although this was prior to the introduction of MDA and did not take these extra selective pressures into account. Successful modelling work in the future will require both a better knowledge of the transmission dynamics of schistosomes under MDA pressure, taking into account factors such as hybridization and zoonotic hosts, along with a better understanding of the genetic loci under positive selection for PZQ resistance.

Although not based on parasitic life cycles, models on the origin of species by sympatric speciation (Dieckmann and Doebeli, 1999), with explicit description of genetic determinism, may act as starting points for the development of models which incorporate hybridization and/or drug-resistant alleles. Dieckmann and Doebeli (1999) show that sympatric speciation is a likely outcome of competition for resources, where individuals mate preferentially with like individuals, a situation exacerbated in infections where parasites surviving treatment are more likely to mate with others also surviving either due to potentially harbouring resistant alleles, or increased resistance through hybrid vigour. Such scenarios could theoretically lead to increased drug resistance despite coinfections with susceptible parasites, and may ultimately lead to isolation of drug-resistant strains within individuals harbouring only those parasites which survive treatment.

Research into adaptive evolution may also act as a useful springboard for future models discussed here. Models on plant parasites, for instance, have indicated that evolutionary divergence of parasite phenotypes can be driven by seasonal transmission and associated fitness trade-offs (Hamelin et al., 2011), such as may also occur with annual treatments and reduced density-dependent factors. One such recent model describes how competition explains intra-host diversification of parasites (Rascalou and Gourbiere, 2014). They show that parasite adaptive evolution is faster in highly fragmented parasite populations and for weakly aggregated and virulent parasites, all factors which could be affected by drug-selective pressures, hybridizations and associated trade-offs.

Genomics has the potential to shine a light on these discoveries to elucidate the timing and exact nature of these hybridization events. An exemplary study on the plant pathogenic fungus *Zymoseptoria pseudotritici* (Stukenbrock et al., 2012) used the nuclear genomes of five individuals to investigate a recent hybridization event. This species shows an unusual pattern of genome-wide diversity, with the genome broken up into small (5.8 kb) stretches of very high haplotype diversity interspersed between equally long sections showing no differentiation within the population. This mosaic pattern of haplotypes is indicative of a recent hybridization event followed by a population bottleneck. Estimates of recombination rate and mutation rate in this species allowed two different, approximately congruent estimates of the timing of this hybridization event, from data on the size of the haplotype blocks and the number of point mutations occurring in the genome. It is important to note that both understanding the pattern of hybridization, and dating the founding event for these hybrids was only possible through a genome-level analysis: if only individual genes had been isolated, then the global picture that this study has unearthed would have been lost. Through applying these kinds of methods to WGS data from hybrid populations it should be possible to gain a more complete picture of the nature and timing of introgression between *Schistosoma* species. Understanding how often new populations of hybrids are founded, how long they persist and how their population genetics differs from parental populations will be important in understanding how these populations are likely to respond to MDA and the scope for PZQ resistance to spread within and between populations.

4. CONCLUSIONS

Accurately parameterized mathematical models incorporating genetic or genomic data can, for example, inform on rates of changes of schistosome phenotypes or genotypes associated with drug resistance, so that monitoring and evaluation studies can understand what they need to monitor, how and when is best to monitor, and advise on optimal treatment strategies to maximize the gains from limited resources. Although the cost of NGS has fallen exponentially in recent years, the outlay of sequencing all 380 Mb of the schistosome genome remains currently beyond the financial capacity of most institutions, although RAD-seq or exon-capture may represent a more cost-effective alternative (Gilabert and Wasmuth, 2013). When dealing with a disease that affects the very poorest in the world, the most useful public health interventions are those which cost the least and are broadly sustainable; therefore it is important that this technology is targeted to answer the most relevant public health questions. These NGS technologies, to complement population genetic data available from methodologies such as microsatellites, can thus be combined with mathematical models, and we anticipate that the next few years will represent a highly exciting and important era in such trans-disciplinary research of profound theoretical and applied importance.

ACKNOWLEDGMENTS

We are very grateful to two anonymous referees for comments on the manuscript. PHLL is funded through an Imperial College Junior Research Fellowship; TC is funded through a Medical Research Council (MRC DPT PhD) studentship (supervised by JPW and JAC); JAC is funded by the Wellcome Trust through their support of the Wellcome Trust Sanger Institute (Grant 098051); JPW holds a Chair in Parasitic Diseases. RVC manuscript reference number PPB_00915.

REFERENCES

Allendorf, F.W., Hohenlohe, P.A., Luikart, G., 2010. Genomics and the future of conservation genetics. Nat. Rev. Genet. 11, 697–709.

Anderson, R.M., May, R.M., 1991. Infectious Diseases of Humans: Dynamics and Control. Oxford University Press, New York.

Anderson, R.M., May, R.M., Gupta, S., 1989. Non-linear phenomena in host-parasite interactions. Parasitology 99 (Suppl.), S59–S79.

Anderson, R.M., Truscott, J.E., Pullan, R.L., Brooker, S.J., Hollingsworth, T.D., 2013. How effective is school-based deworming for the community-wide control of soil-transmitted helminths? PLoS Negl. Trop. Dis. 7, e2027.

Anderson, T., 2009. Mapping the spread of malaria drug resistance. PLoS Med. 6, e1000054.

Anderson, T.J., Haubold, B., Williams, J.T., Estrada-Franco, J.G., Richardson, L., Mollinedo, R., Bockarie, M., Mokili, J., Mharakurwa, S., French, N., Whitworth, J., Velez, I.D., Brockman, A.H., Nosten, F., Ferreira, M.U., Day, K.P., 2000. Microsatellite markers reveal a spectrum of population structures in the malaria parasite *Plasmodium falciparum*. Mol. Biol. Evol. 17, 1467–1482.

Anderson, T.J., Romero-Abal, M.E., Jaenike, J., 1993. Genetic structure and epidemiology of *Ascaris* populations: patterns of host affiliation in Guatemala. Parasitology 107 (Pt 3), 319–334.

Archie, E.A., Luikart, G., Ezenwa, V.O., 2009. Infecting epidemiology with genetics: a new frontier in disease ecology. Trends Ecol. Evol. 24, 21–30.

Ariey, F., Witkowski, B., Amaratunga, C., Beghain, J., Langlois, A.C., Khim, N., Kim, S., Duru, V., Bouchier, C., MA, L., Lim, P., Leang, R., Duong, S., Sreng, S., Suon, S., Chuor, C.M., Bout, D.M., Menard, S., Rogers, W.O., Genton, B., Fandeur, T., Miotto, O., Ringwald, P., Le Bras, J., Berry, A., Barale, J.C., Fairhurst, R.M., Benoit-Vical, F., Mercereau-Puijalon, O., Menard, D., 2014. A molecular marker of artemisinin-resistant *Plasmodium falciparum* malaria. Nature 505, 50–55.

Assefa, L.M., Crellen, T., Kepha, S., Kihara, J.H., Njenga, S.M., Pullan, R.L., Brooker, S.J., 2014. Diagnostic accuracy and cost-effectiveness of alternative methods for detection of soil-transmitted helminths in a post-treatment setting in western Kenya. PLoS Negl. Trop. Dis. 8, e2843.

Balloux, F., Lehmann, L., 2012. Substitution rates at neutral genes depend on population size under fluctuating demography and overlapping generations. Evol. Int. J. Org. Evol. 66, 605–611.

Basáñez, M.G., Mccarthy, J.S., French, M.D., Yang, G.J., Walker, M., Gambhir, M., Prichard, R.K., Churcher, T.S., 2012. A research agenda for helminth diseases of humans: modelling for control and elimination. PLoS Negl. Trop. Dis. 6, e1548.

Basch, P.F., Basch, N., 1984. Intergeneric reproductive stimulation and parthenogenesis in *Schistosoma mansoni*. Parasitology 89 (Pt 2), 369–376.

Berriman, M., Haas, B.J., Loverde, P.T., Wilson, R.A., Dillon, G.P., Cerqueira, G.C., Mashiyama, S.T., Al-Lazikani, B., Andrade, L.F., Ashton, P.D., Aslett, M.A., Bartholomeu, D.C., Blandin, G., Caffrey, C.R., Coghlan, A., Coulson, R., Day, T.A., Delcher, A., Demarco, R., Djikeng, A., Eyre, T., Gamble, J.A., Ghedin, E., Gu, Y., Hertz-Fowler, C., Hirai, H., Hirai, Y., Houston, R., Ivens, A., Johnston, D.A., Lacerda, D., Macedo, C.D., Mcveigh, P., Ning, Z., Oliveira, G., Overington, J.P., Parkhill, J., Pertea, M., Pierce, R.J., Protasio, A.V., Quail, M.A., Rajandream, M.A., Rogers, J., Sajid, M., Salzberg, S.L., Stanke, M., Tivey, A.R., White, O., Williams, D.L., Wortman, J., Wu, W., Zamanian, M., Zerlotini, A., Fraser-Liggett, C.M., Barrell, B.G., El-Sayed, N.M., 2009. The genome of the blood fluke *Schistosoma mansoni*. Nature 460, 352–358.

Blake, I.M., Burton, M.J., Bailey, R.L., Solomon, A.W., West, S., Munoz, B., Holland, M.J., Mabey, D.C., Gambhir, M., Basanez, M.G., Grassly, N.C., 2009. Estimating household and community transmission of ocular *Chlamydia trachomatis*. PLoS Negl. Trop. Dis. 3, e401.

Blanton, R.E., Blank, W.A., Costa, J.M., Carmo, T.M., Reis, E.A., Silva, L.K., Barbosa, L.M., Test, M.R., Reis, M.G., 2011. *Schistosoma mansoni* population structure and persistence after praziquantel treatment in two villages of Bahia, Brazil. Int. J. Parasitol. 41, 1093–1099.

Blaxter, M., Kumar, S., Kaur, G., Koutsovoulos, G., Elsworth, B., 2012. Genomics and transcriptomics across the diversity of the Nematoda. Parasite Immunol. 34, 108–120.

Blouin, M.S., 2003. DNA-based methods for pedigree reconstruction and kinship analysis in natural populations. Trends Ecol. Evol. 18, 503–511.

Bockarie, M.J., Kelly-Hope, L.A., Rebollo, M., Molyneux, D.H., 2013. Preventive chemotherapy as a strategy for elimination of neglected tropical parasitic diseases: endgame challenges. Philos. Trans. R. Soc. Lond. Ser. B, Biol. Sci. 368, 20120144.

Borrmann, S., Straimer, J., Mwai, L., Abdi, A., Rippert, A., Okombo, J., Muriithi, S., Sasi, P., Kortok, M.M., Lowe, B., Campino, S., Assefa, S., Auburn, S., Manske, M., Maslen, G., Peshu, N., Kwiatkowski, D.P., Marsh, K., Nzila, A., Clark, T.G., 2013. Genome-wide screen identifies new candidate genes associated with artemisinin susceptibility in *Plasmodium falciparum* in Kenya. Sci. Rep. 3, 3318.

Bright, A.T., Winzeler, E.A., 2013. Epidemiology: resistance mapping in malaria. Nature 498, 446–447.

Castillo-Chavez, C., Feng, Z., Xu, D., 2008. A schistosomiasis model with mating structure and time delay. Math. Biosci. 211, 333–341.

Chan, J.D., Zarowiecki, M., Marchant, J.S., 2013. Ca(2)(+) channels and praziquantel: a view from the free world. Parasitol. Int. 62, 619–628.

Chan, M.S., Guyatt, H.L., Bundy, D.A., Medley, G.F., 1994. The development and validation of an age-structured model for the evaluation of disease control strategies for intestinal helminths. Parasitology 109 (Pt 3), 389–396.

Cheeseman, I.H., Miller, B.A., Nair, S., Nkhoma, S., Tan, A., Tan, J.C., Al Saai, S., Phyo, A.P., Moo, C.L., Lwin, K.M., Mcgready, R., Ashley, E., Imwong, M., Stepniewska, K., Yi, P., Dondorp, A.M., Mayxay, M., Newton, P.N., White, N.J., Nosten, F., Ferdig, M.T., Anderson, T.J., 2012. A major genome region underlying artemisinin resistance in malaria. Science 336, 79–82.

Chin, C.S., Alexander, D.H., Marks, P., Klammer, A.A., Drake, J., Heiner, C., Clum, A., Copeland, A., Huddleston, J., Eichler, E.E., Turner, S.W., Korlach, J., 2013. Nonhybrid, finished microbial genome assemblies from long-read SMRT sequencing data. Nat. Methods 10, 563–569.

Churcher, T.S., Basáñez, M.G., 2008. Density dependence and the spread of anthelmintic resistance. Evolution 62, 528–537.
Churcher, T.S., Basáñez, M.G., 2009. Sampling strategies to detect anthelmintic resistance: the perspective of human onchocerciasis. Trends Parasitol. 25, 11–17.
Cole, S.T., Brosch, R., Parkhill, J., Garnier, T., Churcher, C., Harris, D., Gordon, S.V., Eiglmeier, K., Gas, S., Barry 3rd, C.E., Tekaia, F., Badcock, K., Basham, D., Brown, D., Chillingworth, T., Connor, R., Davies, R., Devlin, K., Feltwell, T., Gentles, S., Hamlin, N., Holroyd, S., Hornsby, T., Jagels, K., Krogh, A., Mclean, J., Moule, S., Murphy, L., Oliver, K., Osborne, J., Quail, M.A., Rajandream, M.A., Rogers, J., Rutter, S., Seeger, K., Skelton, J., Squares, R., Squares, S., Sulston, J.E., Taylor, K., Whitehead, S., Barrell, B.G., 1998. Deciphering the biology of *Mycobacterium tuberculosis* from the complete genome sequence. Nature 393, 537–544.
Coudeville, L., Garnett, G.P., 2012. Transmission dynamics of the four dengue serotypes in southern Vietnam and the potential impact of vaccination. PLoS One 7, e51244.
Criscione, C.D., Anderson, J.D., Sudimack, D., Peng, W., Jha, B., Williams-Blangero, S., Anderson, T.J., 2007. Disentangling hybridization and host colonization in parasitic roundworms of humans and pigs. Proc. R. Soc. Biol. Sci. 274, 2669–2677.
Criscione, C.D., Anderson, J.D., Sudimack, D., Subedi, J., Upadhayay, R.P., Jha, B., Williams, K.D., Williams-Blangero, S., Anderson, T.J., 2010. Landscape genetics reveals focal transmission of a human macroparasite. PLoS Negl. Trop. Dis. 4, e665.
Criscione, C.D., Blouin, M.S., 2005. Effective sizes of macroparasite populations: a conceptual model. Trends Parasitol. 21, 212–217.
Criscione, C.D., Poulin, R., Blouin, M.S., 2005. Molecular ecology of parasites: elucidating ecological and microevolutionary processes. Mol. Ecol. 14, 2247–2257.
Criscione, C.D., Valentim, C.L., Hirai, H., Loverde, P.T., Anderson, T.J., 2009. Genomic linkage map of the human blood fluke *Schistosoma mansoni*. Genome Biol. 10, R71.
Daniels, R., Chang, H.H., Sene, P.D., Park, D.C., Neafsey, D.E., Schaffner, S.F., Hamilton, E.J., Lukens, A.K., Van Tyne, D., Mboup, S., Sabeti, P.C., Ndiaye, D., Wirth, D.F., Hartl, D.L., Volkman, S.K., 2013. Genetic surveillance detects both clonal and epidemic transmission of malaria following enhanced intervention in Senegal. PLoS One 8, e60780.
Dieckmann, U., Doebeli, M., 1999. On the origin of species by sympatric speciation. Nature 400, 354–357.
Doenhoff, M.J., Cioli, D., Utzinger, J., 2008. Praziquantel: mechanisms of action, resistance and new derivatives for schistosomiasis. Curr. Opin. Infect. Dis. 21, 659–667.
Dong, Y., Xie, M., Jiang, Y., Xiao, N., Du, X., Zhang, W., Tosser-Klopp, G., Wang, J., Yang, S., Liang, J., Chen, W., Chen, J., Zeng, P., Hou, Y., BIAN, C., Pan, S., Li, Y., Liu, X., Wang, W., Servin, B., Sayre, B., Zhu, B., Sweeney, D., Moore, R., Nie, W., Shen, Y., Zhao, R., Zhang, G., Li, J., Faraut, T., Womack, J., Zhang, Y., Kijas, J., Cockett, N., Xu, X., Zhao, S., Wang, J., Wang, W., 2013. Sequencing and automated whole-genome optical mapping of the genome of a domestic goat (*Capra hircus*). Nat. Biotechnol. 31, 135–141.
Downing, T., Imamura, H., Decuypere, S., Clark, T.G., Coombs, G.H., Cotton, J.A., Hilley, J.D., De Doncker, S., Maes, I., Mottram, J.C., Quail, M.A., Rijal, S., Sanders, M., Schonian, G., Stark, O., Sundar, S., Vanaerschot, M., Hertz-Fowler, C., Dujardin, J.C., Berriman, M., 2011. Whole genome sequencing of multiple *Leishmania donovani* clinical isolates provides insights into population structure and mechanisms of drug resistance. Genome Res. 21, 2143–2156.
Farhat, M.R., Shapiro, B.J., Kieser, K.J., Sultana, R., Jacobson, K.R., Victor, T.C., Warren, R.M., Streicher, E.M., Calver, A., Sloutsky, A., Kaur, D., Posey, J.E., Plikaytis, B., Oggioni, M.R., Gardy, J.L., Johnston, J.C., Rodrigues, M.,

Tang, P.K., Kato-Maeda, M., Borowsky, M.L., Muddukrishna, B., Kreiswirth, B.N., Kurepina, N., Galagan, J., Gagneux, S., Birren, B., Rubin, E.J., Lander, E.S., Sabeti, P.C., Murray, M., 2013. Genomic analysis identifies targets of convergent positive selection in drug-resistant *Mycobacterium tuberculosis*. Nat. Genet. 45, 1183–1189.

Feng, Z., Curtis, J., Minchella, D.J., 2001. The influence of drug treatment on the maintenance of schistosome genetic diversity. J. Math. Biol. 43, 52–68.

Fenwick, A., Webster, J.P., Bosque-Oliva, E., Blair, L., Fleming, F.M., Zhang, Y., Garba, A., Stothard, J.R., Gabrielli, A.F., Clements, A.C., Kabatereine, N.B., Toure, S., Dembele, R., Nyandindi, U., Mwansa, J., Koukounari, A., 2009. The Schistosomiasis Control Initiative (SCI): rationale, development and implementation from 2002–2008. Parasitology 136, 1719–1730.

Ferdig, M.T., Cooper, R.A., Mu, J., Deng, B., Joy, D.A., Su, X.Z., Wellems, T.E., 2004. Dissecting the loci of low-level quinine resistance in malaria parasites. Mol. Microbiol. 52, 985–997.

French, M.D., Churcher, T.S., Basanez, M.G., Norton, A.J., Lwambo, N.J., Webster, J.P., 2012. Reductions in genetic diversity of *Schistosoma mansoni* populations under chemotherapeutic pressure: the effect of sampling approach and parasite population definition. Acta Trop 128 (2), 196–205.

French, M.D., Churcher, T.S., Gambhir, M., Fenwick, A., Webster, J.P., Kabatereine, N.B., Basáñez, M.G., 2010. Observed reductions in *Schistosoma mansoni* transmission from large-scale administration of praziquantel in Uganda: a mathematical modelling study. PLoS Negl. Trop. Dis. 4, e897.

Gardner, M.J., Hall, N., Fung, E., White, O., Berriman, M., Hyman, R.W., Carlton, J.M., Pain, A., Nelson, K.E., Bowman, S., Paulsen, I.T., James, K., Eisen, J.A., Rutherford, K., Salzberg, S.L., Craig, A., Kyes, S., Chan, M.S., Nene, V., Shallom, S.J., Suh, B., Peterson, J., Angiuoli, S., Pertea, M., Allen, J., Selengut, J., Haft, D., Mather, M.W., Vaidya, A.B., Martin, D.M., Fairlamb, A.H., Fraunholz, M.J., Roos, D.S., Ralph, S.A., Mcfadden, G.I., Cummings, L.M., Subramanian, G.M., Mungall, C., Venter, J.C., Carucci, D.J., Hoffman, S.L., Newbold, C., Davis, R.W., Fraser, C.M., Barrell, B., 2002. Genome sequence of the human malaria parasite *Plasmodium falciparum*. Nature 419, 498–511.

Gilabert, A., Wasmuth, J.D., 2013. Unravelling parasitic nematode natural history using population genetics. Trends Parasitol. 29, 438–448.

Gower, C.M., Gouvras, A.N., Lamberton, P.H., Deol, A., Shrivastava, J., Mutombo, P.N., Mbuh, J.V., Norton, A.J., Webster, B.L., Stothard, J.R., Garba, A., Lamine, M.S., Kariuki, C., Lange, C.N., Mkoji, G.M., Kabatereine, N.B., Gabrielli, A.F., Rudge, J.W., Fenwick, A., Sacko, M., Dembele, R., Lwambo, N.J.T., Tchuente, L.A., Rollinson, D., Webster, J.P., 2013. Population genetic structure of *Schistosoma mansoni* and *Schistosoma haematobium* from across six sub-Saharan African countries: implications for epidemiology, evolution and control. Acta Trop. 128, 261–274.

Gower, C.M., Shrivastava, J., Lamberton, P.H.L., Rollinson, D., Webster, B.L., Emery, A., Kabatereine, N.B., Webster, J.P., 2007. Development and application of an ethically and epidemiologically advantageous assay for the multi-locus microsatellite analysis of *Schistosoma mansoni*. Parasitology 134, 523–536.

Greenberg, R.M., 2013. New approaches for understanding mechanisms of drug resistance in schistosomes. Parasitology 140, 1534–1546.

Grossman, S.R., Shlyakhter, I., Karlsson, E.K., Byrne, E.H., Morales, S., Frieden, G., Hostetter, E., Angelino, E., Garber, M., Zuk, O., Lander, E.S., Schaffner, S.F., Sabeti, P.C., 2010. A composite of multiple signals distinguishes causal variants in regions of positive selection. Science 327, 883–886.

Gryseels, B., Polman, K., Clerinx, J., Kestens, L., 2006. Human schistosomiasis. Lancet 368, 1106–1118.

Gupta, S., Maiden, M.C., Feavers, I.M., Nee, S., May, R.M., Anderson, R.M., 1996. The maintenance of strain structure in populations of recombining infectious agents. Nat. Med. 2, 437–442.

Gurarie, D., King, C.H., 2005. Heterogeneous model of schistosomiasis transmission and long-term control: the combined influence of spatial variation and age-dependent factors on optimal allocation of drug therapy. Parasitology 130, 49–65.

Hamelin, F.M., Castel, M., Poggi, S., Andrivon, D., Mailleret, L., 2011. Seasonality and the evolutionary divergence of plant parasites. Ecology 92, 2159–2166.

Harris, S.R., Feil, E.J., Holden, M.T., Quail, M.A., Nickerson, E.K., Chantratita, N., Gardete, S., Tavares, A., Day, N., Lindsay, J.A., Edgeworth, J.D., De Lencastre, H., Parkhill, J., Peacock, S.J., Bentley, S.D., 2010. Evolution of MRSA during hospital transmission and intercontinental spread. Science 327, 469–474.

Hayton, K., Su, X.Z., 2008. Drug resistance and genetic mapping in *Plasmodium falciparum*. Curr. Genet. 54, 223–239.

He, Y.X., Salafsky, B., Ramaswamy, K., 2001. Host–parasite relationships of *Schistosoma japonicum* in mammalian hosts. Trends Parasitol. 17, 320–324.

Holroyd, N., Sanchez-Flores, A., 2012. Producing parasitic helminth reference and draft genomes at the Wellcome Trust Sanger Institute. Parasite Immunol. 34, 100–107.

Hotez, P.J., Pecoul, B., 2010. 'Manifesto' for advancing the control and elimination of neglected tropical diseases. PLoS Negl. Trop. Dis. 4, e718.

Huyse, T., Van den Broeck, F., Jombart, T., Webster, B.L., Diaw, O., Volckaert, F.A., Balloux, F., Rollinson, D., Polman, K., 2013. Regular treatments of praziquantel do not impact on the genetic make-up of *Schistosoma mansoni* in Northern Senegal. Infect. Genet. Evol. 18, 100–105.

Huyse, T., Webster, B.L., Geldof, S., Stothard, J.R., Diaw, O.T., Polman, K., Rollinson, D., 2009. Bidirectional introgressive hybridization between a cattle and human schistosome species. PLoS Pathog. 5, e1000571.

Jones, A.G., Ardren, W.R., 2003. Methods of parentage analysis in natural populations. Mol. Ecol. 12, 2511–2523.

Jost, L.O.U., 2008. GST and its relatives do not measure differentiation. Mol. Ecol. 17, 4015–4026.

Kao, R.R., Haydon, D.T., Lycett, S.J., Murcia, P.R., 2014. Supersize me: how whole-genome sequencing and big data are transforming epidemiology. Trends Microbiol. 22, 282–291.

Kaplan, R.M., Vidyashankar, A.N., 2012. An inconvenient truth: global worming and anthelmintic resistance. Veterinary Parasitol. 186, 70–78.

Katz, N., Chaves, A., Pellegrino, J., 1972. A simple device for quantitative stool thick-smear technique in Schistosomiasis mansoni. Rev. do Inst. Med. Trop. São Paulo 14, 397–400.

Keeling, M.J., Rand, D.A., Morris, A.J., 1997. Correlation models for childhood epidemics. Proc. Biol. Sci./R. Soc. 264, 1149–1156.

Kim, Y., Escalante, A.A., Schneider, K.A., 2014. A population genetic model for the initial spread of partially resistant malaria parasites under anti-malarial combination therapy and weak intrahost competition. PLoS One 9, e101601.

Klepac, P., Metcalf, C.J., Mclean, A.R., Hampson, K., 2013. Towards the endgame and beyond: complexities and challenges for the elimination of infectious diseases. Philos. Trans. R. Soc. Lond. Ser. B, Biol. Sci. 368, 20120137.

Koukounari, A., Donnelly, C.A., Moustaki, I., Tukahebwa, E.M., Kabatereine, N.B., Wilson, S., Webster, J.P., Deelder, A.M., Vennervald, B.J., Van Dam, G.J., 2013. A latent Markov modelling approach to the evaluation of circulating cathodic antigen strips for schistosomiasis diagnosis pre- and post-praziquantel treatment in Uganda. PLoS Comput. Biol. 9, e1003402.

Kruger, F.J., 1990. Frequency and possible consequences of hybridization between *Schistosoma haematobium* and *S. mattheei* in the Eastern Transvaal Lowveld. J. Helminthol. 64, 333–336.
Lamberton, P.H., Kabatereine, N.B., Oguttu, D.W., Fenwick, A., Webster, J.P., 2014. Sensitivity and specificity of multiple Kato-Katz thick smears and a circulating cathodic antigen test for *Schistosoma mansoni* diagnosis pre- and post-repeated-praziquantel treatment. PLoS Negl. Trop. Dis. 8, e3139.
Lamberton, P.H.L., Hogan, S.C., Kabatereine, N.B., Fenwick, A., Webster, J.P., 2010. In vitro praziquantel test capable of detecting reduced in vivo efficacy in *Schistosoma mansoni* human infections. Am. J. Trop. Med. Hyg. 83, 1340–1347.
Lammie, P.J., Fenwick, A., Utzinger, J., 2006. A blueprint for success: integration of neglected tropical disease control programmes. Trends Parasitol. 22, 313–321.
Laxminarayan, R., Klugman, K.P., 2011. Communicating trends in resistance using a drug resistance index. BMJ Open 1, e000135.
Lelo, A.E., Mburu, D.N., Magoma, G.N., Mungai, B.N., Kihara, J.H., Mwangi, I.N., Maina, G.M., Kinuthia, J.M., Mutuku, M.W., Loker, E.S., Mkoji, G.M., Steinauer, M.L., 2014. No apparent reduction in schistosome burden or genetic diversity following four years of school-based mass drug administration in mwea, central kenya, a heavy transmission area. PLoS Negl. Trop. Dis. 8, e3221.
Levin, S.A., 1992. Mathematics and Biology: The Interface. Lawrence Berkeley Laboratory, University of California, Berkeley, CA. Available: http://www.bio.vu.nl/nvtb/Contents.html (accessed 30.10.14).
Levin, S.A., Grenfell, B., Hastings, A., Perelson, A.S., 1997. Mathematical and computational challenges in population biology and ecosystems science. Science 275, 334–343.
Liang, Y.S., Coles, G.C., Doenhoff, M.J., Southgate, V.R., 2001. In vitro responses of praziquantel-resistant and -susceptible *Schistosoma mansoni* to praziquantel. Int. J. Parasitol. 31, 1227–1235.
Lietman, T.M., Gebre, T., Ayele, B., Ray, K.J., Maher, M.C., See, C.W., Emerson, P.M., Porco, T.C., Group, T.S., 2011. The epidemiological dynamics of infectious trachoma may facilitate elimination. Epidemics 3, 119–124.
Lu, D.B., Rudge, J.W., Wang, T.P., Donnelly, C.A., Fang, G.R., Webster, J.P., 2010. Transmission of *Schistosoma japonicum* in marshland and hilly regions of China: parasite population genetic and sibship structure. PLoS Negl. Trop. Dis. 4, e781.
Luikart, G., England, P.R., Tallmon, D., Jordan, S., Taberlet, P., 2003. The power and promise of population genomics: from genotyping to genome typing. Nat. Rev. Genet. 4, 981–994.
Luz, P.M., Vanni, T., Medlock, J., Paltiel, A.D., Galvani, A.P., 2011. Dengue vector control strategies in an urban setting: an economic modelling assessment. Lancet 377, 1673–1680.
Lynch, C., Roper, C., 2011. The transit phase of migration: circulation of malaria and its multidrug-resistant forms in Africa. PLoS Med. 8, e1001040.
Mather, A.E., Reid, S.W., Maskell, D.J., Parkhill, J., Fookes, M.C., Harris, S.R., Brown, D.J., Coia, J.E., Mulvey, M.R., Gilmour, M.W., Petrovska, L., De Pinna, E., Kuroda, M., Akiba, M., Izumiya, H., Connor, T.R., Suchard, M.A., Lemey, P., Mellor, D.J., Haydon, D.T., Thomson, N.R., 2013. Distinguishable epidemics of multidrug-resistant *Salmonella typhimurium* DT104 in different hosts. Science 341, 1514–1517.
Mcmanus, D.P., Gray, D.J., Li, Y., Feng, Z., Williams, G.M., Stewart, D., Rey-Ladino, J., Ross, A.G., 2010. Schistosomiasis in the People's Republic of China: the era of the Three Gorges Dam. Clin. Microbiol. Rev. 23, 442–466.
Medley, G., Anderson, R.M., 1985. Density-dependent fecundity in *Schistosoma mansoni* infections in man. Trans. R. Soc. Trop. Med. Hyg. 79, 532–534.
Nair, S., Willaims, J.T., Brockman, A., Paiphum, L., Mayxay, M., Newton, P.N., Guthmann, J.-P., Smithuis, F.M., Hien, T.T., White, N.J., Nosten, F.,

Anderson, T.J.C., 2003. A selective sweep driven by pyrimethamine treatment in southeast asian malaria parasites. Mol. Biol. Evol. 20, 1526–1536.
Nee, S., Holmes, E.C., May, R.M., Harvey, P.H., 1994. Extinction rates can be estimated from molecular phylogenies. Philos. Trans. R. Soc. Lond. Ser. B, Biol. Sci. 344, 77–82.
Nkhoma, S.C., Nair, S., Al-Saai, S., Ashley, E., Mcgready, R., Phyo, A.P., Nosten, F., Anderson, T.J., 2013. Population genetic correlates of declining transmission in a human pathogen. Mol. Ecol. 22, 273–285.
Norton, A.J., Gower, C.M., Lamberton, P.H., Webster, B.L., Lwambo, N.J., Blair, L., Fenwick, A., Webster, J.P., 2010. Genetic consequences of mass human chemotherapy for *Schistosoma mansoni*: population structure pre- and post-praziquantel treatment in Tanzania. Am. J. Trop. Med. Hyg. 83, 951–957.
Ntd_Modelling_Consortium. Manta Ray Media. Available: http://www.ntdmodelling.org/ (accessed 30.10.14).
Oleksyk, T.K., Smith, M.W., O'brien, S.J., 2010. Genome-wide scans for footprints of natural selection. Philos. Trans. R. Soc. Lond. Ser. B, Biol. Sci. 365, 185–205.
Ou, C.Y., Ciesielski, C.A., Myers, G., Bandea, C.I., Luo, C.C., Korber, B.T., Mullins, J.I., Schochetman, G., Berkelman, R.L., Economou, A.N., Et, A.L., 1992. Molecular epidemiology of HIV transmission in a dental practice. Science 256, 1165–1171.
Pampoulie, C., Ruzzante, D.E., Chosson, V., Jörundsdóttir, T.D., Taylor, L., Thorsteinsson, V., Daníelsdóttir, A.K., Marteinsdóttir, G., 2006. The genetic structure of Atlantic cod (*Gadus morhua*) around Iceland: insight from microsatellites, the PanI locus, and tagging experiments. Can. J. Fish. Aquat. Sci. 63, 2660–2674.
Parkhill, J., 2013. What has high-throughput sequencing ever done for us? Nat. Rev. Microbiol. 11, 664–665.
Paterson, S., Viney, M.E., 2000. The interface between epidemiology and population genetics. Parasitol. Today 16, 528–532.
Pearce, R.J., Pota, H., Evehe, M.S., Ba El, H., Mombo-Ngoma, G., Malisa, A.L., Ord, R., Inojosa, W., Matondo, A., Diallo, D.A., Mbacham, W., Van Den Broek, I.V., Swarthout, T.D., Getachew, A., Dejene, S., Grobusch, M.P., Njie, F., Dunyo, S., Kweku, M., Owusu-Agyei, S., Chandramohan, D., Bonnet, M., Guthmann, J.P., Clarke, S., Barnes, K.I., Streat, E., Katokele, S.T., Uusiku, P., Agboghoroma, C.O., Elegba, O.Y., Cisse, B., Ie, A.E., Giha, H.A., Kachur, S.P., Lynch, C., Rwakimari, J.B., Chanda, P., Hawela, M., Sharp, B., Naidoo, I., Roper, C., 2009. Multiple origins and regional dispersal of resistant dhps in African *Plasmodium falciparum* malaria. PLoS Med. 6, e1000055.
Pindolia, D.K., Garcia, A.J., Huang, Z., Smith, D.L., Alegana, V.A., Noor, A.M., Snow, R.W., Tatem, A.J., 2013. The demographics of human and malaria movement and migration patterns in East Africa. Malar. J. 12, 397.
Plaisier, A.P., Alley, E.S., Van Oortmarssen, G.J., Boatin, B.A., Habbema, J.D.F., 1997. Required duration of combined annual ivermectin treatment and vector control in the Onchocerciasis Control Programme in West Africa. Bull. World Health Organ. 75, 237–245.
Protasio, A.V., Dunne, D.W., Berriman, M., 2013. Comparative study of transcriptome profiles of mechanical- and skin-transformed *Schistosoma mansoni* schistosomula. PLoS Negl. Trop. Dis. 7, e2091.
Protasio, A.V., Tsai, I.J., Babbage, A., Nichol, S., Hunt, M., Aslett, M.A., De Silva, N., Velarde, G.S., Anderson, T.J., Clark, R.C., Davidson, C., Dillon, G.P., Holroyd, N.E., Loverde, P.T., Lloyd, C., Mcquillan, J., Oliveira, G., Otto, T.D., Parker-Manuel, S.J., Quail, M.A., Wilson, R.A., Zerlotini, A., Dunne, D.W., Berriman, M., 2012. A systematically improved high quality genome and

transcriptome of the human blood fluke *Schistosoma mansoni*. PLoS Negl. Trop. Dis. 6, e1455.

Prugnolle, F., de Meeûs, T., Durand, P., Sire, C., Théron, A., 2002. Sex-specific genetic structure in *Schistosoma mansoni*: evolutionary and epidemiological implications. Mol. Ecol. 11, 1231–1238.

Prugnolle, F., Liu, H., de Meeûs, T., Balloux, F., 2005a. Population genetics of complex life-cycle parasites: an illustration with trematodes. Int. J. Parasitol. 35, 255–263.

Prugnolle, F., Roze, D., Théron, A., de Meeûs, T., 2005b. F-statistics under alternation of sexual and asexual reproduction: a model and data from schistosomes (platyhelminth parasites). Mol. Ecol. 14, 1355–1365.

Prugnolle, F., Théron, A., Pointier, J.P., Jabbour-Zahab, R., Jarne, P., Durand, P., de Meeûs, T., 2005c. Dispersal in a parasitic worm and its two hosts: consequence for local adaptation. Evolution 59, 296–303.

Rascalou, G., Gourbiere, S., 2014. Competition, virulence, host body mass and the diversification of macro-parasites. J. R. Soc. Interface 11, 20131108.

Resik, S., Lemey, P., Ping, L.H., Kouri, V., Joanes, J., Perez, J., Vandamme, A.M., Swanstrom, R., 2007. Limitations to contact tracing and phylogenetic analysis in establishing HIV type 1 transmission networks in Cuba. AIDS Res. Hum. Retroviruses 23, 347–356.

Riley, S., Carabin, H., Belisle, P., Joseph, L., Tallo, V., Balolong, E., Willingham, A.L., Fernandez, T.J., Gonzales, R.O., Olveda, R., Mcgarvey, S.T., 2008. Multi-host transmission dynamics of *Schistosoma japonicum* in Samar province, the Philippines. PLoS Med. 5, e18.

Rivers, C.M., Lofgren, E.T., Marathe, M., Eubank, S., Lewis, B.L., 2014. Modeling the impact of interventions on an epidemic of Ebola in Sierra Leone and Liberia. PLoS Curr. Outbreaks 1.

Rollinson, D., Knopp, S., Levitz, S., Stothard, J.R., Tchuem Tchuente, L.A., Garba, A., Mohammed, K.A., Schur, N., Person, B., Colley, D.G., Utzinger, J., 2013. Time to set the agenda for schistosomiasis elimination. Acta Trop. 128, 423–440.

Rollinson, D., Webster, J.P., Webster, B.L., Nyakaana, S., Jørgensen, A., Stothard, J.R., 2009. Genetic diversity of schistosomes and snails: implications for control. Parasitology 136, 1801–1811.

Rudge, J.W., Lu, D.B., Fang, G.R., Wang, T.P., Basanez, M.G., Webster, J.P., 2009. Parasite genetic differentiation by habitat type and host species: molecular epidemiology of *Schistosoma japonicum* in hilly and marshland areas of Anhui Province, China. Mol. Ecol. 18, 2134–2147.

Rudge, J.W., Webster, J.P., Lu, D.B., Wang, T.P., Fang, G.R., Basáñez, M.G., 2013. Identifying host species driving transmission of schistosomiasis japonica, a multi-host parasite system, in China. Proc. Natl. Acad. Sci. U.S.A. 110, 11457–11462.

Sabeti, P.C., Varilly, P., Fry, B., Lohmueller, J., Hostetter, E., Cotsapas, C., Xie, X., Byrne, E.H., Mccarroll, S.A., Gaudet, R., Schaffner, S.F., Lander, E.S., International Hapmap, C., Frazer, K.A., Ballinger, D.G., Cox, D.R., Hinds, D.A., Stuve, L.L., Gibbs, R.A., Belmont, J.W., Boudreau, A., Hardenbol, P., Leal, S.M., Pasternak, S., Wheeler, D.A., Willis, T.D., Yu, F., Yang, H., Zeng, C., Gao, Y., Hu, H., Hu, W., Li, C., Lin, W., Liu, S., Pan, H., Tang, X., Wang, J., Wang, W., Yu, J., Zhang, B., Zhang, Q., Zhao, H., Zhao, H., Zhou, J., Gabriel, S.B., Barry, R., Blumenstiel, B., Camargo, A., Defelice, M., Faggart, M., Goyette, M., Gupta, S., Moore, J., Nguyen, H., Onofrio, R.C., Parkin, M., Roy, J., Stahl, E., Winchester, E., Ziaugra, L., Altshuler, D., Shen, Y., Yao, Z., Huang, W., Chu, X., He, Y., Jin, L., Liu, Y., Shen, Y., Sun, W., Wang, H., Wang, Y., Wang, Y., Xiong, X., Xu, L., Waye, M.M., Tsui, S.K., Xue, H., Wong, J.T., Galver, L.M., Fan, J.B., Gunderson, K., Murray, S.S., Oliphant, A.R., Chee, M.S., Montpetit, A.,

Chagnon, F., Ferretti, V., Leboeuf, M., Olivier, J.F., Phillips, M.S., Roumy, S., Sallee, C., Verner, A., Hudson, T.J., Kwok, P.Y., Cai, D., Koboldt, D.C., Miller, R.D., et al., 2007. Genome-wide detection and characterization of positive selection in human populations. Nature 449, 913–918.

Sanchez, C.P., Mayer, S., Nurhasanah, A., Stein, W.D., Lanzer, M., 2011. Genetic linkage analyses redefine the roles of PfCRT and PfMDR1 in drug accumulation and susceptibility in *Plasmodium falciparum*. Mol. Microbiol. 82, 865–878.

Schistosoma Japonicum Genome, S. & Functional Analysis, C, 2009. The *Schistosoma japonicum* genome reveals features of host-parasite interplay. Nature 460, 345–351.

Smith, G., Grenfell, B.T., Isham, V., Cornell, S., 1999. Anthelmintic resistance revisited: under-dosing, chemoprophylactic strategies, and mating probabilities. Int. J. Parasitol. 29, 77–91 discussion 93–94.

Smith, J.M., Haigh, J., 1974. The hitch-hiking effect of a favourable gene. Genet. Res. 23, 23–35.

Stack, J.C., Murcia, P.R., Grenfell, B.T., Wood, J.L., Holmes, E.C., 2013. Inferring the inter-host transmission of influenza A virus using patterns of intra-host genetic variation. Proc. R. Soc. B. Biol. Sci. 280, 20122173.

Steinauer, M.L., Blouin, M.S., Criscione, C.D., 2010. Applying evolutionary genetics to schistosome epidemiology. Infect. Genet. Evol. 10, 433–443.

Steinauer, M.L., Christie, M.R., Blouin, M.S., Agola, L.E., Mwangi, I.N., Maina, G.M., Mutuku, M.W., Kinuthia, J.M., Mkoji, G.M., Loker, E.S., 2013. Non-invasive sampling of schistosomes from humans requires correcting for family structure. PLoS Negl. Trop. Dis. 7, e2456.

Steinauer, M.L., Hanelt, B., Mwangi, I.N., Maina, G.M., Agola, L.E., Kinuthia, J.M., Mutuku, M.W., Mungai, B.N., Wilson, W.D., Mkoji, G.M., Loker, E.S., 2008. Introgressive hybridization of human and rodent schistosome parasites in western Kenya. Mol. Ecol. 17, 5062–5074.

Steinmann, P., Keiser, J., Bos, R., Tanner, M., Utzinger, J., 2006. Schistosomiasis and water resources development: systematic review, meta-analysis, and estimates of people at risk. Lancet Infect. Dis. 6, 411–425.

Stukenbrock, E.H., Christiansen, F.B., Hansen, T.T., Dutheil, J.Y., Schierup, M.H., 2012. Fusion of two divergent fungal individuals led to the recent emergence of a unique widespread pathogen species. Proc. Natl. Acad. Sci. U.S.A. 109, 10954–10959.

Takala-Harrison, S., Clark, T.G., Jacob, C.G., Cummings, M.P., Miotto, O., Dondorp, A.M., Fukuda, M.M., Nosten, F., Noedl, H., Imwong, M., Bethell, D., Se, Y., Lon, C., Tyner, S.D., Saunders, D.L., Socheat, D., Ariey, F., Phyo, A.P., Starzengruber, P., Fuehrer, H.P., Swoboda, P., Stepniewska, K., Flegg, J., Arze, C., Cerqueira, G.C., Silva, J.C., Ricklefs, S.M., Porcella, S.F., Stephens, R.M., Adams, M., Kenefic, L.J., Campino, S., Auburn, S., Macinnis, B., Kwiatkowski, D.P., Su, X.Z., White, N.J., Ringwald, P., Plowe, C.V., 2013. Genetic loci associated with delayed clearance of *Plasmodium falciparum* following artemisinin treatment in Southeast Asia. Proc. Natl. Acad. Sci. U.S.A. 110, 240–245.

Taylor, L.H., Latham, S.M., Woolhouse, M.E., 2001. Risk factors for human disease emergence. Philos. Trans. R. Soc. Lond. Ser. B, Biol. Sci. 356, 983–989.

Tchuem Tchuenté, L.A., Southgate, V.R., Combes, C., Jourdane, J., 1996. Mating behaviour in Schistosomes: are paired worms always faithful? Parasitol. Today 12, 231–236.

Tchuem Tchuenté, L.A., Southgate, V.R., Imbert-Establet, D., Jourdane, J., 1995. Change of mate and mating competition between males of *Schistosoma intercalatum* and *S. mansoni*. Parasitology 110 (Pt 1), 45–52.

Tsai, I.J., Zarowiecki, M., Holroyd, N., Garciarrubio, A., Sanchez-Flores, A., Brooks, K.L., Tracey, A., Bobes, R.J., Fragoso, G., Sciutto, E., Aslett, M., Beasley, H., Bennett, H.M.,

Cai, J., Camicia, F., Clark, R., Cucher, M., De Silva, N., Day, T.A., Deplazes, P., Estrada, K., Fernandez, C., Holland, P.W., Hou, J., Hu, S., Huckvale, T., Hung, S.S., Kamenetzky, L., Keane, J.A., Kiss, F., Koziol, U., Lambert, O., Liu, K., Luo, X., Luo, Y., Macchiaroli, N., Nichol, S., Paps, J., Parkinson, J., Pouchkina-Stantcheva, N., Riddiford, N., Rosenzvit, M., Salinas, G., Wasmuth, J.D., Zamanian, M., Zheng, Y., Taenia Solium Genome, C., Cai, X., Soberon, X., Olson, P.D., Laclette, J.P., Brehm, K., Berriman, M., 2013. The genomes of four tapeworm species reveal adaptations to parasitism. Nature 496, 57–63.

Turner, H.C., Churcher, T.S., Walker, M., Osei-Atweneboana, M.Y., Prichard, R.K., Basáñez, M.G., 2013. Uncertainty surrounding projections of the long-term impact of ivermectin treatment on human onchocerciasis. PLoS Negl. Trop. Dis. 7, e2169.

Turner, H.C., Walker, M., Churcher, T.S., Basanez, M.G., 2014. Modelling the impact of ivermectin on River Blindness and its burden of morbidity and mortality in African Savannah: EpiOncho projections. Parasit. Vectors 7, 241.

Valentim, C.L., Cioli, D., Chevalier, F.D., Cao, X., Taylor, A.B., Holloway, S.P., Pica-Mattoccia, L., Guidi, A., Basso, A., Tsai, I.J., Berriman, M., Carvalho-Queiroz, C., Almeida, M., Aguilar, H., Frantz, D.E., Hart, P.J., Loverde, P.T., Anderson, T.J., 2013. Genetic and molecular basis of drug resistance and species-specific drug action in schistosome parasites. Science 342, 1385–1389.

Vitti, J.J., Grossman, S.R., Sabeti, P.C., 2013. Detecting natural selection in genomic data. Annu. Rev. Genet. 47, 97–120.

Volkman, S.K., Neafsey, D.E., Schaffner, S.F., Park, D.J., Wirth, D.F., 2012. Harnessing genomics and genome biology to understand malaria biology. Nat. Rev. Genet. 13, 315–328.

Wang, L., Utzinger, J., Zhou, X.N., 2008. Schistosomiasis control: experiences and lessons from China. Lancet 372, 1793–1795.

Webster, B.L., Diaw, O.T., Seye, M.M., Webster, J.P., Rollinson, D., 2013. Introgressive hybridization of *Schistosoma haematobium* group species in Senegal: species barrier break down between ruminant and human schistosomes. PLoS Negl. Trop. Dis. 7, e2110.

Webster, J.P., Molyneux, D., Hotez, P., Fenwick, A., 2014. The contribution of mass drug administration to global health – past, present and future. Philos. Trans. R. Soc. Lond. Ser. B, Biol. Sci. 369, 1471–2970.

Webster, J.P., Norton, A.J., Gower, C.M., 2008. Evolutionary concepts in predicting and evaluating the impact of mass chemotherapy schistosomiasis control programmes on parasites and their hosts. Evol. Appl. 1, 66–83.

Webster, J.P., Shrivastava, J., Johnson, P.J., Blair, L., 2007. Is host-schistosome coevolution going anywhere? BMC Evol. Biol. 7, 91.

Weir, B.S., Cockerham, C.C., 1984. Estimating F-statistics for the analysis of population structure. Evolution 38, 1358–1370.

WHO, 2012. Accelarating Work to Overcome the Global Impact of Neglected Tropical Diseases – A Roadmap for Implementation (Online). http://www.who.int/neglected_diseases/resources/en/index.html.

World Health Organization, 2013. Schistosomiasis. Progress report 2001–2011 and strategic plan 2012–2020. Available: www.who.int/iris/bitstream/10665/78074/1/9789241503174_eng.pdf, accessed: 15 January 2015.

Wolstenholme, A.J., Fairweather, I., Prichard, R., Von Samson-Himmelstjerna, G., Sangster, N.C., 2004. Drug resistance in veterinary helminths. Trends Parasitol. 20, 469–476.

Wootton, J.C., Feng, X., Ferdig, M.T., Cooper, R.A., Mu, J., Baruch, D.I., Magill, A.J., Su, X.Z., 2002. Genetic diversity and chloroquine selective sweeps in *Plasmodium falciparum*. Nature 418, 320–323.

WTSI. http://www.sanger.ac.uk/research/initiatives/globalhealth/research/helminthgenomes/.

Xu, D., Curtis, J., Feng, Z., Minchella, D.J., 2006. On the role of schistosome mating structure in the maintenance of drug-resistant strains. Bull. Math. Biol. 68, 209–229.
Young, N.D., Jex, A.R., Li, B., Liu, S., Yang, L., Xiong, Z., Li, Y., Cantacessi, C., Hall, R.S., Xu, X., Chen, F., Wu, X., Zerlotini, A., Oliveira, G., Hofmann, A., Zhang, G., Fang, X., Kang, Y., Campbell, B.E., Loukas, A., Ranganathan, S., Rollinson, D., Rinaldi, G., Brindley, P.J., Yang, H., Wang, J., Wang, J., Gasser, R.B., 2012. Whole-genome sequence of *Schistosoma haematobium*. Nat. Genet. 44, 221–225.
Zerlotini, A., Aguiar, E.R., Yu, F., Xu, H., Li, Y., Young, N.D., Gasser, R.B., Protasio, A.V., Berriman, M., Roos, D.S., Kissinger, J.C., Oliveira, G., 2013. SchistoDB: an updated genome resource for the three key schistosomes of humans. Nucleic Acids Res. 41, D728–D731.
Zhang, H., Li, D., Zhao, L., Fleming, J., Lin, N., Wang, T., Liu, Z., Li, C., Galwey, N., Deng, J., Zhou, Y., Zhu, Y., Gao, Y., Wang, T., Wang, S., Huang, Y., Wang, M., Zhong, Q., Zhou, L., Chen, T., Zhou, J., Yang, R., Zhu, G., Hang, H., Zhang, J., Li, F., Wan, K., Wang, J., Zhang, X.E., Bi, L., 2013. Genome sequencing of 161 *Mycobacterium tuberculosis* isolates from China identifies genes and intergenic regions associated with drug resistance. Nat. Genet. 45, 1255–1260.
Zhang, J., Lietman, T., Olinger, L., Miao, Y., Stephens, R.S., 2004. Genetic diversity of *Chlamydia trachomatis* and the prevalence of trachoma. Pediatr. Infect. Dis. J. 23, 217–220.

CHAPTER EIGHT

Economic and Financial Evaluation of Neglected Tropical Diseases

Bruce Y. Lee[1], Sarah M. Bartsch, Katrin M. Gorham
Public Health Computational and Operations Research (PHICOR) and International Vaccine Access Center (IVAC), Johns Hopkins Bloomberg School of Public Health, Baltimore, MD, USA
[1]Corresponding author: E-mail: brucelee@jhu.edu

Contents

Advances in Parasitology, Volume 87
ISSN 0065-308X
http://dx.doi.org/10.1016/bs.apar.2015.01.002

Abstract

Economic and financing studies are particularly important for decision-making when resources are scarce or considerably limited. This is the case for neglected tropical diseases (NTDs). In fact, the definition of NTDs is an economic one. The shortage of resources for NTD control may be due in large part to the fact that the burden of NTDs and economic value of control measures have not been fully characterized. A number of economic study methodologies are available: cost of illness can quantify the extent, magnitude, and change of a problem; cost of intervention studies can outline the feasibility and guide the design of a policy or intervention; and cost-benefit, cost-effectiveness, and return-on-investment studies can determine the potential value of different interventions and policies. NTDs have unique characteristics that require special consideration in such analyses. Hence, approaches used for other diseases may need modifications to capture the full impact of NTDs. While the existing literature has made important findings, there is a need for substantially more work, as many NTDs and their associated interventions and policies require more evaluation. With increasing work in this area, NTDs may not be as 'neglected' in the future as they are now.

1. THE IMPORTANCE OF ECONOMICS IN DECISION-MAKING

Economics is essentially the science of decision-making when resources are limited, which is usually the case in medical, public health, or policy decision-making. If resources were unlimited, then decision-making would be considerably easier (e.g. policy makers would not have to choose among programmes or interventions and manufacturers could simply produce whatever was feasible and desired), and behaviour would be substantially different (e.g. many people may switch their current jobs and responsibilities and simple altruism or hedonism could play larger roles). However, in reality, resources are never unlimited and are usually scarce. Even in rare situations in which funding, materials, personnel or fuel seem abundant for a certain set of people or organizations, others invariably face constraints. In actuality, no resource is unlimited. Even things that people in high-income countries may take for granted such breathable air and potable water are proving to be more limited than previously believed.

Since decision-making without considering resource constraints is usually impractical, economics can help to translate policies and interventions originating in other sciences to the real world. There are two broad types of economics: macroeconomics and microeconomics. Macroeconomics looks at aggregate economics or the whole economy. It focusses on the performance, structure, behaviour, and decision-making of the economy as a whole, taking into consideration factors such as national income, unemployment, rate of growth, inflation, etc. Microeconomics takes a more focused, detailed look at portions (e.g. individuals, organizations, or groups) of the overall economy. It focusses on decision-making based on resource allocation and evaluates concepts such as supply and demand, price and output, and how much to produce and charge. The focus of this chapter is on microeconomics in the context of the neglected tropical diseases (NTDs). Economic evaluations are widely used in virtually every field and profession for this purpose. Medicine and public health are no exception. For instance, many agree that vastly improving a low-income country's infrastructure and sanitation could reduce or even eliminate hookworm. However, doing so could be prohibitively expensive in many low-income countries, making it unfeasible to rely solely on such a solution. Therefore, other solutions are currently being implemented or developed. Mass drug administration (MDA) is being used in locations primarily in schoolchildren since MDA in a wider population could prove to be too expensive (in addition to concerns about the hookworm populations developing anthelmintic resistance). A possible approach would be to test everyone in the population annually or even more frequently for hookworm infection and treat those who test positive, but the cost of such mass screening may make such a strategy impractical. Test kits cost money to produce, ship and administer. Personnel may not be readily available and require compensation to administer and interpret such tests. Testing may require the tested person to miss work or school which could cause financial hardship. Monitoring and advertising the testing consumes resources as well. New interventions such as hookworm vaccines are under development but funding to continue their development would be required for them to reach the market, and funders must choose among many different competing public health needs. Moreover, once any new intervention reaches the market, the costs of producing, shipping and administering help to govern whom should receive the intervention as well as how and where the intervention should be employed.

Table 1 shows some major decision-makers and stakeholders involved in the study and control of parasitic infections and how each may use economic studies. Some of these questions crossmultiple decision-makers. For example, a funder, researcher, manufacturer, employer, policy maker and health care worker may be very interested in how much an intervention may cost. So the assignment of different questions to different decision-makers is actually

Table 1 Key stakeholders in public health, questions that need addressing and how economic evaluations can help

Stakeholder	Questions or decisions	How economic evaluation can help
Researcher	How much will a study cost? What is the value of collecting additional information? What is the potential impact of research findings? What are the high-impact questions to examine?	Provides information about costs, effectiveness, and impact of potential research. What type of information is key or makes a difference in results and should be collected.
Product developer/ manufacturer	How much will developing and manufacturing the product cost? What product price will provide sufficient return on investment? Should we continue to develop and manufacture this intervention? Will it be worth the time, effort and cost? What can we claim about our intervention versus other interventions? How much should we charge for the intervention? How much will payers cover the intervention?	Provides guidance for product development and optimization strategies.

Table 1 Key stakeholders in public health, questions that need addressing and how economic evaluations can help—cont'd

Stakeholder	Questions or decisions	How economic evaluation can help
Health care worker	Which intervention is appropriate for a particular patient? Which makes the best use of financing resources? To whom should we provide particular interventions? What is the quality—quantity trade-off and how can we use and apply this knowledge?	Provides data to help workers efficiently treat patients.
Third-party payers	How can costs be minimized? Should we provide coverage for specific interventions? How will covering particular interventions affect our business?	Provides cost-effectiveness and other data to inform coverage decisions.
Employer	How can health care costs be minimized without sacrificing worker productivity? Which conditions reduce productivity and what interventions are available to curb this? Should we encourage (or discourage) third-party coverage for certain interventions? What is the quality—quantity trade-off and how can we use and apply this knowledge?	Provides information that can optimize productivity and worker well-being and how much can invest to sponsor a programme.
Nonprofit funder	How should limited resources be allocated to grantees? Are particular interventions worth investing in? What will the rewards be (to us and society) for investing our resources?	Provides information to determine the best use of funder resources.
Disease control official	How can infections be controlled given the constraints on governmental resources? Should we institute a mass intervention? Should we commit funds to develop a particular product?	Provides data to inform disease control strategies.

quite fluid. In fact, economic studies can help various decision-makers see each other's points of view and bring stakeholders together. A policy maker may need to determine how much funders need to contribute to executing an intervention or advocate an intervention to funders to achieve their cooperation. Economics is the language that many decision-makers such as funders and employers routinely speak, so a discussion without economic data may be fruitless. Even in rare incidences when a decision-maker has no economic constraints or considerations, the decision-maker invariably works among others who have economic considerations.

Table 1 also points to the importance of **perspective** in economic studies. The costs and effects of a disease and/or intervention are borne to different degrees by different stakeholders. An employer may see productivity losses from its employees missing work but not be responsible for the employee's health care costs. Health care costs may be the responsibility of the government or other third-party payer. A health care system or hospital may only see the costs that are not reimbursed by a third-party payer. A particular funder may only contribute to one type of cost. All costs are born by society. Other frequently used perspectives are those of individual patients/families, health service providers (e.g. clinics and hospitals), health systems (i.e. all health care provider costs) and funders. Since there may be substantial discrepancy in the resulting costs and effects, every economic evaluation should clearly establish its perspective, the most common being the societal perspective, third-party perspective and the health care system perspective. Once the perspective is established, the study should be consistent in tabulating only those costs and effects that are directly relevant to that perspective.

While some general economic principles and methods apply widely to all fields, unique aspects of medicine and public health necessitate some tailored approaches. As will be seen in the descriptions of the health economics methodologies later in this Chapter, quantifying the impact of diseases and interventions, integrating these with costs and comparing and ranking possibilities have their challenges. Decision-making can be much easier in situations when one can focus on a single goal such as profit-making and has overarching decision-making authority. These are certainly not the case in medicine and public health. Many stakeholders in these arenas must balance multiple considerations and goals (e.g. a hospital cannot solely focus on profit-making and cannot even focus on treating one disease). Many of these considerations entail agonizing trade-offs (e.g. focussing on one disease or population means neglecting other diseases or populations, leading to dire effects and even death). Some elements such as ethical concerns can be difficult to quantify and potentially highly subjective. For instance, if one has to choose between treating two different types of individuals, the choice may differ depending on the person making the decision. All of this is further complicated by the fact that situations in medicine and public health frequently involve many diverse decision-makers, who cannot exert absolute authority over one another. Instead these decision-makers must negotiate and cooperate. (For example, a country's Ministry of Health can rarely implement an intervention without getting buy-in from multiple other stakeholders.) Additional complexities in medicine and public health abound. Finally, the stakes in medicine and public health can be very high. Decisions can affect millions and even billions of lives for many years.

Economics is particularly important in decision-making for the NTDs, which face considerable resource constraints. These diseases are more common in the lowest income and most disadvantaged populations in the world. These populations often lack the infrastructure, materials, personnel or money to implement proper control or treatment measures. Additionally, compared with other higher profile diseases, such as human immunodeficiency virus (HIV)/AIDS, far less funding is available for the study and control of NTDs.

In many ways the definition of NTDs is economic. NTDs are considered 'neglected' because they may not be receiving enough attention and resources for adequate control. The combined global spending on HIV/AIDS, malaria and tuberculosis control exceeds USD 22 billion annually: an estimated USD 16.8 billion for HIV/AIDS programmes, USD 4.5 billion for malaria control initiatives and USD 1.25 billion for tuberculosis control (Report on the global AIDS epidemic, 2012; Malaria finding & resource utilization, 2010; Global tuberculosis control, 2007). By contrast, total global spending for all 17 NTDs prioritized by the World Health Organization (WHO) combined (trachoma, Chagas disease, guinea worm, human African trypanosomiasis, lymphatic filariasis, leprosy, onchocerciasis, schistosomiasis, soil-transmitted helminthiases (STH), visceral leishmaniasis, buruli ulcer, dengue, echinococcosis, food-borne trematodiases, rabies, taeniasis/cysticercosis and yaws) will likely fall below USD 600 million in 2014 – less than 5% of the total spending on HIV/AIDS, malaria and tuberculosis (Uniting to Combat Neglected Tropical Diseases, 2014).

The reasons that NTDs are 'neglected' also arise in large part from economics. NTDs tend to affect the poorest and most marginalized populations of the world and may in fact be contributing to these populations' poverty. These diseases inflict life-long disability, disfigurement, reduced economic productivity, and social stigma, impairing child health and development, education, pregnancy and worker productivity (Hotez et al., 2006; King, 2010). However, the commitment of resources towards NTDs has historically remained low. Perhaps this may arise from a combination of a relative dearth of economic studies, existing studies not fully capturing the economic impact of NTDs and their control measures, and economic studies not fully translating into decision-making. Therefore, well-designed economic studies and their proper translation into decision-making can be crucial for NTD understanding, advocacy and control.

2. HEALTH AND ECONOMIC BURDEN: COST OF ILLNESS (COI) STUDIES

An important first step in practically all decision-making is to understand and quantify the extent and nature of the problem. Quantifying a problem is particularly useful since subjective terms such as 'substantial' and 'large' may have varying interpretations and do not lend themselves to financial analyses. Typical questions regardless of the problem include the following:

- *What is the magnitude of the problem?* Greater problems warrant expending more time, effort and resources to deal with the problem.
- *Is the problem growing and if so by how much and over what period of time?* Even though a problem may not be substantial today, significant anticipated growth can make the problem worthwhile to tackle soon before it becomes bigger. Quantifying the growth of a problem can help schedule policies, interventions and countermeasures.
- *When will the impact of the problem manifest?* Knowing whether effects of a problem will appear soon or in the future can help with anticipating problems and planning the timing of policies and interventions. When effects take a while to manifest, decision-makers may underestimate the impact *(a feature typical of many NTDs, in which the disease sequelae take time to manifest as a result of cumulative infection).*
- *How will different locations, populations and stakeholders be affected by the problem?* The answers to these questions can help stakeholders to understand how the problem will affect each of them so that they may prepare and plan. This information can also help to target policies and interventions geospatially and coordinate and organize stakeholders.

While noneconomic measures such as disease incidence can help to address some of these questions, they fall short in many ways. First, disease-specific measures such as disease incidence and prevalence are difficult to compare across diseases *(and incidence, the number of new infections per unit time, may be very difficult to quantify in the chronic and insidious NTDs)*. Many stakeholders must make decisions across different diseases. For example, how does one case of hookworm infection compare with one case of leishmaniasis infection and one case of Chagas disease and one case of schistosomiasis infection? Second, even within a given disease, considerable heterogeneity exists. For example, leishmaniasis infections can have different possible outcomes, each having a different level of impact. Third, a number of interventions work across different diseases (e.g. improving sanitation will improve many parasitic and nonparasitic diseases). Fourth, disease-specific measures do not capture all of the positive and negative effects of a disease, as will be explained later. Finally, decision-makers may have to compare health interventions with nonhealth interventions (e.g. a government may have to choose between improving roads or sanitation).

Expanded health effect measures overcome some of the above mentioned limitations but also do not completely characterize the impact of disease and interventions. Common health effect measures include hospitalizations, mortality, various types of morbidity, and other disease outcomes. Some of these may be still difficult to compare across diseases as not all diseases have the same type of outcomes. Also, focussing on a specific set of outcomes may disadvantage those diseases whose common outcomes are not being considered. For example, using mortality or a case fatality ratio as the primary outcome favours those diseases that tend to cause death and does not account for the full impact of chronic ailments that do not cause death immediately (if ever) but can result in substantial suffering (e.g. lymphatic filariasis). Moreover, the effects of diseases go well beyond health effects alone. For example, anaemia from hookworm may not result in hospitalization but could lead to decreased energy and thus hinder productivity at work and advancement in school. Finally, many stakeholders are concerned with impact beyond health effects and rely on economic measures more universally than health effects. For example, telling an employer that a disease will cost her firm millions of dollars may be more relevant than telling her that the disease will result in so many cases of anaemia.

2.1 Costs

One way of quantifying economic burden is to convert all effects into costs, which makes it easier to compare across health and nonhealth problems. Costs are a common measure that a wide variety of decision-makers in many industries and professions understand. Two major sources of costs typically included are,

- *Health care costs*: These include the costs of clinic visits, hospitalizations and various procedures.
- *Productivity losses*: These include time missed from school, work or other productive endeavours from disease symptoms and while receiving health care and/or treatment. Also, premature death results in losing the productivity the person would have had between the age of his or her death and the person's life expectancy (e.g. a person who passes away at 20 years of age and had a life expectance of 75 years old would lose 55 years of productivity).

Depending on where one wants to draw the boundaries, others costs can be included. Potential examples include the costs of developmental and educational delays from illness. However, the causal links with such effects can be harder to draw and quantifying secondary and tertiary effects can be difficult.

A cost of illness (COI) study will typically report the aggregate costs of a disease or a cost per case (or per other epidemiological measure or health outcome) at local, national, regional and global levels. Aggregate costs help to show the overall magnitude of the burden. Cost per case, cost per hospitalization or cost per other epidemiological measure or health outcome provide a unit measurement that helps to extrapolate costs to different locations, times and circumstances. This is particularly helpful when the incidence or prevalence of disease is changing over time. The COI will include all of the direct costs associated with the disease (i.e. costs completely attributable to the disease), such as the materials and personnel time involved in diagnosing and treating the disease. Personnel costs can be a combination of professional charges or the salary needed to pay the health care worker for the time spent diagnosing, treating or caring for the person with the disease. Similarly, costs of clinic visits and hospitalizations can either be derived from charges or aggregating all of the costs associated with the clinic visit or hospital stay (e.g. rent, electricity, amortization of equipment). Note that in many countries, a health care worker, facility or organization often does not receive the same amount charged for a clinic visit or hospitalization. In locations where third-party payers (e.g. government or insurance companies) cover at least some of the costs, how much these third-party payers eventually pay is often the result of negotiations. Therefore, a historic cost-to-charge ratio can convert charges to actual costs. The historic **cost-to-charge ratio** is what proportion of charges tends to be paid over time.

For example, a cost-to-charge ratio of 0.33 suggests that for every USD 1 charged only 33 cents will be paid.

The COI also may include indirect costs, i.e. costs result from the symptoms of the disease (and its treatment) beyond the diagnosis and treatment of the disease. A large source of indirect costs is **productivity losses**. A common proxy for productivity losses is a person's salary. For example, a person infected with hookworm may miss work for medical visits and may suffer anaemia causing reduced work capabilities (anaemia weakens him/her and then makes him/her less effective at work). This may generate productivity losses for absenteeism and reduced work capabilities: for every hour of work a person misses, the lost productivity is equal to the salary that the person would have earned during that hour; if a person operates at 30% effectiveness during an hour of work, the lost productivity is 70% of the person's salary for that hour. Death results in losing that person's remaining expected productivity or the value of his or her expected remaining lifetime earnings. Productivity losses may extend beyond the person afflicted with the disease, by affecting those around them, such as caregivers for children or older adults who make miss work and suffer productivity losses. COI studies are increasingly including other indirect costs such as those emerging from developmental and educational delays and pain and suffering caused by the disease. Indirect costs, such as productivity losses, often are substantially greater than direct costs, so excluding them from the COI can greatly underestimate the economic impact of a disease or its treatments.

The **time horizon** (i.e. the length of time over which the costs and other measures are tabulated) of an economic study is an important consideration. For example, since a disease such as Chagas disease takes years to manifest any symptoms, a COI study would have to have a time horizon well beyond one year to reflect its costs. Of course, the further into the future one must consider, the greater uncertainty is involved (predicting events 20 years from now is certainly more difficult than predicting events for next year). Many uncertain events could occur, including currency fluctuations, changes in treatment and even changes in disease manifestations. Therefore, choosing the time horizon is a balance among the typical natural history of the disease, the data available and the decisions that need to be addressed. Typically, NTDs require longer time horizons than more acute, self-limited diseases such as influenza. This is because NTDs can escape detection for a while and have more insidious chronic effects.

A COI or other economic study that has a time horizon beyond a year should consider **discounting** future costs. Discounting accounts for people's 'time preferences' regarding money – the fact that people prefer to have money now rather than later. In other words, a dollar, pound, euro, yen or rupee 10 years from now is not worth the same as one today. People would rather have the amount now than the same amount years from now. A concrete example is the basis of credit cards, loans and the interest charged. People are frequently willing to eventually pay more (factoring in the interest added) for a product so that they can have it immediately. Essentially, our time preferences keep lenders in business. Discounting is the process of adjusting future costs to an earlier point in time (inflation is the opposite – adjusting costs to future costs). The **net present value** (NPV) is all relevant future costs discounted and past costs inflated to the present or current time. Calculating the NPV brings all costs to a single point in time, allowing us to compare fairly and analyze costs incurred at different points in time. The **discount rate** is the percentage difference in costs from one year to the next year. So if USD 1.03 in 2013 is worth USD 1 in 2012 then the discount rate is 3% or,

$$\text{Discount rate} = (\text{value in year n} + 1 - \text{value in year n})/(\text{value in year n})$$

The discount rate may vary from year-to-year as evidenced by the fact that the consumer price index (CPI), the price level of a sample set of consumer goods and services, fluctuates significantly from year-to-year. Often, economic studies will make the simplifying assumption of using the same discount rate from year-to-year. The most commonly used discount rates for health economic studies have ranged from 3% to 5% (Murray et al., 2012), which have been the average annual increase of measures of cost and price such as the CPI. The following formula adjusts a cost in year n + 1 to a cost in year n,

$$\text{Cost in year n} = (\text{cost in year n} + 1)/(1 + \text{discount rate})$$

So for example, if the discount rate is 3% then a cost of USD 1 in 2014 would be,

$$\text{USD } 1/1.03 = \text{USD } 0.97 \text{ in } 2013$$

Rearranging the formula above allows one to inflate the cost from one year to another,

$$\text{Cost in year n} + 1 = (\text{cost in year n}) \times (1 + \text{discount rate})$$

So, a cost of USD 1 in 2014 would be,

$$\text{USD } 1 \times 1.03 = \text{USD } 1.03 \text{ in } 2015.$$

Discounting over multiple years entails reapplying the formula above for multiple years.

$$\text{Cost in year n} + 1 = (\text{cost in year n} + 2)/(1 + \text{discount rate})$$

$$\text{Cost in year n} = (\text{cost in year n} + 1)/(1 + \text{discount rate})$$

Therefore,

$$\text{Cost in year n} = (\text{cost in year n} + 2)/(1 + \text{discount rate})^2$$

Extending this to any number of years results in the following formula, where n = the number of years in the future a particular cost is incurred,

$$\text{year n} = (\text{cost in year n} + \text{k})\Big/(1 + \text{discount rate})^{\text{k}}$$

In the example of a 3% discount rate, if the cost of the illness in today's dollars is USD 100, its cost in one year will be USD 100/1.03, or USD 97 in today's dollars; the following year, the same costs will equal approximately USD 94 (or USD $100/1.03^2$) in today's dollars.

Determining the total NPV entails summing the NPV of each future cost. The following formula performs this calculation,

$$NPV = \sum_{i=1}^{T} \frac{Ci}{(1+r)^i}$$

where T is the time frame, i is year $1-i$, C is the cost in year i and r is the discount rate. Continuing the example above, if the illness is chronic and runs the course over that third year period (this year, in one year, and in two years), the NPV will be the sum of the discounted costs or USD 191 (USD 100 + USD 97 + USD 94).

It must be kept in mind that applying a standard discount rate across multiple years is an approximation. The discount rate includes the inherent time value of money as well as currency value fluctuations. The latter can vary significantly from year to year. In some years, currency can depreciate. Fluctuations can be especially vast in politically unstable environments. Even regional differences can occur within a country. The standard discount rate mirrors the long-term average change in the time value of money and purchasing power of currency. Depending on the purpose of the study and the precision required, other more specific discount rates may apply.

When the same costs continue indefinitely into the future (i.e. the time horizon is forever and has no limit) we have what is called a **perpetuity**. A perpetuity is a series of periodic payments of equal value (i.e. identical cash flows) with no end. Although perpetual cash flows never end and their total, unadjusted value cannot be calculated, the present value of a perpetuity can be determined because the discounted value of payments far into the future reduces considerably and reaches close to zero. Thus, the future cash flows, after a certain point in time, are eventually presently worth USD 0 and therefore add no value to the NPV of a series of cash flows (i.e. the present value is finite). Therefore, we can estimate the present value of a cost that continues to repeat itself each year forever by considering cash flows up to that point. This is done using the following formula,

$$\text{Present Value} = A/r$$

or the fixed periodic payment (A) divided by the interest rate or discount rate per compounding period (r). As an example, let us assume that NTD control cost USD 1 million annually. If the time horizon of interest is 20 years, we would calculate the NPV using the standard NPV formula above. However, 20 years may be arbitrary. If we want the time horizon to be infinity, we can calculate the present value of the perpetuity. In this example, the present value would be USD 33.3 million (USD 1 million/3%).

2.2 Health effects

Some disease burden studies separate costs from health effects and express these health effects by particular commonly used measures. As mentioned previously, epidemiological measures such as number of cases may be used but do not facilitate cross-disease comparisons as it does not account for potentially vast differences among diseases. Additionally, such epidemiological measures do not account for the wide spectrum of possible clinical outcomes from a disease, and choosing one measure such as deaths alone would unduly favour those diseases that tend to exhibit that outcome (e.g. high-mortality diseases) and disadvantage those that do not (e.g. while Ebola viral infections lead to high mortality, it does not necessarily have a higher burden than STH or many other parasitic diseases that have lower mortality rates). Since the non-life-threatening consequences of diseases vary widely in nature, severity and duration, other measures are needed to compare health effects across diseases. Two frequently employed health effect measures are **quality-adjusted life-years (QALY)** and **disability-adjusted life-years (DALY)**. QALYs and DALYs both measure the quality and quantity of life either of individuals or a population and include the concepts of time, illness, disease and health treatments. QALYs are a measure of years lived in perfect health gained (health gain in quality of life), whereas DALYs are a measure of years of perfect health lost (health loss in the quality of life). Additionally, QALYs focus on health states, while DALYs focus on disease. Although they measure health in different ways, both represent a year of life with a number between 0 and 1. For QALYs, 1 QALY is a completely healthy year of life and 0 is death. Anything in between represents a year of life that is lived in a state of less than perfect health. For instance, 0.75 QALY is a year in life in which the person is 75% functional compared with a fully healthy year of life. By contrast, a DALY is the opposite: how much of a given year is 'lost' to the condition; 1 DALY indicates that the entire year of life was lost, and 0 DALY means that nothing was lost (i.e. the person was completely healthy). QALYs are typically used to generate the estimated number of years that can be added to life for a given intervention or treatment. QALYs are calculated using weights that are related to individual experiences of health, not linked to a particular disease, condition or disability (but often are or can be). These weights are designated to health states based on how people make trade-offs between different dimensions of health (Gold et al., 2002). DALYs measure mortality and morbidity, i.e. years lost due to the poor quality of life due to illness and disability or lifetime lost due to an early death. Therefore, DALYs measure the gap between current health and perfect health (current vs ideal situation) (Gold et al., 2002). DALYs consist of two components, one to measure morbidity and the other mortality: the years lost due to disability (YLD) and years of life lost (YLL) to a particular disease or condition. The disability weight (between 0 and 1) associated with the particular disease/condition multiplied by the duration of and number of persons with that disease/condition determines the YLDs. The number of deaths due to the disease/condition multiplied by the standard life expectancy of the individuals determines the YLL. The disability weight used in DALY calculations quantifies the social preferences for different health states in relation to the societal idea of good health (they do not represent the lived experience of any state or imply societal value for the person in that state). Just as costs are discounted, QALYs and DALYs can be discounted. This reflects the social preference of a healthy year now rather than in the future. As such, the value of a year of life is generally decreased annually by a fixed percentage. It is also important to note that with QALYs it is possible to describe combinations of illness (e.g. comorbidities) as they are linked to health states; this is not possible with DALYs, as they are linked to disease (Gold et al., 2002).

Although these measures are useful, they both face criticisms. For example, defining and measuring weighting factors for health states used in QALYs is subjective and controversial and standard life expectancies may overestimate DALYs saved if the actual life expectancy is shorter. In the context of parasitic diseases, many argue that the disability weight used to assess the magnitude of disease-caused disability is subjective and does not accurately reflect the true burden of disability caused by chronic conditions that can result (van der Werf et al., 2003a; King et al., 2005; AbouZahr and Vaughan, 2000; Reidpath et al., 2003; Jia et al., 2011, 2007). Thus, the actual global burden of NTDs may be higher than even these measures suggest (King and Bertino, 2008).

A challenge is that the impact of NTDs may be more complex to measure. NTDs do not tend to have very acute, salient and severe health effects such as death that can readily draw the attention of stakeholders. For example, mortality for NTDs is not as high as some other higher profile diseases (e.g. HIV/AIDS and malaria). The global burden of disease estimates that NTDs result in 152,300 deaths (2.2/100,000 persons) globally (Lozano et al., 2012). Instead, many of the health ramifications of NTDs are chronic and more insidious. People infected may still be able to perform many of the daily basic functions of life and not require extensive hospitalization. However, their functioning may not be as high as it normally would be. In fact, those infected may not even fully realize that they are infected. For example, STH infection can cause chronic anaemia, resulting in extreme fatigue, loss of physical strength and even cognitive effects, which can in turn decrease current productivity as well as future productivity by impeding educational and cognitive development. Over time such losses can accrue to be substantial (Basta et al., 1979). An additional complication is that there can be a lag time between infection and the manifestation of symptoms. When health effects are more subtle, insidious and in some cases delayed, it can be more difficult to attribute the health effects to the infection. Other conditions could be causing similar health effects (e.g. iron deficiency and malaria can cause anaemia), and infected persons may not seek medical care and therefore never be properly diagnosed. As a result, considerable underreporting of NTDs may be occurring.

Additionally, the strong association of NTDs with poverty makes it challenging to disentangle cause and effect. The effects of NTDs may greatly impair a population's ability to overcome poverty. A population continuously contending with chronic symptoms may not have the wherewithal or reserve to improve their conditions. At the same time, an environment of poverty (including poor sanitation and hygiene, inadequate housing and lack of access to medical care) encourages transmission, exacerbates symptoms and the resulting impact, impedes monitoring and delays diagnosis and proper treatment (WHO, 2006). Since the majority of those infected with NTDs live in regions with ineffective or nonexistent reporting mechanisms, many cases, deaths and costs go unreported. Even in places where surveillance infrastructure exists, many cases go unreported because people focussed on the daily challenges of a low-income environment may not have the time or motivation to seek treatment for general, chronic symptoms. Moreover, low-income environments have many other competing diseases and conditions that draw the attention of the population and health care workers. For example, in malaria-endemic locations, a health care worker may be more likely to ascribe symptoms to malaria or focus on treating those with malaria rather than treating NTDs.

Another issue is the major overlap and interactions among different NTDs and NTDs with other diseases such as HIV, hepatitis C and malaria. Since many parasitic diseases have similar or identical vectors for transmission and NTDs tend to be closely associated with poverty, there is considerable overlap in the populations and locations affected by different NTDs, making it difficult to separate the impact of one NTD from that of another NTD. An estimated 74% of countries affected by NTDs, must combat two or more NTDs (WHO, 2006). Additionally, a number of NTDs may have strong interactions with other diseases, such as HIV/AIDS, hepatitis and malaria. The presence of one disease can increase the risk of developing another disease. One disease can worsen the symptoms and clinical outcomes of another disease (e.g. HIV infection can lead to greater disfigurement caused by cutaneous leishmaniasis (Kruchten et al., 2014)) and thus alter the resulting epidemiological, clinical and economic impact (Abuhab et al., 2013; Fincham et al., 2003; Spiegel et al., 2003).

2.3 Example of a COI model: Chagas disease

To illustrate how one would develop a COI study, let us use as an example a published study on the cost and health burden of Chagas disease (Lee et al., 2013). Development of this study and model proceeded in the following seven steps:

Step 1: Determine the question of interest

The goal of the study was to provide stakeholders such as policy makers, funders, disease control officials and intervention developers with a better understanding of the burden of Chagas disease and how this burden varied geographically. The study entailed developing a stochastic computational economic model of all the countries that had reported cases of Chagas disease. Since Chagas disease is a chronic disease with a lengthy latent period (frequently over 10 years), estimating the burden of Chagas disease can be challenging without the help of a computational model. Most of the effects accrue years after initial infection. Clinical courses can be quite varied with considerable implications regarding costs and health effects. Profiling the burden of Chagas would help decision-makers to know where Chagas should fall on their priority lists, which countries to target for policies and interventions, how much to invest in Chagas disease prevention and country and who will be affected and to what degree.

Step 2: Establish the perspective and time horizon

As indicated before, the perspective(s) and the time horizon of the study/model can substantially impact the design and results of the model and study. For an initial global burden study such as this was, the societal perspective is most common. Understanding how a disease may affect all of society helps prioritize the disease among the many different disease, conditions and problems that policy makers need to address. As mentioned above, the societal perspective incorporates both direct health care costs and indirect costs such as productivity losses. Another common perspective for such burden studies is the third-party payer perspective, which focusses exclusively on direct health care costs. For Chagas disease, the third-party payer perspective would give governments, countries and insurance companies insight into how much this problem is taxing their health care systems.

The lifetime of individuals infected seemed to be a reasonable time frame for this study. Any shorter time frame would not capture all of the major relevant costs and health effects. In fact, since the latent period is so long, a time frame of less than two decades could potentially miss all of the costs and health effects for many infected individuals.

Step 3: Choose/design the appropriate model structure

For this study, a **Markov model** could readily represent the long, many-year and varied potential clinical course of Chagas disease. Markov models are particularly useful in decision problems when risk of particular events is continuous or ongoing over time, timing of events is important, the timing at which events occur is uncertain and events may happen more than once (Sonnenberg and Beck, 1993). In a Markov model, a person is always in one of a finite number of discrete health states (Markov states), and the events of interest are transitions from one state to another.

Figure 1 represents the structure of the Chagas model and its various Markov states through which individuals could cycle. The person stays in a given health state for the duration of a Markov cycle. At the end of each cycle, the person has probabilities of either remaining in the same state or moving ('transition') into another state. The arrows in Figure 1 show how the possible paths that an individual can take each cycle. The length of a cycle should be short enough to account for the different clinically meaningful outcomes that may occur and paths that a person may take, but not so short to make the model unwieldy or require data that are not readily available (Sonnenberg and Beck, 1993). Remaining in a given state for the duration of a cycle may accrue certain costs and health effects. Similarly, transitioning from one state to another may also accrue costs and health effects. As can be seen in Figure 1, the Chagas disease model consisted of five disease states: acute disease, intermediate disease, cardiomyopathy with or without congestive heart failure (CHF), megaviscera (enlargement of the oesophagus or colon) and death. The death state was an absorptive state, meaning that once an individual transitioned into this state, the individual could no longer move to other states. A cycle length of 1 year seemed short enough to capture the heterogeneity in an individual's possible pathways (e.g. a person is less likely to move amongst multiple states in less than a year) but long enough to have data to be reasonably parameterized. A two-year cycle may have been too long. For instance, during two years, one individual

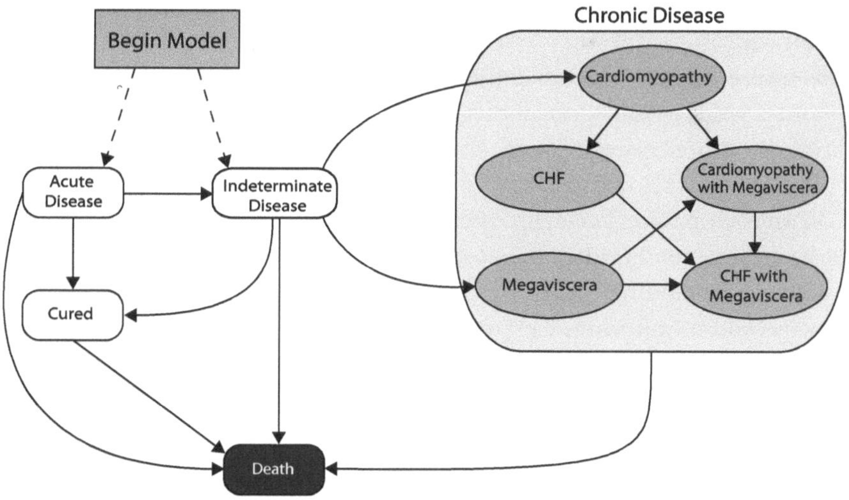

Figure 1 Chagas disease model structure and Markov states. *Reprint from The Lancet, 13, B.Y. Lee, K.M. Bacon, M.E. Bottazzi, P.J. Hotez, Global economic burden of Chagas disease: a computational simulation model, 342–348, Copyright (2013), with permission from Elsevier.*

could have had CHF for the first year and passed away near the end of that first year while another individual could have had CHF for almost the entire two years and passed away at the end of the second year. Even though a model with a two-year cycle length would have handled both individuals the same way, their costs and health effects would have been quite different. Finding data for a model with a 1-month cycle length would have been challenging. There are not enough studies available that show an infected person's monthly risk of developing different clinical outcomes. Most available studies stratify risk on an annual basis.

In the Chagas disease model, all individuals began as infected in either the acute disease or indeterminate states. Upon initial infection with *Trypanosoma cruzi*, 5% of newly infected individuals began in the acute disease state, characterized by minor symptoms (e.g. fever, rash, swelling and nausea) or more serious outcomes (e.g. myocarditis or meningoencephalitis) that could lead to death. Individuals surviving acute infection underwent treatment that included either benznidazole or nifurtimox (and persons accrued the associated costs) and then became symptom-free within 1 year (i.e. 1 cycle). The remaining 95% of infected individuals began in the asymptomatic intermediate disease state, remaining in this state for at least 9 years or 9 cycles (based on the reported time from acute illness to cardiac symptoms). After this 9-year latent period, an individual had a probability each cycle of transitioning to one of the chronic disease states (cardiomyopathy or megaviscera). As can be seen in Figure 1, an individual entering one of these chronic disease states then would have probabilities in progressing to more severe disease states in subsequent cycles. Combinations of different outcomes were possible. An individual with cardiomyopathy had a probability of developing megaviscera, and one with megaviscera had probabilities of developing cardiomyopathy and CHF.

With each passing cycle, an individual in a given state could accrue costs and health effects. Health care costs came from treatment, diagnostics and monitoring, including, where applicable, costs for initial consults, regular checkups and monitoring. Certain states had probabilities of specific clinical outcomes that could bring additional costs. For example, an individual in the cardiomyopathy states with or without CHF had probabilities of undergoing pacemaker implantation (which reduced mortality risk) and an annual probability of developing cardiomyopathy or CHF. Each individual could have up to two pacemaker implantations, with the second being less likely than the first. Megaviscera affected either the colon or oesophagus and could require surgery with an accompanying risk of death. Overall, because therapeutics for Chagas disease are relatively ineffective in chronic disease states, individuals with chronic disease had no possibility of cure. As described above, death is an absorptive state (i.e. individuals in this state leave the model). Individuals could enter this state from any state (Figure 1) and could result from either Chagas disease-related causes or other causes per age-adjusted mortality (this accounts for the probability of death from non-Chagas-related causes throughout the course of their lifetime).

Step 4: Establish the model outcome measures

The main outcome measures for the Chagas disease model were both direct and indirect costs as measured by currency and health effects as measured by DALYs. Therefore, each event or outcome in the model accrued certain quantities of these measures. In this study, DALYs were converted into productivity losses (i.e. wages adjusted by DALYs) to generate indirect costs. Death resulted in years of life lost (YLL) based on a person's age and the average life-expectancy.

Step 5: Identify appropriate data sources and populate/calibrate the model

Table 2 shows the disparate sources and the input values for the major model parameters. Costs for acute disease included initial consultation, general and specific diagnostic tests and treatment. Costs for indeterminate disease included periodic medical checkups with laboratory tests and X-rays. To represent variability and uncertainty, many of the parameters draw from distributions rather than single values. Even when fairly precise data are available, distributions account for person-to-person, event-to-event or circumstance-to-circumstance variation. As is often the case with NTDs, distributions also account for the lack of precise, quality or generalizable data. For instance, for the Chagas model, a total of 33 countries across four regions had readily available data. For some of the remaining countries, extrapolation from countries with existing data was necessary and based on a country's gross domestic product (GDP) per person. Sensitivity analyses helped to account for inaccuracy in extrapolation. The model used a 3% discount rate to adjust all costs to present-day values.

Step 6: Run baseline scenario(s)

The baseline scenario assumed the parameters presented in Table 2. This study ran the baseline scenario for each of the four GDP quartiles. Each simulation involved sending 1,000 individuals through the model 1,000 times (resulting in 1 million total iterations). Thus, each simulation run pulled a different value from the parameter-specific distribution (described above); this is known as **Monte Carlo simulation** or **probabilistic sensitivity analyses**. The simulation then resulted in a distribution of outcomes

Table 2 Input parameters and values for the global burden of Chagas disease study[a]

Parameter	Country income quartile	Value	Standard deviation or range	Source
Diagnosis, treatment and monitoring costs (2012 USD)				
Acute disease	Low income	257	–	Literature
	Low middle income	785	–	Literature
	High middle income	2,613	–	Literature
	High income	4,829	–	Literature
Indeterminate disease	Low income	72	–	Literature
	Low middle income	221	–	Literature
	High middle income	1,603	–	Literature
	High income	3,280	1,230	Literature
Chronic disease (cardiomyopathy)	Low income	612	–	Literature
	Low middle income	805	75	Literature
	High middle income	4,463	9,238	Literature
	High income	6,006	–	Literature
Chronic disease (CHF)	Low income	1,609	–	Literature
	Low middle income	2,701	2,561	Literature
	High middle income	5,224	3,073	Literature
	High income	9,495	15,952	Literature
Chronic disease (megaviscera)	Low income	705	–	Literature
	Low middle income	1,044	–	Literature
	High middle income	2,217	–	Literature
	High income	3,640	–	Literature

Chronic disease (surgery for megacolon)	Low income	36	–	Literature
	Low middle income	111	–	Literature
	High middle income	1,204	–	Literature
	High income	2,527	2,610	Literature
Chronic disease (pacemaker implant)	Low income	649	–	Literature
	Low middle income	1,986	–	Literature
	High middle income	6,608	–	Literature
	High income	31,252	0.23–3.34%	Literature
Duration of absenteeism (days)				
Acute disease	–	11	7–15	Literature
Indeterminate disease	–	5	–	Literature
Chronic disease (cardiomyopathy or CHF)	–	15	–	Literature
Chronic disease (megaviscera)	–	11	7–15	Literature
Megacolon surgery	–	135	–	Literature
Risk (given* Trypanosoma cruzi *infection) (in percent)				
Acute disease	–	5.0	–	Literature
Chronic disease (cardiomyopathy, annual, endemic)	–	2.0	–	Literature
Chronic disease (CHF, annual)	–	4.0	–	Literature
Chronic disease (megaviscera, annual)	–	0·0225	–	Literature
Probabilities of seeking treatment (in percent)				
Acute disease	–	34.0	–	Literature
Indeterminate disease	–	50.0	–	Literature

(*Continued*)

Table 2 Input parameters and values for the global burden of Chagas disease study[a]—cont'd

Parameter	Country income quartile	Value	Standard deviation or range	Source
Chronic disease (cardiomyopathy or CHF)	—	78.0	—	Literature
Chronic disease (megaviscera)	—	78.0	—	Assumed same as cardiac disease
Megaviscera surgery ($\leq$2 surgeries)	—	5.0	—	Literature
First pacemaker implant	—	3.5	0.4–6.67	Literature
Second pacemaker implant	—	1.75	0.23–3.34	Literature
Cure, acute disease	—	65.0	50–80	Literature
Cure, indeterminate disease	—	8.0	6–10	Literature
Treatment side effects	—	5.0	—	Literature
Probabilities of mortality				
Acute disease (annual)	—	5.0	—	Literature
Indeterminate disease (annual)	—	0.18	0.17–0.19	Literature
Chronic disease (cardiomyopathy, annual)	—	4.2	—	Literature
Chronic disease (CHF, annual)	—	30.0	—	Literature
Megaviscera surgery	—	2.25	—	Literature

CHF, congestive heart failure.
[a]Literature, as cited in Lee et al., 2013.

of which we can report the mean, median, distribution type, spread or other outcome measure to describe the variability of outcomes. Multiplying the costs (direct and indirect) and DALYs per case by the cases per location yields the burden estimates for countries, regions (Latin America, Europe, the USA and Canada and Australia) and the world.

Step 7: Conduct sensitivity analyses

In addition to the probabilistic sensitivity analyses (described above) that were performed in each simulation, one-way sensitivity analysis explored the effects of varying individual parameters. These parameters included the treatment seeking probability (±10% for nonsurgical treatment and ±1% for surgical treatment), treatment costs (±5%), duration of absenteeism (±5%), and age at initial infection (0–50 years), and annual cardiomyopathy risk (0.47–2.0%). Each of these parameters may vary from country to country and have an impact on the burden of Chagas disease. The initial age of infection was varied to account for all forms of transmission, such as congenital, vector-borne, and transfusion-related. The annual risk of cardiomyopathy was varied based on the literature for endemic and nonendemic countries, as it is reportedly lower in nonendemic countries.

Table 3 summarizes results from this study. On average, an individual with chronic infection incurred USD 474 in health care costs (range: USD 222–USD 914) and 0.51 DALYs (range: 0.38–0.60) annually, resulting in an average NPV total lifetime of USD 3,456 (range: USD 2,623–USD 4,060). The global burden of Chagas disease exceeded that of cholera (USD 5.43 annually) and rotavirus (USD 2.0 billion), two diseases that until the time of the study had received greater media exposure. The majority (~87%) of Chagas disease costs (USD 164 billion of a total USD 188.8 billion annually) stem from productivity losses so that concentrating on direct health care costs will grossly underestimate the burden of Chagas disease. Although Chagas disease is endemic in Latin America, the burden in other locations not traditionally associated with Chagas disease (e.g. USA and Canada and Europe) was not insignificant.

2.4 COI studies in the scientific literature

The Chagas disease study is one example of an NTD economic burden study. Table 4 lists some major NTD COI studies. Not all NTDs have comprehensive COI studies. Of the ones that do, there is considerable variation in the methodology, data sources and scope, making them difficult to compare. Many of the COI studies centre on a specific location, which is helpful for that location but may limit generalizability. Some are even

Table 3 Global health and economic burden of Chagas disease (from Lee et al., 2013)

	Latin America	USA and Canada	Europe	Australia	Global
Annual burden					
Health care cost	492 (179–983)	119 (6–597)	17 (2–58)	0.3 (0.07–0.9)	627 (186–1,639)
Total cost	6,182 (3,782–8,095)	865 (65–2,871)	140 (19–311)	3 (1–4)	7,189 (3,867–22,281)
DALYs	772,404 (396,255–1,063,932)	27,687 (1,823–99,384)	6,093 (742–14,690)	85 (22–153)	806,170 (398,840–1,178,159)
Net present value of currently infected cases					
Health care cost	17,116 (13,301–19,213)	6,769 (1,354–16,263)	828 (312–1,368)	21 (16–25)	24,733 (14,984–36,843)
Total cost	152,098 (94,480–524,778)	31,751 (3,888–103,715)	4,849 (1,071–10,853)	98 (46–159)	188,797 (99,485–382,089)
DALYs	28,017,511 (9,392,622–44,2466305)	1,123,552 (83,648–4,470,747)	240,731 (32,485–652,598)	3,456 (987–6,868)	29,385,250 (9,509,737–49,396,520)

Note: costs are USD in millions, range is variation from sensitivity analysis.
DALYs, disability-adjusted life-years.

Table 4 Summary of studies on the health and economic burden of parasitic neglected tropical diseases

Disease	Setting	Health and/or economic burden
Chagas	Latin America	USD 6.2 billion and 772,404 DALYs annually (Lee et al., 2013)
Chagas	USA and Canada	USD 865 million and 27,687 DALYs annually (Lee et al., 2013)
Chagas	Europe	USD 140 million and 6,093 DALYs annually (Lee et al., 2013)
Chagas	Australia	USD 3 million and 85 DALYs annually (Lee et al., 2013)
Chagas	Mexico	USD 10,160 lifetime cost for timely diagnoses and treated case USD 11,877 lifetime cost for undiagnosed case (Ramsey et al., 2014)
Chagas (chronic cardiomyopathy with and without congestive heart failure)	Colombia	USD 1,028 annual treatment cost per patient; USD 11,619 lifetime treatment cost per patient (Castillo-Riquelme et al., 2008)
Cystic echinococcosis	Global/worldwide	No under-reporting: USD 763,980,979 total costs and 285,407 DALYs lost Adjusting for under-reporting: USD 1,918,318,955 total costs and 1,009,662 DALYs lost (Budke et al., 2006)
Echinococcosis	Ningxia Hui Autonomous Region, People's Republic of China	Cystic echinococcosis: Mean USD 220 (USD 30–USD 4,189) Alveolar echinococcosis: Mean USD 243 (USD 30–USD 1,500) (Yang et al., 2006)
Cystic echinococcosis	Iran	USD 93.39 million (95% CI: USD 6.1–USD 222.7 million) annually (direct and indirect costs) USD 1,539 per surgical case (Fasihi Harandi et al., 2012)
Cystic echinococcosis	Spain	EUR133.4 million (95% CI: EUR6.7–EUR13.4 million) direct and indirect costs in 2005 (Benner et al., 2010)
Echinococcosis	Tunisia	USD 6.3 million in total costs and USD 0.81 million in lost productivity due to disability (Majorowski et al., 2005)

(Continued)

Table 4 Summary of studies on the health and economic burden of parasitic neglected tropical diseases—cont'd

Disease	Setting	Health and/or economic burden
Cystic echinococcosis	Peru	USD 2.4 million (95% CI: USD 1.1−USD 4.8 million) in total costs; 1,139 DALYs due to surgical cases (Moro et al., 2011)
Cystic echinococcosis	China	USD 1,493 per person in hospitals costs; USD 1,436 per person in indirect costs; 1.03 DALYs per person (Wang et al., 2012)
Cystic echinococcosis	India	USD 8.75 million (Singh et al., 2014)
Human African trypanosomiasis	Serere, Uganda (population 46,455)	USD 147 per patient (treatment costs); 1,430 age-weighted DALYs (17 DALYs per case) (Fevre et al., 2008)
Visceral leishmaniasis	Italy	EUR1,370,228 in direct costs for ordinary and day-hospital admissions (Mannocci et al., 2007)
Visceral leishmaniasis	Sudan	USD 450 median total cost per episode (IQR: USD 387−USD 544); USD 211 median cost per episode for health facilities; USD 18 in direct costs and USD 22 in indirect costs per episode for households (Meheus et al., 2013)
Visceral leishmaniasis	Bihar, India	USD 131 total median treatment costs per person; USD 210 total median treatment and income loss costs to households (Sarnoff et al., 2010)
Visceral leishmaniasis	Bihar, India	USD 127 median cost per patient, USD 83 of which was medical care (Sundar et al., 2010)
Visceral leishmaniasis	Nepal	USD 165 median total cost per episode for household (Uranw et al., 2013)
Schistosomiasis	Ghana	EUR1.81−EUR2.13 per person in health care costs (van der Werf et al., 2003b)

DALYs, disability-adjusted life-years; IQR, interquartile range.

further centred on particular situations or types of costs. The following subsections summarize the studies in Table 4 by NTD.

2.4.1 Chagas disease

One modelling study of Chagas disease in Mexico concentrated on the COI that would arise with earlier and more timely diagnosis and if undiagnosed. The study found that early and timely diagnosis resulted in lower lifetime costs than an undiagnosed case, and a majority of costs resulted from lost working days (i.e. productivity losses) (Ramsey et al., 2014). This study utilized a Markov model (states included: acute phase, indeterminate asymptomatic phase, symptomatic chronic phase, no progression and death) simulating the lifetime of a Chagas disease cohort to evaluate three scenarios: (1) all individuals detected and treated early (acute phase of disease), (2) all individuals detected, but only 80% treated (assumes patient refusal or those not clinically capable of receiving treatment) and (3) no diagnosis or treatment. Future costs were appropriately discounted (5% rate used) and included direct and indirect COI. Results were stratified by disease phase and cost category, resulting in a total lifetime cost of USD 10,160 for timely diagnosis and treatment (compared with USD 11,877 with no diagnosis). Although the study performed Monte Carlo simulations for parameters with distributions, additional sensitivity analyses were not performed. Additionally, the authors did not include productivity losses during the acute phase of illness (since the cohort age is 10 years old), however most modelling studies would still consider productivity losses, regardless of age, as everyone contributes to society, even if unemployed.

Another Chagas study concentrated specifically on treatment costs. This study used a model to estimate the treatment costs of chronic Chagas patients with cardiomyopathy (with and without CHF) in Colombia from the payer perspective (Castillo-Riquelme et al., 2008). Data came from a retrospective review of treatment costs (63 patient records from 3 different hospitals). Treatment costs (from patient records) were from payer perspective, were stratified by levels of care (basic, intermediate, specialized), and included diagnostic testing, surgical procedures, rehabilitation, ambulatory care and hospitalization; future costs were discounted. The mean cost per patient was USD 1,028 annually and USD 11,619 over a lifetime (estimates weighted by care seeking and utilization patterns). Cardiomyopathy with CHF cost more, as did seeking care at higher levels. This study provides detailed estimates of the cost of chronic Chagas at different levels of care and for those with and without CHF; it also provides cost details for the services utilized. However, this study did not appear to perform a sensitivity analysis around any estimates, only using average values.

2.4.2 Human African trypanosomiasis

One COI study evaluated human African trypanosomiasis (HAT), but focussed on a specific outbreak. This study explored the health system costs and age-specific and total DALY burden of HAT based on real data from an outbreak in a village in Uganda (Fevre et al., 2008). This study included all patients seeking treatment during the outbreak, used health record data, accounted for under-reporting, specifically evaluated *Trypanosoma brucei rhodesiense*, and evaluated uncertainty in parameter estimates. Results showed that under-reporting accounts for 93% of the DALY estimate; 1,431 total DALYs with under-reporting compared with 101 DALYs with no under-reporting. Assuming no under-reporting, the highest DALY burden was among the 5- to 14-year olds; however, with under-reporting, the highest burden was among the 15- to 29-year olds. The total cost of treatment (drugs and hospital costs) was USD 11,961 to the health system (USD 147/patient). While this study highlights the burden of HAT, the true burden may be even more substantial as this study did not include productivity losses and costs to patients.

2.4.3 Echinococcosis

A few COI studies for echinococcosis evaluated the burden in both human and animal population for different countries (Benner et al., 2010; Fasihi Harandi et al., 2012; Majorowski et al., 2005; Moro et al., 2011; Singh et al., 2014; Wang et al., 2012; Budke et al., 2006). A substantial proportion of human symptomatic cases go unreported, leaving the true prevalence of disease unknown. Thus, many report on the cost of surgical cases, while, some have attempted to estimate the number of undiagnosed or asymptomatic cases. Therefore, many existing COI studies for echinococcosis probably underestimate the true economic burden.

Studies calculating the direct and indirect costs of echinococcosis for the human and livestock populations concentrate on Tunisia, India, Iran and Spain (Benner et al., 2010; Fasihi Harandi et al., 2012; Majorowski et al., 2005; Singh et al., 2014). All of these are modelling studies, but employ different model types. While two of these studies are limited to surgical cases (Fasihi Harandi et al., 2012; Majorowski et al., 2005), the other two estimated the number of asymptomatic or undiagnosed cases following a similar extrapolation method (Benner et al., 2010; Singh et al., 2014). Data for these studies came from the literature, local reports, hospital databases and other sources. Direct costs included hospitalization, drug treatment, surgery, diagnostic costs and clinical testing. These studies described indirect costs as a percentage of reduced productivity during the given time span; additionally, some adjusted for factors such as age, gender and employment status. All of these studies accounted for uncertainty in estimates by modelling distributions (mostly by Monte Carlo simulation) and/or performing sensitivity analyses. In Iran, surgical cases of echinococcosis cost an estimated USD 93.39 million, with USD 1.09 million in direct costs (1.2% of total) and USD 92.34 million in indirect costs (Fasihi Harandi et al., 2012). In India, total costs were estimated at USD 8.75 million (95% confidence interval: USD 61–USD 13.6 million) (Singh et al., 2014). In Spain, echinococcosis estimated to cost EUR603,671 in direct costs and EUR132,795,199 in indirect costs (when including asymptomatic or undiagnosed cases) or EUR274,643 in indirect costs (excluding asymptomatic or undiagnosed cases) (Benner et al., 2010). These studies show that lost productivity can be substantial. They are estimated at USD 0.81 million in Tunisia (Majorowski et al., 2005) and cost an estimated EUR132,795,199 (99.5% of the total human population costs and 89.1% of the total burden) in Spain (in 2005) (Benner et al., 2010). These studies not only highlight the large burden resulting from productivity losses but the overall contribution of infection in the human population to the total economic burden (with the remaining burden incurred by livestock). Costs in the human population contributed to approximately 40% of the total burden in Iran (Fasihi Harandi et al., 2012). In India, the human population represented 4.2% of the total burden (USD 8.75 million of USD 212.4 million) (Singh et al., 2014). In Spain, the human population constituted 90% of the total burden when productivity losses for asymptomatic or undiagnosed cases are included, but only 5.3% of the total burden when these losses were not included (Benner et al., 2010).

Two other studies also estimated the health burden of echinococcosis. One of them was a medical record review of 2,018 patients from 5 hospitals in China (between 2004 and 2008) to estimate the direct and indirect costs of cystic echinococcosis (Wang et al., 2012). Investigators used the human capital method (i.e. using the gross national product (GNP), DALY and productivity weight) to estimate indirect costs. Because currently there is no disability weight estimate for echinococcosis, the authors used that of liver cancer as the clinical symptoms are similar between the two. The overall median hospitalization cost was USD 1,231 per patient (medication was highest cost component) and the indirect cost was USD 1,436 per person, with 1.03 DALYs lost per case. Age-specific results showed 20- to 25-year olds incurred the highest DALY losses (1.37 for males and 1.52 for females), while 30- to 35-year olds incurred the largest economic burden (USD 1721/person). The study stratified the burden by age group, showing how DALYs and costs may correlate with age; however since the study was restricted to five hospitals in China, the generalizability is unclear. The second study estimated the burden of cystic echinococcosis in Peru (Moro et al., 2011). This study assigned costs and DALYs to only certain types of cases (i.e. surgical cases), while assigning productivity losses to surgical cases, outpatients and asymptomatic cases. In Peru, cystic echinococcosis surgical treatment cost an estimated USD 836,064 annually and resulted in 1,139 DALYs per year. Productivity losses cost USD 1,592,764 annually, for a total burden estimate of USD 2.4 billion per year. The economic burden may be potentially underestimated as the costs of outpatient cases do not appear to have been considered.

One study modelled the global economic and DALY burden of cystic echinococcosis (Budke et al., 2006). The incidence estimates came from the annual number of detected cases per susceptible population for each country with reported echinococcosis cases. Additionally, the authors assumed ~10% of annual cases are undiagnosed and do not receive treatment. The study considered direct costs only for surgical cases (including diagnostics, surgery, hospitalization and postoperative costs), but indirect costs (e.g. wage losses) and DALYs for surgical and undiagnosed cases. Disability weights derived from liver cancer weights (as the outcomes are similar and echinococcosis-specific weights are not available). The analysis estimated the global burden with and without an under-reporting factor (a fourfold increase to account for under-reporting of patients who received treatment). Globally, echinococcosis cost USD 193,529,740 in direct and indirect costs and results in 285,407 DALYs annually, assuming there is no under-reporting. When accounting for under-reporting, direct and indirect costs totalled USD 763,980,979 and DALYs totalled 1,009,662 annually. While this study used Monte Carlo techniques, quantity and quality of data determined the probability distributions for each parameter, rather than variation in reported values. Additional assumptions surrounding these distributions may lead to wide variation in results (e.g. assumed normal distribution for incidence and uniform distributions over a wide range for other parameters) and a set retirement age of 65 estimated lost productivity for mortality, which may not be the same in each country nor accurately reflect variation between countries. Additionally, this study did not include alveolar echinococcosis cases and may underestimate the true burden.

2.4.4 Leishmaniasis

Five studies evaluated the burden of leishmaniasis (Mannocci et al., 2007; Meheus et al., 2013; Sarnoff et al., 2010; Sundar et al., 2010; Uranw et al., 2013), all of which were done by medical record review or surveys, with most reporting costs incurred by the household. In most cases, visceral leishmaniasis cases need to be hospitalized for the full duration of treatment. In Italy, using data from discharged hospitalized cases from 1993 to 1998, direct costs totalled EUR1,370,228 in 2003 (Mannocci et al., 2007). In Sudan, the median total cost of inpatient care per episode of visceral leishmaniasis ranged from USD 117 to USD 366 per patient (provider perspective), varying with the cost of a hospital day (Meheus et al., 2013). The median medical cost was USD 45 per patient, with the cost of drugs representing 91%. Only inpatient care was considered in this study (as patients are hospitalized for the full duration of treatment) and costs included hospitalization (hotel unit cost, which included recurrent expenditures and capital costs), drugs, diagnosis, medical supplies and laboratory investigation that came from medical record reviews or health facility records.

Studies also reported the burden leishmaniasis poses on households. Two studies evaluated the household cost of visceral leishmaniasis illness in Bihar, India, as both of these studies were surveys and are subject to recall bias (Sarnoff et al., 2010; Sundar et al., 2010). One study estimated the average overall annual household expenditures were USD 1,312 and medical care compromised most of the cost; households could not afford treatment as 87% had to take out loans (Sarnoff et al., 2010). The other estimated a median USD 127 per patient, which is more than 3 months of income (Sundar et al., 2010). The direct household costs in Sudan averaged USD 212 per episode; households bore 53% of total costs (despite free drugs being available), representing 40% of the annual household income, with more than 75% incurring high out-of-pocket expenses (Meheus et al., 2013). In Nepal, the total cost (direct medical, nonmedical and indirect costs) of visceral leishmaniasis was an estimated USD 165 (direct medical costs included out-of-pocket expenses only), this represents 11% of annual household income; 56% of households take out loans to cover these expenses (Uranw et al., 2013).

2.4.5 Schistosomiasis

One study evaluated the functioning of the Ghanaian health system with regards to diagnostics and treatment of schistosomiasis (van der Werf et al., 2003a). This study interviewed health care workers (interviews conducted in 2000) in different regions of Ghana to determine the cost of schistosomiasis treatment. Interviews with health care workers describing how they would treat a patient with schistosomiasis determined the costs that the patient would accrue. For example, the cost of diagnosis was included only if the health care worker reported that it would be utilized. Costs were specific to the level of care, based on referral responses of the health care worker. Besides finding that many cases are unrecognized and do not receive adequate treatment, the study concluded the drug praziquantel was not available in many clinics, the overall cost of treatment ranged from EUR1.81 to EUR2.13 per patient (depending on *Schistosoma* species), with drug costs representing approximately 40% of the total cost. The results of this study may not be generalizable outside this region.

2.4.6 Summary of COI studies

Differences among many of these studies' methodologies and scope hinder comparisons. Studies evaluated the COI from a range of perspectives (societal, provider, households), each of which has different cost inclusions. In addition to these, the types of costs included vary from study to study, even when the studies took the same perspective. Some only included household cost for treatment and income loss; others considered food and transport costs, hospitalization costs, outpatient, laboratory tests, annual screening, clinical procedures, medicines and productivity losses. In contrast, others calculated hospital 'hotel' unit costs including data for recurrent expenditures (e.g. administration costs, building maintenance, utilities) and capital costs (e.g. buildings, equipment, vehicles, furniture) (Meheus et al., 2013). Estimates for echinococcosis used treatment costs only for those who are diagnosed or undergo surgery but include productivity losses for the undiagnosed (Budke et al., 2006). Additionally, these costs may not be generalizable to other areas or countries, as in-country variation exists. Thus care should be taken when making comparisons between NTDs and even between studies evaluating the NTD.

The methods for calculating productivity losses also varied: lost work days, DALYs, productivity loss percentage or productivity loss weight — making it difficult to compare these estimates. Although the methods differ, studies report on the proportion of the economic burden represented by productivity losses. For example, lymphatic filariasis results in 0.81 h lost per work day, with 88% of patients reporting absenteeism (Keating et al., 2014). For Chagas disease, most costs are due to missed working days, between 44% and 75% depending on timing of diagnosis and treatment (Ramsey et al., 2014). Productivity loss estimates for visceral leishmaniasis range from USD 22 to USD 210 per episode for a household (Meheus et al., 2013; Sarnoff et al., 2010), representing 53% of household costs (Uranw et al., 2013) and are an estimated USD 1,436 per person for echinococcosis (Wang et al., 2012).

While these studies initially start to paint the picture and landscape of the burden of parasitic NTDs, existing gaps in the literature leave much work to be done. For some NTDs, a few studies have estimated their burden (e.g. visceral leishmaniasis and echinococcosis), although they are highly variable and may not include all the effects. Burden estimates for others (dracunculiasis, HAT, lymphatic filariasis, schistosomiasis, and the STHs) are severely lacking.

3. COST OF INTERVENTIONS

Knowing the cost of an intervention and each of its components can help to guide its design, choice and implementation. Cost of intervention studies quantify the overall cost of an intervention and its various components. A cost of intervention study can determine whether the intervention's overall cost may preclude its use. For instance, testing an entire country's population each month for the presence of hookworm and then giving everyone who tests positive medications may be very effective in controlling hookworm but much too expensive to employ. The personnel, facilities, transportation and time required to operationalize these efforts for an entire country every month may be prohibitively expensive. Thus, an alternative less resource-intensive intervention may be necessary. Specialized subsets of cost of intervention studies are cost of diagnosis, cost of vaccination, cost of vector control, costs of prevention and cost of treatment studies. All are similar but focus on different types of interventions. Their specific cost components may be different (e.g. the cost of vector control may not include the cost of health care workers).

Cost of intervention studies also may show how costs may change when different components of the intervention are altered, which can, in turn, help to design and implement the intervention. In essence, the study profiles what aspects of an intervention are more resource intensive than others (e.g. if the cost of delivering goods to a particular region are found to be very high, research may focus on how to deliver the goods to the region in a cheaper manner). Different scenarios and sensitivity analyses can identify the main cost drivers for the intervention.

Costs can be fixed or variable. Fixed costs do not depend on the volume of products or services produced or provided. By contrast, variable costs change based on volume. An example is the cost of a building that is purchased or rented to serve as a storage location for vaccines. The cost of the building will not change based on the number of vaccines stored each day in the building. More vaccines delivered do not mean higher building cost. The cost of vaccines in an immunization programme, however, is variable. If more vaccines are required, delivered and used, the costs of the programme increase. A certain component can have both fixed and variable costs. For example, in producing vaccines, the costs of initially setting up a manufacturing line may be fixed. The cost of purchasing equipment may be the same if you are producing 100 vaccines versus 10,000. However, the cost of materials to produce the vaccines will depend directly on the number of vaccines produced.

Similar to the COI, the cost of an intervention may include direct and indirect costs. Direct costs include the producing, delivering and administering the goods required for the intervention. Material, personnel, storage and transport costs factor prominently. Common indirect costs are productivity losses and travel costs for people to receive the intervention (e.g. a person missing time from work to receive a treatment). Intangible costs can be key indirect costs as well. For some interventions, the indirect costs can be considerable. Some interventions may cause psychological stress or social stigma (e.g. some individuals may find it embarrassing to be treated for a parasitic disease). Another cost to consider is that emerging from the potential side effects of an intervention, ranging from minor side effects such as diarrhoea that may result in productivity loss, to allergic reactions that merit treatment, to serious side effects such as encephalopathy from treatment with diethylcarbamazine. Calculating the costs from death via an intervention side effect is similar to calculating the costs of death from a disease.

One challenge is determining how to handle capital expenditures for something that will be used over a long period of time. Assigning all of the costs to the year of purchase may unduly punish that year and favour the other years even though the asset is used for many years. (For example, a truck, for instance, purchased in 2014 may be used for the next decade). **Amortization** is the process of spreading the cost of an asset over the lifetime of its use. One method of amortization, often called straight-line amortization, is simply to divide the total cost of the asset by the number of expected years of its use (e.g. a truck that costs USD 30,000 and has an expected lifetime of 10 years will incur a cost of USD 3,000 each year). The drawback of this method is that it does not account for differing intensity of use over the asset's lifetime (e.g. a truck may be used more frequently during certain years). Thus, another method of amortization is to spread the cost over a measure of the asset's use (e.g. for a truck that measure may be miles driven). So in the course of a truck's lifetime, it may be used for 30,000 miles. Therefore each mile used, the truck would incur USD 30,000/30,000 miles or USD 1). This method also accounts for the fact that an asset's lifetime may be shortened if used more heavily and lengthened if used more lightly. Assets whose costs are usually amortized include buildings, vehicles and other equipment. When an asset has remaining value after the end of its lifetime (i.e. residual value), the cost that is amortized is the purchase cost minus the residual value (e.g. if after 10 years, the USD 30,000 truck can be sold for USD 1,000, the cost that is amortized is USD 30,000−USD 1,000 or USD 29,000).

A given NTD intervention may affect multiple NTDs at once, further complicating the economics of NTDs and their prevention and control measures. Vector control measures such as the use of insecticides and bed nets may affect multiple NTDs and some treatments such as broad-spectrum benzimidazoles for the treatment of STHs work against multiple NTDs. Different control measures and treatments can also interact, potentially enhancing or inhibiting each other. Furthermore, the efficacy and application of control measures can change as the epidemiology of the target NTDs change. Although parasitic diseases are diverse and can be controlled through a variety of measures, as they are often interrelated in their transmission and treatment, and generally affect populations with limited resources, there is great interest in and debate over how those limited resources should be allocated for their control and treatment. For example, conventional MDA of chemotherapies for many NTDs have been the 'gold standard' of treatment in resource-limited areas for decades, but increasingly there is debate over whether alternative strategies for determining which populations to administer chemotherapies to (via techniques such as geographic information system (GIS) mapping, morbidity questionnaires and environmental sampling), over the development of more effective drugs and potential vaccines, and even over whether certain NTDs are worth treating at all (Hotez and Ferris, 2006; Modabber, 2010; Brooker et al., 2009; Kreimer et al., 2010).

The complex and close relationship between poverty and parasitic diseases, their widespread distribution in resource-limited settings, the historical neglect of the problem and the various options for prevention and treatment, mean that parasitic disease control is increasingly regarded as an economic investment in human capital and poverty reduction, rather than just a health intervention (Hotez and Ferris, 2006; Engels and Savioli, 2006; Alvar et al., 2006; Adhikari et al., 2010). For these reasons, economic analysis is a critical component of assessing the burden of NTDs and formulating appropriate disease prevention and control strategies for allocating these scare resources. All of the complexities and nuances of parasitic infections make economic analysis of NTDs complicated, but also underscore the need for robust economic evaluations of the diseases and their control programmes. These types of analyses are crucial to delineate the true burden of these illnesses and in determining the difference between different interventions and implementing policy, as they can help to delineate which might be the least costly, which may be the most effective, which may be the most cost-effective and which may have the biggest impact.

3.1 Cost of intervention studies in the scientific literature

Table 5 lists some published cost of interventions studies. A majority of published NTD economic studies are indeed cost of intervention studies (although the number of cost of intervention studies still pales in comparison to those for other more 'prominent' diseases such as HIV). Many of the cost of intervention studies are pathogen specific (e.g. for Chagas disease, leishmaniasis, onchocerciasis, schistosomiasis, and STH interventions) but some evaluate interventions across different NTD pathogens. Most of the existing cost of intervention studies concentrate on specific programmes under specific situations, and for particular geographic regions. This may be helpful for those regions, but may limit generalizability. Below, these studies are summarized by NTD pathogen.

3.1.1 Chagas disease

Two studies focussed on indoor residual spraying against triatomine bugs, the Chagas disease vector. A retrospective review of vector control (spraying) costs entailed interviewing vector control programme staff (Castillo-Riquelme et al., 2008). The study considered both direct (e.g. field workers, supplies, transport) costs and overhead costs. The average cost per sprayed house was USD 27 with the insecticide cost representing on average 57% of this cost. However, these costs varied substantially among villages studied. This was in part due to the low cost of procurement if purchased through the central government, as central level negotiates prices based

Table 5 Summary of the literature on the cost of interventions

Disease	Setting	Intervention	Cost (USD)
Chagas disease	Colombia	Indoor residual spraying (IRS)	27 per house (Castillo-Riquelme et al., 2008)
Chagas disease	Guatemala	Surveys, followed by IRS, and community-based surveillance with selective spraying	921,815 (total) (Hashimoto et al., 2012)
Visceral leishmaniasis	India	Active case detection for new cases	50–106 per case (Hirve et al., 2010)
Visceral leishmaniasis	Bangladesh	Active case detection for new cases	172 per case (Hirve et al., 2010)
Visceral leishmaniasis	Nepal	Active case detection for new cases	262 per case (Hirve et al., 2010)
Visceral leishmaniasis	Nepal	IRS	4.4 per household
		Long-lasting insecticidal nets (LLIN)	4.35 per household
		Ecological vector management	5.25 per household (Das et al., 2008)
Visceral leishmaniasis	Bangladesh	IRS	11.7 per household
		LLIN	3.5 per household
		Ecological vector management	9.3 per household (Das et al., 2008)
Visceral leishmaniasis	India	IRS	2.4 per household
		LLIN	5.1 per household
		Ecological vector management	14.0 per household (Das et al., 2008)
Visceral leishmaniasis	India	Periodic random surveys compared with health facility-based routine monitoring for drug effectiveness	19.65 per community health centre; 15.52 per patient (Malaviya et al., 2011)

(Continued)

Table 5 Summary of the literature on the cost of interventions—cont'd

Disease	Setting	Intervention	Cost (USD)
Visceral leishmaniasis	Bangladesh	Active case detection	22 per case detected (Huda et al., 2012)
Visceral leishmaniasis	Nepal	Active case detection	199 per case detected (Huda et al., 2012)
Visceral leishmaniasis	India	Active case detection	320 per case detected (Huda et al., 2012)
Visceral leishmaniasis	Sudan	Insecticide-treated bed net distribution	2,286,195 (total) (Ritmeijer et al., 2007)
Visceral leishmaniasis	India, Nepal and Bangladesh	Active case detection via different approaches:	Cost per new case detected by approach
		Camp	274–376
		Index case	180–212
		Incentive based	229
		Blanket	377–473 (Singh et al., 2011)
Lymphatic filariasis	Burkina Faso, Dominican Republic, Egypt, Ghana, Haiti, Philippines, Tanzania	Mass drug administration (MDA)	0.40–5.87 per person treated, varying with newest of programme, volunteer use and population treated (Goldman et al., 2007)
Onchocerciasis	Ghana	Annual and biannual mass community-directed ivermectin treatment	Annual treatment: 0.39–0.45 per person treated Biannual treatment: 0.58–0.73 per person treated (Turner et al., 2013)

Schistosomiasis	Plateau and Nasarawa states, Nigeria	Screening of *Schistosoma haematobium* only with all-age MDA;	18,637
		screening of *S. haematobium* and *Schistosoma mansoni* with all-age MDA;	36,816
		MDA to school-age children without screening;	15,510
		MDA to all ages without screening	68,610 (Gutman et al., 2009)
Schistosomiasis	China	Health education and promotion	0.88 Yuan per case (Hu et al., 2005)
Soil-transmitted helminthiases (STH)	Vietnam	School-based deworming with cost containment measures aimed to reduce total costs	0.03 per treated child (Montresor et al., 2007)
STH	Myanmar	School-based deworming	0.05 per treated child (Montresor et al,. 2004)
STH	Uganda	School-based delivery of albendazole	0.063–0.105 per schoolchild treated
			0.04–0.08 delivery cost per schoolchild treated (Kabatereine et al., 2005)
STH	Cambodia	Deworming with mebendazole and health education to schoolchildren twice a year	0.057–0.122 per treated child (Sinuon et al., 2005)
Schistosomiasis and STH	Burkina Faso	Community- and school-based MDA	1,067,284 total; 0.32 per child treated (Gabrielli et al., 2006)
Schistosomiasis and STH	Niger	School-based and community-distributed MDA	0.58 per treatment (456,718 total for two years) (Leslie et al., 2011)

on the previous year's budget year, but when a department runs out it must be purchased directly from private providers, usually at a higher price. Another study estimated for the costs of an indoor residual spraying (IRS) programme that included initial spraying followed by community-based surveillance and then with selective respraying in 317 villages in Guatemala (Hashimoto et al., 2012). The average number of units sprayed and surveyed multiplied by the unit costs of each determined the total cost; the spraying cost per house included insecticide, labour and transport, while surveillance was estimated at 80% of spray costs, and entomological survey costs were estimated. Overall, direct costs for spraying during the initial universal spraying phase totalled USD 720,890. The follow-up surveillance with selective spraying cost USD 53,243 to yield a total of USD 774,133 for the entire spraying programme. Surveys conducted throughout the time period to evaluate the impact of the IRS cost an estimated USD 147,682. Although this programme took place from 2000 to 2008, different villages started at different years (from 2000 to 2003) and not all villages were surveyed each year. Additionally, cost estimates are from varying sources and were not directly calculated from the programme, therefore they may not capture all costs or be an accurate representation.

3.1.2 Leishmaniasis

Four studies evaluated different interventions for visceral leishmaniasis in India, Nepal and Bangladesh (Das et al., 2008; Hirve et al., 2010; Huda et al., 2012; Singh et al., 2011). Three of the studies evaluated case detection and the other looked at three interventions (IRS, insecticide-treated nets and ecological vector management); all of them are observational studies. Of the case detection studies, each evaluated two or more of the following different active case detection approaches: camp-based approach (mobile teams visit target villages), index case (search in neighbourhood of known case), incentive-based (active detection by village health care workers who received incentive) and blanket screening (house to house). These studies evaluated direct costs and included the costs of items such as training, training materials, diagnostic kits, per diems, and transport/travel and personnel (staff time). An evaluation of all four methods found that total cost ranged from USD 889—USD 3,307 for the camp approach, USD 309—USD 800 for the index case approach, USD 380—USD 2,173 for the incentive-based approach and USD 3,187—USD 9,281 for the blanket approach and varied by country. The camp approach and incentive-based approach tended to be the least costly methods per new visceral leishmaniasis case detected (Singh et al., 2011). In another study, active case detection with blanket screening cost an additional USD 50–USD 262 per new case detected (USD 934–USD 2,071 total direct costs) compared with passive detection (i.e. routine surveillance) (Hirve et al., 2010). The other study evaluated the camp and index case method (Huda et al., 2012). The cost per camp and cost per new case detected, respectively, was USD 283 and USD 105 in Bangladesh, USD 349 and USD 1,046 in India and USD 688 and USD 459 in Nepal by the camp method. By the index case search method, the cost was USD 97 in India (USD 1,311 per new case detected) and USD 91 in Nepal. Overall, the camp search strategy was better. The index case search strategy could not be evaluated for Bangladesh. While these three studies are helpful for programme planning and evaluation, they may not be generalizable outside these regions. Additionally, they may not yield the same benefits (e.g. number new cases detected) in areas with differing epidemiological conditions.

Another observational study evaluating visceral leishmaniasis control determined the total programme and component costs for the mass distribution of impregnated bed nets in Sudan (Ritmeijer et al., 2007). This study evaluated a programme from 1996 to 2002 and while it focussed on evaluating the efficacy and coverage of bed nets, calculated the total documented costs (including the three distribution phases) and split them into net costs and additional costs (staff, transport and storage). The total cost was USD 2,286,195 USD 6.40 per net distributed), of which 79.7% was the cost of the nets.

3.1.3 Lymphatic filariasis

One study evaluated the economic and financial costs of lymphatic filariasis MDA programmes in seven countries from the national programme perspective (Goldman et al., 2007). The study defined economic costs as the value of all resources used (includes donated items) and financial costs as actual cash disbursements. This study used both retrospective and prospective data collection (interviews, questionnaires, programme record reviews) and included costs for personnel, supplies, drugs and capital and recurrent costs for equipment, transportation and facilities (annualizing capital costs). Overall, the financial costs per person treated ranged from USD 0.06 to USD 2.23, while economic costs ranged from USD 0.40 to USD 5.87. The economic cost per person treated for each of the seven countries in the most recent year reported was as follows: Burkina Faso USD 4.82; Ghana USD 4.88; Tanzania USD 4.53; Dominican Republic USD 1.56; Egypt USD 1.34; Philippines USD 0.40; Haiti's economic cost was not available, but financial cost was USD 1.30. Limitations of this study include challenges in estimating the proportion of time and money allocated specifically to lymphatic filariasis (vs other ministry of health programmes) and differing definitions of certain activity categories.

3.1.4 Onchocerciasis

One study quantified costs associated with annual versus biannual mass community-directed treatment with ivermectin (CDTI) in Ghana from the health care provider perspective (Turner et al., 2013). Data came from interviews and national records and centred on four districts where CDTI is used (one annually, two biannually and one where in 55% of communities it is annual and in 45% it is biannual). This study included several costs and types of costs: drug distribution chain (in-country distribution costs), mobilization and sensitization (promotion, information dissemination and advocacy), training of volunteers (community drug distributors), other training, reporting, surveillance and evaluation (disease and treatment), administration and other project activities not included elsewhere. Resource types of costs included: capital and recurrent costs of transportation, personnel, per diems, capital and recurrent supplies and equipment costs, overheads, and volunteer community drug distributor time. It did not include costs of drug manufacture and transport to country. For annual CDTI, the financial cost was USD 0.39 per person treated and the economic cost USD 0.45 per person treated per year. For biannual CDTI, financial costs ranged from USD 0.58 to USD 0.62 per person treated and economic costs ranged from USD 0.69 to USD 0.73 per person. Within the district using a mix of annual and biannual CDTI, financial costs were USD 0.40 per person treated, while economic costs were USD 0.50 per person treated. Personnel costs contributed most to total costs. As most control programmes in Ghana are integrated, obtaining accurate costs for a single disease intervention can be difficult; additionally, the data used were subject to some degree of recall bias.

3.1.5 Schistosomiasis

Two studies evaluated the cost of schistosomiasis control. One, a modelling study, assessed four different MDA methods in Nigeria (Gutman et al., 2009). The strategies evaluated were: 1) village-by-village screening for *Schistosoma haematobium* with MDA given to villages where 20% of school-aged children are infected; 2) screening for both species with MDA if 20% prevalence of *S. haematobium* and 10% of *Schistosoma mansoni*; 3) presumptive annual treatment of all school-aged children; 4) presumptive annual treatment of all eligible adults and children. Total costs were higher in year one, but over 5 years *S. haematobium* screening of 30,000 people with

MDA cost USD 18,673, screening for *S. haematobium* and *S. mansoni* with MDA cost USD 36,816, presumptive treatment of children cost USD 15,510, and presumptive treatment of adults and children cost USD 68,610. The importance of community-wide treatment in highly endemic areas should be considered, even though treating school-aged children was the least costly. These results may not be generalizable to other areas.

The other study, an observational study in the Poyang Lake region of the People's Republic of China, determined the cost of health education and promotion for *schistosomiasis japonica* (Hu et al., 2005). The total cost of the health education and promotion cost 88 Yuan per 100 persons. The intervention saw compliance increases and the disease prevalence dropped by 83.7% in schoolchildren and 63.4% in women. These costs were based on a pilot study and came from questionnaires, therefore, they are not generalizable to other areas.

3.1.6 Soil-transmitted helminthiases

Several studies of MDA for STH prevention evaluated programmes currently in place. One study determined the district level financial costs for school-based distribution of albendazole in Uganda from the perspective of the control programme (Kabatereine et al., 2005). The study estimated a total cost of USD 0.063—USD 0.105 per schoolchild treated and a delivery cost of USD 0.04—USD 0.08 per child treated. In Cambodia, deworming with mebendazole and health education sought to cover 75% of schoolchildren twice a year (Sinuon et al., 2005). Cost estimates included the cost of training, drug, health education material and monitoring. Total cost was USD 118,969 (USD 0.122 per child treated) in round one and USD 158,851 (USD 0.057/child treated) in round one. This study did not include teacher training or time to administer the drug or distribution to the schools (these were not financial costs to the programme as they were financed/donated). In a school in Myanmar, a crude calculation of the cost of a pilot exercise for MDA with albendazole was estimated at USD 0.05 per child (Montresor et al., 2004). This estimate included the cost to buy and deliver drugs, cost for teacher training and training materials and personnel per diems. Based on estimates for the number of schoolchildren and assumption they should be treated twice a year, the total yearly cost for one treatment would be USD 573,516. The authors point out that these are rough estimates based on one site, before scaling up, and are not discounted or do not account for time spent. While these studies include appropriate costs and methods for determining the total intervention cost for evaluating the programme, they may not be generalizable to other countries, or even other regions within the same country.

Another study evaluated the impact of changing a current programme to reduce costs. This programme aimed to be a cost containment intervention designed to reduce costs of MDA while maintaining coverage (i.e. a cost containment intervention) in Vietnam (Montresor et al., 2007). The following was done in an attempt to contain costs: no baseline survey conducted; estimated amount of drug distributed in packets of 100 tablets reduce repackaging; drug delivery was 'piggybacked' with other drug transport to health stations; health education materials not developed; no incentives for teachers to administer treatment (but given treatment themselves); parasitological evaluation not conducted; and a fixed amount of money was supplied to province to use according to needs. The only costs considered were direct financial costs (drug procurement, quality control, travel for personnel, monitoring, transport fuel). The total financial costs of the programme were USD 81,000, this translates to USD 0.03 per child treated, which is low compared with the cost in previous years (USD 0.71 and USD 0.11 per child treated). Additionally, a higher coverage rate was reached compared with the previous years.

3.1.7 Multiple NTDs

Two studies evaluated school and community-based control measures for schistosomiasis and STHs. The first evaluated a combined school- and community-based programme targeting school-aged children for treatment against schistosomiasis (with praziquantel) and STH (with albendazole) from 2004 to 2005 in Burkina Faso (Gabrielli et al., 2006). The total cost was USD 1,067,284 (69.4% spent on drugs) to treat 3,322,564 children; total cost per treated child was USD 0.32 (USD 0.308 for school-based treatment and USD 0.330 for community-based treatment). Of that, the delivery costs (total campaign cost minus drug costs) was USD 325,936 and drug costs totalled USD 741,348. The study did not report more detail as to which costs were included in the campaign. The second is a retrospective study of a two-year (2004–2006) of school-based and community-distributed MDA in four districts of Niger (Leslie et al., 2011). Data sources included government accounts, receipts, records of activities, and surveys. Cost elements included programme costs (e.g. capital, recurrent and variable costs), opportunity costs (value of contributions for things with no monetary spending), value to personnel time and reporting and technical support. The total economic cost for this programme (including programmatic, national and local government and international support) was USD 456,718 over the two years. This is USD 0.58 per treatment (with the average drug cost USD 0.28 per treatment), school-based treatment cost USD 0.76 and community distribution cost USD 0.46. This study breaks down cost by level and type of cost, finding that 75% of total costs were programme costs, of which 65% was drug cost (49% of total cost).

Furthermore, a few studies have shown the economic benefits of integrating interventions for different NTDs or NTDs and other diseases. A study suggested that in sub-Saharan Africa, integrating preventive chemotherapy programmes for lymphatic filariasis, onchocerciasis, STHs, schistosomiasis, and trachoma could yield cost-savings of up to 47% (USD 52 million) (Brady et al., 2006). An integrated drug delivery strategy would cost between USD 0.53 and USD 0.62 per person treated (saving USD 55 to USD 102 million) for a programme reaching 100% of the target population in sub-Saharan Africa (Brady et al., 2006). A similar study showed that integration of preventive chemotherapy treatment for trachoma, schistosomiasis, lymphatic filariasis and STHs in Niger saved 16–21% in 2008 and 2009 (Leslie et al., 2013).

3.1.8 Summary of cost of intervention studies

Studies of MDA cost demonstrated that drug costs constitute a majority of the costs. A number of drug manufacturers donate drugs to national ministries of health, thereby keeping drug costs fairly low (Fenwick et al., 2005; Molyneux and Nantulya, 2004; Albonico et al., 2006). However, one continuing question is the sustainability of MDA programmes with reliance on drug donations and fragile health care infrastructures. These limitations often mean that only fractions of the population can receive MDA. It is unclear how the cost structures may change if MDA were expanded to cover large swathes of the population and manufacturers were to change their donation practices (Anderson et al., 2013). Globally, annual MDA ivermectin treatment for onchocerciasis cost between USD 0.20 and USD 1.20 per person in Africa (compared with treatment with doxycycline which ranges from USD 1.77 to USD 2.77 per person) (Keating et al., 2014). Implementing annual MDA for STH control programmes across Latin America and the Caribbean cost USD 47 million annually to target USD 78.7 million school-aged children (Colston and Saboya, 2013).

As with the COI studies, the published cost of intervention studies range widely in the interventions evaluated (vaccines to drug treatments to surveillance), settings studied, methodologies used and types of costs included, limiting their comparability and generalizability. While a few studies suggest that combining and integrating NTD prevention and control programmes may save costs, more work is needed in this area. Most of the existing studies evaluate specific individual programmes under specific situations, and for particular geographical regions. Moreover, most existing studies have evaluated interventions in highly endemic areas, which may be quite different from areas with a lower prevalence. The cost structure could change considerably with disease prevalence. For example, for active case detection, blanket-screening approaches are most costly, but may be worthwhile in highly endemic areas where more cases would be detected and treated than in areas of lower prevalence. Additionally, most studies evaluated current programmes without exploring the potential range of implementation scenarios. Again, this limits generalizability and does not accommodate possible changes in the future. Finally, not all interventions have undergone economic evaluation.

4. COST-BENEFIT AND COST-EFFECTIVENESS ANALYSES

While COI studies can profile the nature and size of the problem and cost of intervention studies can identify which interventions may be feasible from a cost standpoint and what factors affect these costs, cost-benefit and effectiveness studies help decision-makers determine the relative trade-offs between costs of an intervention and its effects. In other words, what is the 'bang for the buck'? On rare occasions, the effects or costs between two options are the same, making the choice between the two relatively simple: choosing the cheapest if the effects were the same and the most effective if the costs were the same. However, this is usually the exception and not the rule. Typically, the decision therefore comes down to the trade-offs between the different option's costs and effects.

4.1 Cost-benefit analysis

Cost-benefit analyses (CBA) convert all of the costs and effects into a single unit, typically monetary terms (e.g. dollars, pounds, euros, yen, yuan, pesos, won, rupees, etc.) so that different options can be readily compared with a common unit of measure. So, the following formula determines the cost-benefit of an intervention:

$$\text{Cost-benefit of an intervention} = \text{the benefits of the intervention in monetary terms} - \text{cost of the intervention in monetary terms}$$

where the benefits of the intervention in monetary terms is the difference in costs and health effects (expressed in monetary terms) between the situation with the intervention versus the situation without the intervention. The benefits of the intervention may be the number of cases prevented by the intervention multiplied by the cost of each case, with the product expressed in monetary terms.

When comparing interventions the higher the calculated value (i.e. net benefit), the more favourable the intervention is (e.g. intervention A with a value of USD 20,000 is favourable over intervention B with a value of USD 1,000). A positive cost-benefit result means that the given option will actually save costs (e.g. intervention C with a value of USD 500 means that its net value is positive).

CBA can help to address the following questions:

- What is the cost impact of choosing one intervention or scenario versus another?
- What is the break-even point for a given intervention or programme?
- How effective does an intervention or programme need to be in order for it to garner cost-savings?
- How much can a programme or intervention cost to return savings based on a given effectiveness?

A CBA is fairly straightforward and easy to understand with the outcome measure as one term. However, converting health effects is not always easy. Moreover, the goal is not always simply to minimize costs. Many times, the goal is to maximize health effects (e.g. minimize deaths or other health outcomes) as long as the costs of doing so are not prohibitive. People often are willing to pay to save lives or prevent suffering.

4.2 Cost-effectiveness analysis

Cost-effectiveness analyses (CEA) emerged to help to determine how much it would cost to achieve a desired health effect. Thus, the outcome measure is the cost per unit health effect. Common measures of health effects are DALYs averted or QALYs saved (described above). Other frequently used measures are cases, hospitalizations, clinic visits, deaths or other health outcomes averted. Therefore, common outcome measures for CEAs are cost per DALY averted, cost per QALY, cost per case averted, cost per death averted, etc. Cost-effectiveness studies can answer questions such as:

- Are interventions/programmes worth developing and implementing based on estimated costs?
- What characteristics would a given product or intervention needs to have in order to be cost-effective?
- Which course of action among the available will be best?
- How should scare resources be allocated to garner the best outcomes?

When evaluating two or more competing interventions or programmes, the **incremental cost-effectiveness ratio** (ICER) can be determined. The ICER is calculated by:

$$\text{ICER} = (\text{Cost A} - \text{Cost B})/(\text{Effectiveness A} - \text{Effectiveness B})$$

where A and B represent two interventions or programmes. When one intervention is being evaluated, B can be the current practice which may be no intervention. (Note that when effectiveness is measured in DALYs, the incremental effectiveness must be inverted or the negative value taken.) Figure 2(a) plots the costs and effectiveness of A and B for comparison, where each point represents the intersection of the intervention's costs and effectiveness. The ICER value is the slope between the plotted points, which represents the difference in costs and effectiveness between the two interventions. This ICER value can also be plotted; Figure 2(b) shows the four possibilities for the ICER:

- the new intervention or programme can be more costly and more effective than the current strategy (ICER value);
- the new intervention can be less costly and less effective (ICER value);
- the new intervention can be less costly and more effective (**dominant**); and
- the new intervention can be more costly and less effective (**dominated**).

The dotted line represents the **willingness-to-pay** (WTP) threshold. The intervention's cost-effectiveness results from comparing the ICER value to a given threshold. ICER values below this threshold are considered to be cost-effective, while values above the line are not (grey shading in Figure 2(b)). For developing countries, this is typically the country's GDP per capita. ICER values less than the country's GDP per capita are deemed highly cost-effective, values from 1 to 3 times the GDP per capita are deemed cost-effective, and values more than 3 times the GDP per capita are not considered cost-effective.

4.2.1 Example of a cost-effectiveness model: a hookworm vaccine

To illustrate a CEA evaluating an NTD intervention, we walk through an example in the literature exploring the potential economic value of a hookworm vaccine in Brazil (Lee et al., 2011):

Step 1: Determine the question of interest

The goal of this study was to determine the potential economic value of a human hookworm vaccine, which is currently under development. Understanding a vaccine's potential economic value can assist scientists,

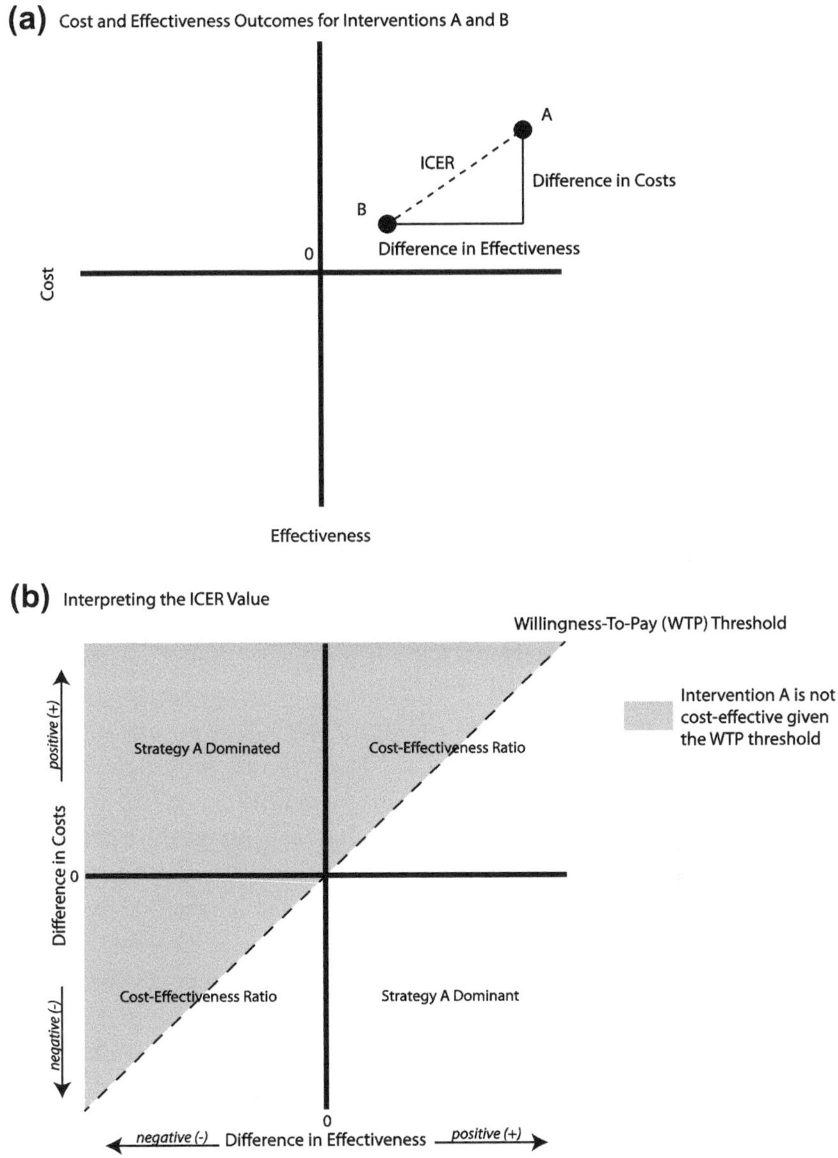

Figure 2 Interpreting cost-effectiveness (a) cost and effectiveness outcomes for interventions A and B; (b) interpreting the incremental cost-effectiveness ratio value.

manufacturers, public health officials and other decision-makers and help to guide clinical development, investment, marketplace positioning and eventual implementation. While, regular mass anthelmintic drug chemotherapy remains the primary method for hookworm treatment and transmission prevention, reports of decreases in drug efficacy over the past two decades raise concerns of emerging drug resistance. Thus, there is a need to develop alternative hookworm control methods such as vaccines. This study entailed constructing a computer simulation model to evaluate the potential economic value of a hookworm vaccine and how it may change with varying vaccine characteristics (prevention, probability of egg reduction and cost) and different environmental conditions such as infection prevalence, severity and drug resistance. Constructing economic models early in a vaccine's development, when vaccine characteristics and market strategy can still be adjusted, may improve a vaccine's chances of success.

Step 2: Establish the perspective and time horizon

The study assumed the societal perspective, and the model's time frame was 20 years, long enough to account conservatively for age-dependent risks of infection, the possibility of reinfection with hookworm, the duration of vaccine protection, and the accrued benefits of the vaccine.

Step 3: Choose/design the appropriate model structure

Figure 3 shows the structure of the model, which consisted of four arms: 1) vaccine plus drug treatment, 2) vaccine only, 3) drug treatment only versus 4) no intervention. Each arm led into a Markov model and its various mutually exclusive states: no hookworm, light intensity infection, moderate/heavy intensity infection and death. Death, as in the Chagas disease Markov model, is an absorptive state. The two different infection states (light intensity infection and moderate/heavy intensity infection) accounted for differences in an individual's risk of different outcomes, intervention efficacy (that may vary by the person's worm burden) and health effects (i.e. DALYs). To mirror what is seen in an endemic area, an individual had a probability of beginning in any of the three nondeath states. In other words, an individual could begin with no hookworm infection or either a light intensity or moderate/heavy intensity hookworm infection. The probability of beginning in each of these states depended on the individual's age-specific probability of infection and heavy intensity infection.

Given that treatment programmes for hookworm in endemic countries is annual, a cycle length of 1 year seemed appropriate. Thus, each year an individual had probabilities of remaining in the same state or moving into another. Therefore, an individual could remain in the same state (e.g. infected) for any number of years up to 18 years (as the model runs for 20 cycles). However, half of infections self-resolved within 5 years. Each year infected individuals had age-specific and infection-intensity-specific probabilities of developing anaemia. An individual with anaemia then received an

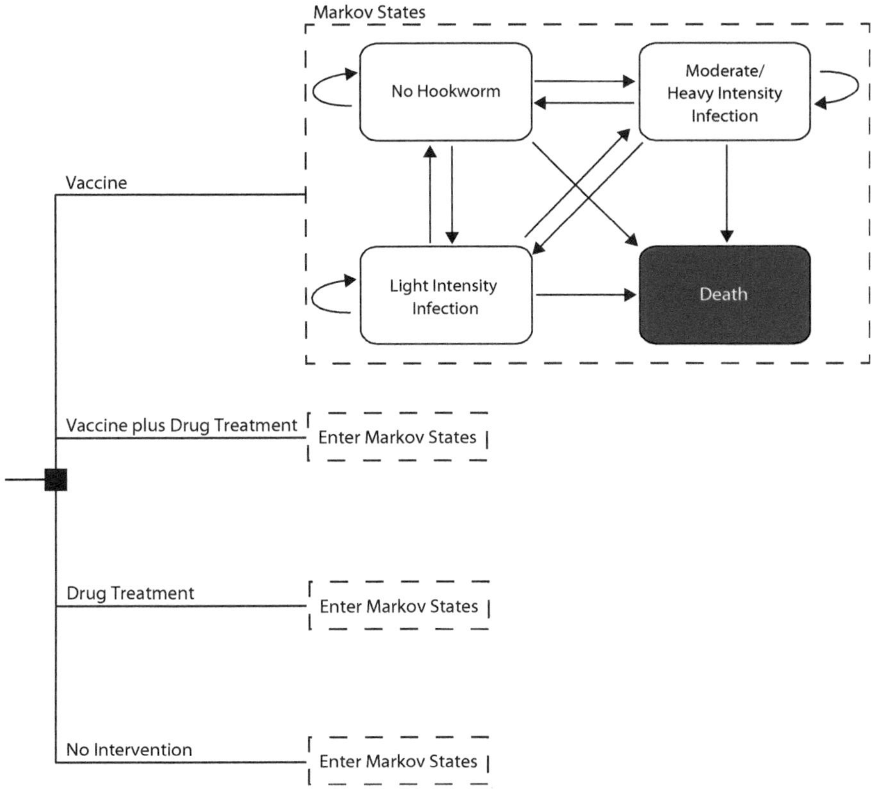

Figure 3 Hookworm vaccine model structure and Markov states. *Reprinted from Vaccine, Vol. 29, B.Y. Lee, K.M. Bacon, R. Bailey, A.E. Wiringa, K.J. Smith, The potential economic value of a hookworm vaccine, pp. 1201–1210, Copyright 2010, with permission from Elsevier.*

assigned anaemic haemoglobin level value. An individual could develop cognitive impairment if infected for at least two years.

The model accounted for how different combinations of vaccine and drug treatment would affect the probabilities of transitioning among the various Markov states and their associated outcomes. Vaccination had a probability (based on the assumed vaccine's efficacy) of decreasing an individual's risk of hookworm infection and risk of higher intensity infection. Treatment (which consisted of the drug albendazole) had probabilities of being curative and preventative. Treatment could clear infection and reinfection did not occur for one year after treatment and it had a probability of decreasing the risk of hookworm infection. An individual travelling through a treatment arm would receive treatment once annually (i.e. per cycle) via a school setting when the individual was of school-age and a community delivery programme when the individual was an adult.

Step 4: Establish model outcome measures

The main outcome of this study was the ICER. The ICER is the difference in costs over the difference in DALYs between scenarios with the vaccine and without the vaccine. Thus, each event in the model accrued direct and indirect costs (measured in currency) and health effects (measured in DALYs). For each simulation the ICER value formula follows:

$$\text{ICER} = (\text{Cost}_{\text{intervention}} - \text{Cost}_{\text{no intervention}})/(\text{DALYs accrued}_{\text{no intervention}} - \text{DALYs accrued}_{\text{intervention}})$$

It is important to note that the DALYs are inverted (i.e. DALYs with no intervention minus DALYs with intervention) as DALYs measure 'bad' health outcomes, with 0 being healthy. Therefore, the closer an intervention is to 0 DALYs accrued, the better it is (as described above).

Step 5: Identify appropriate data sources and populate/calibrate the model

Table 6 shows the data inputs for the model. As can be seen, hookworm outcomes and risks were age-specific. Values for these parameters were specific to Brazil when available. The model did not include costs for clinic visits and sick leave due to a lack of data. A 3% discount rate converted past and future costs into 2010 values. Direct costs included vaccination, drug treatment and MDA costs. Indirect costs include productivity losses, which as mentioned above, can be estimated in different way. Anaemia led to productivity losses, dependent on the person's haemoglobin level (Hb level) via the following formula (Shastry and Weil, 2003):

$$\text{Productivity loss} = 1 - (\text{Hb level}/\text{Hb threshold})^{1.5}$$

where the Hb threshold was the threshold for anaemia and 11.5 g/dL for school-aged children and 12 g/dL for women of child-bearing age and the 1.5 g/dL was the fluctuation in haemoglobin level that would have no additional effect on productivity (Shastry and Weil, 2003).

Step 6: Run baseline scenario(s)

A baseline scenario existed for each of the two target populations. School-aged children had a start age of 7 years and a 60% risk of infection (which changed over time); women of reproductive age had a start age of 13 years and a 76% risk of infection, which changed over time. Baseline scenarios assumed the following vaccine characteristics: a USD 30 vaccination cost, an efficacy of 60% for preventing infection, and an efficacy of 80% for preventing heavy intensity infection. Each simulation run sent 1000 individuals through the model 1,000 times (for 1 million unique trials).

Table 6 Model input parameters, values and sources for the hookworm vaccine model[a]

Variable	Mean	Source
Risk of hookworm infection		
0–9 years old	60%	Literature
10–19 years old	76%	Literature
20–59 years old	73%	Literature
60 years old and older	70%	Literature
Risk of hookworm infection being heavy intensity (given hookworm infection)		
0–2 years old	0%	Literature
3–5 years old	5%	Literature
6–10 years old	8%	Literature
11–15 years old	8%	Literature
16–20 years old	10%	Literature
21–30 years old	13%	Literature
31–40 years old	12%	Literature
41–50 years old	3%	Literature
51 years old and older	23%	Literature
Probability of anaemia (school-age children)	1%	WHO Global Database on Anaemia[b]
Probability of anaemia (women of reproductive age)	3%	Literature
Probability of anaemia from heavy intensity infection (5–19 years old)	17%	Literature
Probability of anaemia from heavy intensity infection (20 years old and older)	9%	Literature
Probability of anaemia from light intensity infection	3%	Literature
Albendazole cure rate	76% (Range: 57–95%)	Literature
Albendazole egg reduction	0.93% (Range: 79–99%)	Literature
Crude mortality rate	1%	Literature
Costs (2010 USD)		
Community	0.63	Literature
School	0.05	Literature
Albendazole	0.06	Literature

(Continued)

Table 6 Model input parameters, values and sources for the hookworm vaccine model[a]—cont'd

Variable	Mean	Source
Disability-adjusted life-year disability weights		
Anaemia	0.024	Global Burden of Disease 2004
Heavy intensity infection	0.006	Global Burden of Disease 2004
Cognitive impairment	0.024	Global Burden of Disease 2004
Haemoglobin cutoffs for anaemia (g/dL)		
Children	11.5	WHO Report, Worldwide Prevalence of Anaemia
Women of reproductive age	12	WHO Report, Worldwide Prevalence of Anaemia

[a]Literature, as cited in Lee et al., 2011.
[b]http://www.who.int/vmnis/database/anaemia/en.

Step 7: Conduct sensitivity analyses

Sensitivity analyses varied key parameters to show how cost-effectiveness and other outcomes vary by vaccine characteristics (by varying drug cure rates, vaccine efficacy for preventing infection and for preventing heavy intensity infection, vaccine cost and the requirement of booster shots) and for epidemiological factors (such as hookworm infection prevalence). Specifically, the risk of infection varied from 25% to 50% of the age-specific rates and drug cure rate and egg reduction rate varied from 50% to 100% of baseline values to model resistance. Vaccine efficacy for preventing infection ranged from 30% to 60% and for preventing heavy intensity infection ranged from 40% to 80%. The vaccine cost varied from USD 1 to USD 100 to represent different vaccination prices, the requirement of booster shots and various administration strategies (e.g. vaccination during child health days). Treatment costs varied by ±50% to account for less expensive medications and additional administration costs.

Additional scenarios varied the duration of vaccine protection (5 years vs 20 years) and compliance with follow-up vaccination when the duration of protection was less than 20 years. Scenarios with a duration of protection less than 20 years assumed that 50% of persons were compliant and received the

Table 7 Incremental cost-effectiveness of hookworm control strategies targeting reproductive-age women and school-age children

		Baseline infection prevalence		50% Infection prevalence	
Vaccine cost (USD)	**Efficacy**	**76% Drug cure rate**	**38% Drug cure rate**	**76% Drug cure rate**	**38% Drug cure rate**
Reproductive-age women					
30	60%	Combination	Combination	Combination	Combination
	30%	Combination	Combination	USD 5,532	Combination
100	60%	USD 2,655	Combination	USD 9,496	USD 1,001
	30%	USD 8,807	USD 510	USD 28,552	Drug/combination
School-age children					
30	60%	Combination	Combination	Combination	Combination
	30%	Combination	Combination	Combination	Combination
100	60%	Combination	Combination	Combination	Combination
	30%	Combination	Combination	USD 7,964	USD 4,648

Combination = the combination treatment strategy was dominant over all other strategies tested.
Drug/Combination = while the drug treatment dominated the no intervention strategy, combination treatment dominated the vaccine only strategy.

booster vaccination; those not compliant lost all previous vaccine protection and had the baseline risk of infection.

Table 7 reports the ICER values by infection prevalence and drug cure rate. Overall, the vaccination plus drug treatment (combination) strategy

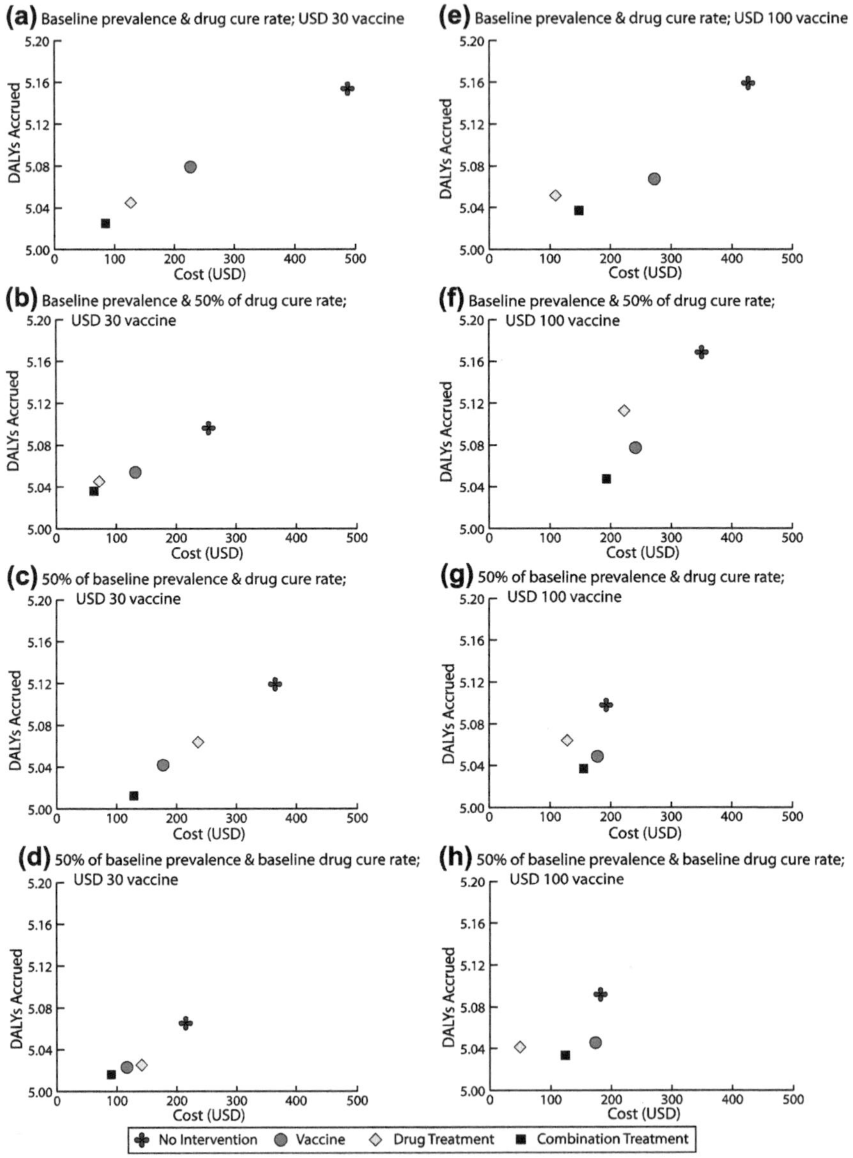

Figure 4 Intervention costs and DALYs accrued for strategies targeting school-age children with optimal vaccine efficacy. *Reprinted from Vaccine, Vol. 29, B.Y. Lee, K.M. Bacon, R. Bailey, A.E. Wiringa, K.J. Smith, The potential economic value of a hookworm vaccine, pp. 1201–1210, Copyright 2010, with permission from Elsevier.*

dominated (i.e. was less costly and more effective) all other explored strategies in a wide range of scenarios in both target populations. This strategy only failed to dominate when the vaccine efficacy was low (≤30% in preventing infection, ≤40% in reducing egg production) and/or the vaccine cost was ≥USD 100. Figure 4 shows the costs and DALYs accrued by strategy targeting school-aged children (assuming the optimal vaccine efficacy: 60% infection prevention, 80% heavy intensity prevention). The difference between any points on a plot represents the incremental (i.e. difference) cost and DALYs accrued, while the slope between them represents the ICER. If a strategy's point is lower and to the left of another, it is dominant over the other strategy. Figure 4 shows a clear hierarchy of strategies with the vaccine plus drug treatment (combination) being the most economically favourable, followed by drug treatment, vaccine alone and no intervention. However, when at 50% of baseline hookworm infection, the combination strategy failed to dominate in all scenarios, but remained highly cost-effective. Among women of child-bearing age, the cost-effectiveness was much less sensitivity to infection prevalence. Vaccine plus drug treatment dominated all scenarios at baseline prevalence for vaccine prices up to USD 100.

4.3 Cost-benefit and cost-effectiveness studies in the scientific literature

Table 8 shows some of the major cost-effectiveness and cost-benefit analyses for NTDs found in the literature. Most focus on interventions, treatment or evaluate specific programmes in place and do not cover many interventions, policies and locations. Some focus on particular situations or types of cost from limited perspectives. The NTD subsections below summarize the studies in Table 8.

4.3.1 Chagas disease

Six studies evaluated on Chagas disease interventions (two vector control, two vaccine, one screening, one blood supply screening) (Agapova et al., 2010; Lee et al., 2010, 2012a; Sicuri et al., 2011; Vazquez-Prokopec et al., 2009; Wilson et al., 2005). Studies found the interventions evaluated were cost-effective in Chagas endemic areas. Of the vector control studies, a retrospective study evaluated the changes in vector control that took place between 1993 and 2004 in Argentina (Vazquez-Prokopec et al., 2009). During this time vector control shifted from a vertical strategy (i.e. insecticide application by qualified personnel) to a horizontal strategy (i.e. surveillance and spraying by community leaders); authors also evaluated a mixed

Table 8 Summary of literature on the cost-benefit and cost-effectiveness of parasitic neglected tropical diseases interventions

Disease	Setting	Intervention	Cost-effectiveness
Chagas	Latin America and Caribbean	Vector control programme, vector control programme plus new drug treatment and no control programme	Both vector control programmes (with and without drug treatment) dominate no vector control; vector control with drug treatment is cost-effective USD 699/QALY compared with vector control alone (Wilson et al., 2005)
Chagas	United States	Screening U.S. blood supply	USD 757,000/QALY test all donors once; USD 29,000/QALY test donors if transfusion patient ≤39 years old (Agapova et al., 2010)
Chagas	Argentina	Surveillance and spraying performed by different mechanisms: vertical strategy (performed qualified personnel) horizontal strategy (performed by community volunteers)mixed strategy (vertical followed by horizontal)	USD 132 per case averted (vertical) USD 45 per case averted (horizontal) USD 82 per case averted (mixed) (Vazquez-Prokopec et al., 2009)
Chagas	Latin American population living in nonendemic areas	Screening newborns and/or pregnant mother and treatment if positive	Screening dominated (over no screening) for both cases (Sicuri et al., 2011)
Chagas	Latin America	Preventative vaccine	Cost-effective (ICERs ≤16,614/DALY) or dominant under most scenarios tested (Lee et al., 2010)

Chagas	Latin America	Therapeutic vaccine	Highly cost-effective (ICERs ≤1,075/DALY) or dominant under most scenarios tested (Lee et al., 2012a)
Human African trypanosomiasis	Angola	Eflornithine vs melarsoprol drug treatment	USD 8,169 per additional life year saved, USD 299 per additional YLL averted (Robays et al., 2008)
Cutaneous leishmaniasis	Latin America	Vaccine	USD 66–USD 140 per case averted (Bacon et al., 2013)
Cutaneous leishmaniasis	Afghanistan	HealthNet TPO standard treatment (sodium stibogluconate (SSG))	USD 1,181/DALY averted (95% CI: USD 761–USD 1,827/DALY) (Reithinger and Coleman, 2007)
Visceral leishmaniasis	India	Vaccine	Dominant (vs no vaccination) when cost ≤USD 30 and efficacy ≥50% Cost-effective (ICER ≤USD 757/DALY) for all scenarios with cost ≤USD 100 and efficacy ≥25% (Lee et al., 2012b)
Visceral leishmaniasis	Endemic countries	Drug regimens: antimonials (SSG), amphotericin B deoxycholate, miltefosine, lipid formulation of amphotericin B	USD 362 per death averted USD 328 per death averted (dominant strategy) USD 457 per death averted USD 1,622 per death averted (Vanlerberghe et al., 2007)
Schistosomiasis	Nigeria	Unqualified haematuria Terminal haematuria Dysuria Chemical reagent strip Visual urine examination	USD 2.16 per correct case diagnosed USD 2.50 per correct case diagnosed USD 3.57 per correct case diagnosed USD 12.91 per correct case diagnosed USD 13.46 per correct case diagnosed (Fatiregun et al., 2009)

(Continued)

Table 8 Summary of literature on the cost-benefit and cost-effectiveness of parasitic neglected tropical diseases interventions—cont'd

Disease	Setting	Intervention	Cost-effectiveness
Schistosomiasis	China	Health education and promotion	3.73 Yuan per 100 persons for 1% reduction in reinfection in women and children (Hu et al., 2005)
Onchocerciasis, lymphatic filariasis, schistosomiasis and soil-transmitted helminthiasis (STH)	Nigeria	Annual triple drug administration (ivermectin, albendazole and praziquantel)	USD 0.06 per patient treated (Evans et al., 2011)
Schistosomiasis and HIV	Zimbabwe	Community-based intervention combined clean water, sanitation, and health education, with praziquantel administration to school-aged children	Very cost-effective when intervention cost ≤USD 300 per person over 20 years and cost-effective when intervention cost ≤USD 875 per person over 20 years (Ndeffo Mbah et al., 2013b)
Schistosomiasis and STH	Niger	School-based and community-distributed mass drug administration	USD 0.78 per schistosomiasis infection averted in children over two years of intervention; USD 4.6 per schistosomiasis infection averted in adults over two years of intervention (in treated and untreated population) (Leslie et al., 2011)
Schistosomiasis and STH	Tanzania	Treatment (albendazole and praziquantel)	USD 6—USD 8 per anaemia case prevented (Guyatt et al., 2001)

QALY, quality-adjusted life-years; ICER, incremental cost-effectiveness ratio; DALY, disability-adjusted life-years; YLL, years of life lost.

strategy (i.e. community leader surveillance with spraying by qualified personnel). This analysis compared the observed cost-effectiveness of the horizontal strategy with the expected value of the vertical and mixed, projected over a 25-year period. Direct costs of the programmes included staff costs, supplies and mobility (e.g. fuel) and indirect programme costs consisted of vehicle maintenance and salaries when not assigned to the field. Effectiveness was measured in number of Chagas cases averted. Overall, the total cost of the horizontal and mixed strategies were lower than the vertical. The horizontal had the lowest cost-effectiveness ratio (USD 83/case averted), which was 1.9—3.3 times lower than the other strategies; however, the number of cases (580 cases, 3,709 averted) was 1.6—4.0 times higher than the other strategies. The mixed strategy would have averted more cases (3,924 averted) than the horizontal strategy and cost USD 149/case averted. Authors conclude that the mixed strategy would have been the model cost-effective. This study showed the economic benefits of getting communities involved in NTD control (as personnel and other overhead costs are reduced); however this came with an increase in opportunity costs and should be weighed carefully. While this study successfully evaluated a programme in Argentina, data came from a high transmission region and therefore may not be generalizable to areas of lower transmission.

The other vector control study utilized a Markov model to evaluate Chagas interventions 19 countries in Latin America and the Caribbean (Wilson et al., 2005). This study explored both an incidence (assuming no disease, e.g. a birth cohort, evaluating disease progression) and a population prevalence (assuming distribution of Chagas outcomes in the population) models and evaluated three strategies: (1) a vector control programme, (2) a vector control programme plus a potential new drug treatment given after the acute disease stage and (3) no vector control. The Markov states consisted of the following: no disease, acute disease, indeterminate disease, chronic disease and death. Chronic outcomes included cardiomyopathy with and without CHF and megaviscera. Effectiveness was measured in QALYs. Direct costs included the intervention costs (either vector control or new drug), and health care costs for Chagas disease and outcomes. For the incidence model, both vector control alone and vector control plus new drug treatment dominated no vector control and vector control plus a new drug was cost-effective (ICER: USD 699/QALY) compared with vector control alone. For the prevalence model, both vector control and vector control plus a new drug dominated no vector control and vector control plus a new drug had an ICER of USD 289/QALY compared with vector control alone. Both strategies for vector control are very cost-effective compared with the gross national income (GNI) per capita from Latin American countries (USD 3,260). This is a very well designed and thorough study, authors appropriately accounted for time preference, performed sensitivity analysis, reports other important outcomes (not discussed here) and adequately point out the limitations of their study and how they may impact results.

Additional studies sought to determine the potential cost-effectiveness of Chagas vaccines and the impact of various vaccine characteristics on its economic value. Two separate studies (by the same authors) utilized Markov models to evaluate Chagas vaccines; one a preventative vaccine in Latin America (Lee et al., 2010) and the other a therapeutic vaccine in Mexico (Lee et al., 2012a). Both are based on the same basic model structure and evaluated vaccines for the desired and minimally acceptable targets. The Markov states included the following: susceptible/well, acute phase, indeterminate phase, chronic cardiomyopathy with and without CHF; the therapeutic vaccine assumed everyone started in the indeterminate state and included megaviscera outcomes. The preventative vaccine simulated children <1 year of age across their lifetime starting in the susceptible/well state, while the therapeutic vaccine simulated infected individuals starting in the indeterminate state across their lifetime. Both models included direct COIs (appropriately discounted over time) and measured effectiveness in DALYs; the preventative vaccine model included indirect costs, modelling the societal perspective. Both vaccines were found to be highly cost-effective or cost-effective, depending on vaccine characteristics and disease risk. The preventative vaccine saved costs in many tested scenarios and was highly cost-effective compared with no vaccination in many scenarios; the vaccine was dominant or had ICERs ≤USD 12,536/DALY averted when the risk of Chagas infection was ≥5% (for all vaccine costs and efficacy values tested). At a 1% risk of infection, the vaccine was cost-effective when ≤USD 30 and ≥50% efficacious. The therapeutic vaccine was also highly cost-effective and economically dominant for many tested scenarios, even a vaccine meeting the minimally acceptable target. Cost-effectiveness varied with key parameters, with the risk of severe adverse events being the strongest driver (i.e. autoimmune response); the number of doses required also had a large impact (as this also increased the risk of adverse events). A therapeutic vaccine was found to be more cost-effective than a preventative vaccine (it can be given to a more focussed proportion of the population who are at higher risk of developing chronic outcomes) (Lee et al., 2012a). Both of these studies are very well designed and thorough, include appropriate outcome measures, performed adequate sensitivity analyses of key variables to determine thresholds for vaccination and vaccine characteristics, and included other outcome measures (e.g. cost per cardiomyopathy case averted, cost of vaccinating an individual). However, they could be more readily compared if they both included indirect costs.

Two studies assessed the cost-effectiveness of screening for Chagas in nonendemic countries. Chagas is a public health concern even in nonendemic countries and is a growing problem due to migration and transmission through blood derivatives, organ transplant and through vertical transmission. One study modelled different screening methods of the U.S. blood supply though surveillance of various target groups (including testing donors from endemic countries) and will not be discussed in detail (Agapova et al., 2010). The other focussed on screening pregnant women from Latin America in nonendemic areas, with subsequent screening of newborns if positive (vertical transmission) (Sicuri et al., 2011). Results showed that testing was dominant (less costly and more effective) than no testing. Testing newborns resulted in EUR22/QALY and for mothers it was EUR96/QALY (both less than no testing). While this model included future costs of treatment, screening, etc., for Chagas outcomes (recurrent and nonrecurrent costs) discounted to the net present value, it is not a Markov model and may not account for variation over time and variation in when costs are accrued. Cost data came only from one hospital. Additionally, the study evaluated effectiveness in QALYs, which were calculated from DALYs. While this study undertook probabilistic sensitivity analyses, much of the variation in parameters was unknown and distributions were estimated with a range of 25% of the mean; one-way analyses sought to determine thresholds.

4.3.2 Human African trypanosomiasis

One study evaluated the cost-effectiveness of drug treatments (eflornithine and melarsoprol) for HAT (Robays et al., 2008). This study assumed the health care system perspective and only included cost of patient care (i.e. direct COIs). It had a sound methodology that included reputable and reliable data sources and adequate sensitivity analyses. The cost inclusions for the perspective evaluated were inclusive and analysis was done with and without drug costs to account for the donation of products (which was assured in the region for a set time period). Outcomes of interest included cost per death avoided, cost per YLL avoided and the increment cost per death (and per YLL) avoided. Eflornithine was found to cost USD 552—USD 845 per patient and USD 560—USD 856 per death averted (depending on drug inclusion costs) and was cost-effective compared with melarosprol (the current treatment) with an ICER of USD 58–USD 299/YLL averted. This study also adequately illustrated the impact of various parameters on the cost-effectiveness, with arsenical encephalopathy and the relapse rate of melarsoprol having the largest impact. Although this study took place in only one area (Caxito, Angola), limiting its generalizability, it shows the potential gains of changing drug treatment.

4.3.3 Leishmaniasis

Four studies evaluated on leishmaniasis (two on visceral and two on cutaneous): two evaluated treatment and two evaluated potential vaccines (Reithinger and Coleman, 2007; Vanlerberghe et al., 2007; Bacon et al., 2013; Lee et al., 2012b). Of the two vaccine studies, one focussed on visceral and the other on cutaneous leishmaniasis. The visceral leishmaniasis vaccine study compared the value of vaccination with two available treatments from the societal perspective in Bihar, India (Lee et al., 2012b). This was evaluated using a Markov model consisting of five states as follows: susceptible/well, immune, visceral leishmaniasis (either symptomatic or asymptomatic), post-kala-azar dermal leishmaniasis (PKDL) and death. Care-seeking behaviour and its associated costs and outcomes were included, as was lost wages for the patient and a family member (assumed for the duration of hospitalization). Individuals entered the model at birth and vaccination was evaluated over their lifetime. Vaccination dominated over no vaccination for all scenarios when the vaccine was ≤USD 5, and when ≤USD 30% and 50% efficacious. Vaccination was cost-effective for all scenarios tested (ICERs ≤USD 757/DALY averted). Vaccination reduced the number of visceral leishmaniasis and PKDL cases and typically resulted in cost-savings. Strengths of this study include the design, adequate sensitivity analyses of key parameters, the inclusion of PKDL and tendency to be conservative about the benefits of vaccination. However, the focus on one state in India reduces generalizability and as the authors note, the potential effects of HIV coinfection were not included.

The cutaneous leishmaniasis vaccine study used a Markov model to evaluate its potential economic value in seven Latin American countries (Bolivia, Brazil, Colombia, Ecuador, Mexico, Peru and Venezuela) simulated a vaccine given at birth and the benefits over their lifetime (Bacon et al., 2013). Markov states included the following: uninfected, cutaneous leishmaniasis, mucocutaneous leishmaniasis, diffuse cutaneous leishmaniasis, prior infection (recently recovered and at risk) and death. Cost-effectiveness was measured in cost per case averted (and evaluated compared with the cost of the recommended treatment of pentavalent antimonials, USD 180). Cost of a case accounted for treatment-seeking behaviour and the associated costs and outcomes. A 2-dose (USD 0.5/dose) 70% efficacious vaccine protecting for 5 years, saved costs to avert a case (compared with treatment) as long as the infection risk was ≥0.1% (averting 14–52 cases per 1,000 individuals vaccinated, 50–93% of mucocutaneous cases and 40–91% of diffuse cutaneous cased, depending on revaccination and compliance). Depending on vaccine efficacy, vaccination cost USD 66–USD 140/case averted when the infection risk was ≥0.1%. Across the scenarios tested, the infection risk needed to be at least 0.03% for the vaccine to cost less to avert a case than treatment. The cost-effectiveness of vaccination was also compared with no vaccination. The methodology used was robust, data came from reputable sources and varied in probabilistic sensitivity analysis and the authors conducted adequate sensitivity analysis of key parameters. While the authors note that the value of vaccination may change from country to country, based on input parameters, their results may be generalizable across these seven countries and for countries with similar local conditions.

Of the treatment studies, one sought to determine the most efficient (most deaths averted) therapeutic approach for visceral leishmaniasis to inform drug policy in India from the health service perspective (Vanlerberghe et al., 2007). Four drug treatments were considered: antimonials, amphotericin B deoxycholate, miltefosine and lipid formulation of amphotericin B. The model started with a person with clinical signs and symptoms; diagnostic testing was performed and those with positive tests received treatment (both true and false positive cases). The COI included diagnostics, drug treatment, possible retreatment episodes, ancillary drugs and patient care during treatment course until cure (included hospitalization, adjuvant treatments, etc.). Results showed (under baseline conditions) that treatment with amphotericin B deoxycholate was the most effective, averting 349 deaths per 1,000 suspected cases (87.2% of all deaths attributable to visceral leishmaniasis with a 40% probability of disease given symptoms). Overall the miltefosine was the dominant strategy (USD 328/death averted), antimonials were dominated by miltefosine, AmBisome was dominated by amphotericin B deoxycholate and amphotericin B deoxycholate cost USD 4,543/death averted. Drug price was the main driver of cost-effectiveness. Although this study evaluated treatment options in India, the conclusions may be applicable to other endemic regions so long as their effectiveness and cost estimates are similar to local figures.

Strengths of this study include the use of one- and two-way sensitivity analyses, robust data from the literature (supplemented with a clinical trial). However, possible delays in correct diagnosis were not considered, PKDL cases were not included, and the model did not capture long-term benefits or risks of treatment strategies (e.g. change in transmission or drug resistance). While deaths averted are valid and useful outcome measures, other quality of life measures, such as DALYs may capture other effects of treatment, such as adverse side effects.

The other drug treatment study evaluated the programme diagnosis and treatment services for cutaneous leishmaniasis patients attending the programmes clinics in Kabul city, Afghanistan (Reithinger and Coleman, 2007). Treatment consisted of intralesional and intramuscular sodium stibogluconate. Programme costs alone were USD 129,820 and medical costs for 16,390 patients totalled USD 102,224 in 2003; 214 DALYs were averted due to successful treatment. The standard treatment regimen cost USD 27/patient treated (95% confidence interval: USD 20—USD 36) and USD 1,181/DALY averted (95% confidence interval: USD 761—USD 1,827/DALY averted), which is not cost-effective (i.e. not less than 3 × GDP of USD 165 in 2004). These estimates were most sensitive to disability weight used and lesion duration, followed by drug compliance and efficacy. While this study performed probabilistic sensitivity analyses, effectiveness and compliance data came from clinical trials, which can lead to overestimates of these parameter values (since clinical trials are controlled this may introduce bias as participants tend to more carefully follow protocols). The authors point out that they did not include benefits to transmission to be conservative about their estimates. As this study is focussed in one city with a high incidence rate, there is a lack of generalizability.

4.3.4 Schistosomiasis

Two studies evaluated the cost-effectiveness of schistosomiasis interventions. One examined the short-term effects of health education and promotion and long-term effects on reinfection patterns in the Poyang Lake region of China (Hu et al., 2005). Cost-effectiveness was secondary to other study outcomes (i.e. this is not primarily a CEA) and measured as the cost for reducing 1% of the infection rate per 100 persons. The total cost of the health education and promotion cost 88 Yuan per 100 persons. Compliance increased among men, women and school-children and the prevalence dropped by 83.7% in school-children and 63.4% in women. This translates to 3.73 Yuan per 100 persons for reducing the reinfection rate in school-children and women by 1%. As cost-effectiveness was not the primary outcome, the methodology and cost data used are not well described. Additionally, the study uses an unusual measure for cost-effectiveness, reports no variation, and as it is based on a pilot study, is not generalizable to other areas. The other study was a comparative cost-effectiveness analysis of four screening methods for urinary schistosomiasis in school-based programmes in Nigeria (Fatiregun et al., 2009). Direct (e.g. production and delivery of materials, and training materials) and indirect (e.g. school time used to train teachers and used for screening) costs of screening were included. The cost per number of cases correctly diagnosed were calculated for each screening method and the method with the lowest value was deemed the most cost-effective. In addition, the effect of varying costs was investigated assuming that the number of missed cases was opportunity cost (i.e. translated to cost per missed case). Unqualified haematuria screening cost the lowest per correctly identified case (USD 2.16 to diagnose a case correctly). Urine examination, the standard, cost USD 13.46 per correctly identified case. With a fixed budget of USD 424 to identify cases, unqualified haematuria was able to diagnose 196 cases correctly (compared with 32 cases with visual examination), suggesting that it is cost-effective than the visual method. It should be noted that this was evaluated in only one school with a small sample size (592 students), thus greatly limiting generalizability of results. The validity of the tests is affected by the prevalence, so results may not be applicable in lower risk groups or lower prevalence settings. Additionally, the ICER was not calculated, which is important when choosing between interventions competing for the same limited resources.

4.3.5 Multiple NTDs

Four studies evaluated the cost-effectiveness of interventions for several NTDs, showing the benefits of combined strategies (Leslie et al., 2011; Evans et al., 2011; Guyatt et al., 2001; Ndeffo Mbah et al., 2013a). All of these studies focussed on African countries; two evaluated real programmes (Leslie et al., 2011; Evans et al., 2011); one modelled the impact on transmission (Ndeffo Mbah et al., 2013b) and one evaluated the effect of deworming on anaemia (Guyatt et al., 2001). One evaluated the cost-savings when switching from the separate delivery of three drugs for onchocerciasis, lymphatic filariasis, schistosomiasis and STH MDA to coadministration of the drugs (triple drug administration), which was found to be clinically safe (Evans et al., 2011). Another evaluated programme costs for schistosomiasis and STH control in Niger; cost-effectiveness was evaluated between the start of the intervention (baseline) and the follow-up period (2 years) across eight schools and four communities (Leslie et al., 2011). Overall, it cost USD 2.5/schistosomiasis infection averted in the treated population of children (USD 1.10/infection averted) and adults (USD 6.5/infection averted) over two rounds of MDA. Including indirect effects (infections averted in treated and untreated persons) it cost USD 1.78/schistosomiasis infection averted. This can result in cost-savings when compared with the productivity losses incurred for cases. Although this study evaluated the cost of MDA for both schistosomiasis and STH, the cost-effectiveness was only evaluated for schistosomiasis infection. While these studies evaluated real programmes and are helpful evaluations, they have limited generalizability. The two remaining studies, although evaluating NTD interventions, focus on other outcomes, such as the reduction in HIV transmission (Ndeffo Mbah et al., 2013b) and the impact on anaemia (Guyatt et al., 2001), and will not be discussed here.

For NTDs like echinococcosis that affect both human and animal populations, especially livestock, cost-benefit studies include the potential cost-savings of an intervention not only for the human population, but for animals as well (e.g. sheep production). For example, echinococcosis in humans and sheep (Jimenez et al., 2002) and schistosomiasis in humans and cattle (Zhou et al., 2005). The latter is cost-benefit study (a retrospective evaluation of six representative counties in China) finding a net benefit-cost ratio was 6.20, meaning that for every dollar spent, the national schistosomiasis control programme (from 1992 to 2000) gained USD 6.20. While these studies are helpful and have a rounded approach to evaluating an intervention or programme (by including other populations), care should be taken when comparing across diseases and control programmes making sure they evaluate similar measure or outcomes to appropriately weigh options for resource allocation.

4.3.6 Summary of CBA and CEA studies

While the existing cost-benefit and cost-effectiveness studies have provided important information, the current literature does not cover many interventions, policies and locations, including some regions where NTD burden is the highest. As with other economic studies of NTDs, many existing studies are specific to particular locations and vary in the types of costs and health effects included, restricting their generalizability. They vary in whether they incorporate costs such as those for drug treatment, diagnostics, indirect costs of transportation and/or food, recurrent episodes, patient care until cure and productivity losses. Different studies have different time horizons, ranging from a year or less so that only the relatively immediately effects are captured to the person's lifetime.

5. DETERMINING RATES OF RETURN: RETURN-ON-INVESTMENT AND INTERNAL RATE OF RETURN ANALYSES

Many decision-makers are interested in specifically how much the investment of funds will yield. Return-on-investment (ROI) and internal rate of return (IRR) calculations quantify the economic return as a percentage of the economic investment for a particular programme and thus both can be used to compare different investments' potential profitability over a period of time and can aid in finical planning and operationalization.

5.1 Return-on-investment

The following formula determines the return on investment:

$$\text{ROI} = (\text{gain from investment} - \text{cost of investment})/\text{cost of investment}$$

The gain from investment minus the cost of investment is the net return. In most cases, the ROI indicates the financial return on a given financial investment as both gains and costs are expressed in currency. The resulting ROI fraction is most often expressed as a percentage and must consider the time period being evaluated. Thus ROIs can be normalized to a common time period when comparing the ROIs of two different programmes. For example, if a programme to prevent an insect-borne parasitic disease through the distribution of bed nets and insecticide spraying has an estimated cost of USD 1 million for a given country, and will result in economic gains of USD 1.5 million over a 3-year period, the total ROI over that 3-year period would be 50%:

$$\text{ROI} = (\text{USD } 1.5 \text{ million} - \text{USD } 1 \text{ million})/\text{USD } 1 \text{ million}.$$

How does this compare to mass chemotherapy administration, which we will say in this example has a 25% ROI over a period of 2 years? The annual rate of return (total ROI divided by time frame) would be 17% and 13% for bed nets/insecticides and chemotherapy, respectively.

As the total cost and investment figures necessary to determine the ROI may be discounted and/or adjusted for inflation, may include direct, indirect or intangible costs and benefits, or some combination of each, and may vary in the currency, arriving at ROI estimates can be very involved and challenging.

5.1.1 Example of ROI study: a Chagas disease vaccine

To illustrate an ROI study, we will look at a study estimating the ROI for a therapeutic vaccine for Chagas in Mexico (Lee et al., 2012a):

Step 1: Determine the investment and amount of investment

This study focussed on the following question: what is the value of investing in the clinical development of a therapeutic Chagas vaccine, across different vaccine coverage scenarios. This involved determining the minimum target population coverage required during the vaccine's first year on the market to yield a positive ROI. Information from this study could guide donors and funders as well as vaccine developers. Different scenarios considered the NPV of clinical development costsas USD 100 million, USD 150 million, USD 200 million and USD 400 million.

Step 2: Establish the return

The value of the therapeutic Chagas vaccine is the costs averted from treating Chagas disease, which a Markov model was used to estimate. Figure 5 shows the structure of the model. The time horizon was an infected person's remaining lifetime (with a 1-year cycle length). The perspective was the third-party payer. Patients entered the model in the indeterminate state and had probabilities of staying in the same state or moving to another state (e.g. megaviscera, cardiomyopathy with and without CHF, or combination of the two) until transitioning to the death state (either Chagas or unrelated causes). Vaccination attenuated the probability of megaviscera, CHF, or disease-associated death and had an associated probability of a severe adverse event (vaccine-induced cardiomyopathy).

The cost-savings from vaccination (i.e. disease costs averted for each successfully vaccinated person) would depending on the vaccine characteristics such as cost, efficacy and mode of action. This study used target product profiles (TPP) for therapeutic vaccines currently under development. The aimed for TPP was as follows: the vaccine would prevent cardiomyopathy, cost USD 46, have a target population of children over 2 years of age, require only one dose, confer lifetime protection and have a 3% probability of severe adverse effects. The minimally acceptable TPP was as follows: the vaccine would delay the onset of cardiomyopathy by 10 years, cost USD 200, have a target population of adults 16 years of age and older, require two doses, confer lifetime protection and have a 3% probability of severe adverse effects.

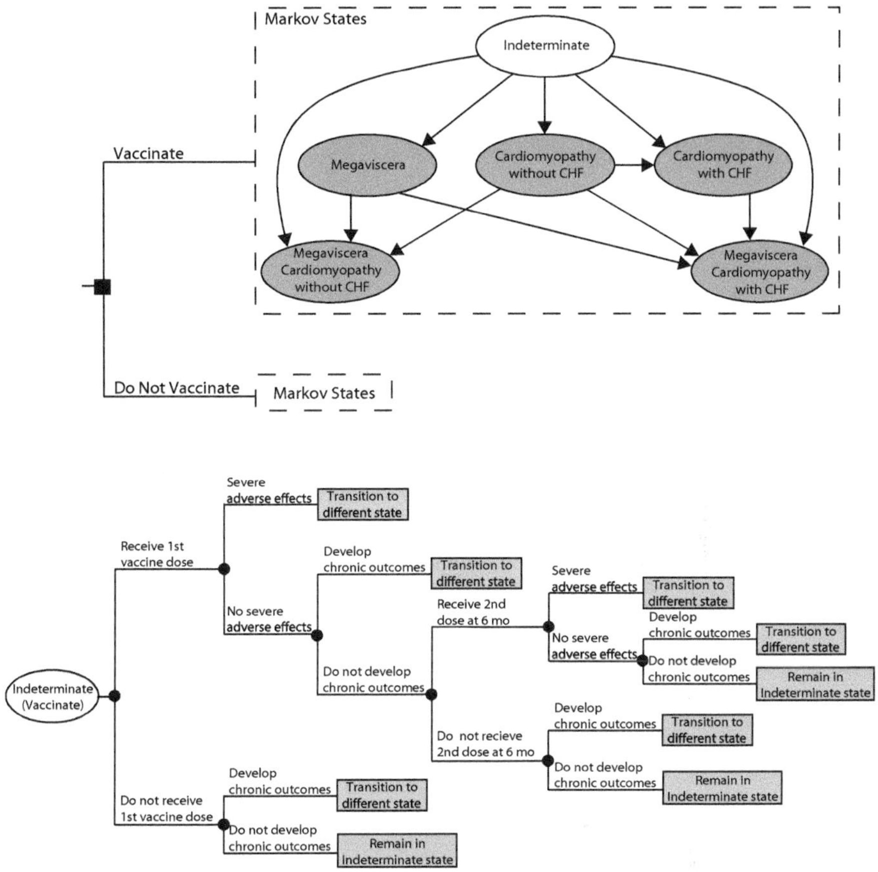

Figure 5 Chagas therapeutic vaccine model structure and Markov states. *Modeling the economic value of a Chagas' disease therapeutic vaccine, B.Y. Lee, K.M. Bacon, A.R. Wateska, et al., Human Vaccines & Immunotherapeutics, Sep 16, 2012 Taylor & Francis, reprinted with permission.*

Step 3: Choose the time course for economic value realized

It can be challenging to determine the time course over which the results of investment will be realized, especially when investing in the development of an intervention. Considerable uncertainty can exist regarding when the intervention may reach the market. Clinical development of a vaccine can proceed over many time courses and encounter unexpected obstacles. Therefore, this study varied the length of clinical development and resulting licensure from 15 to 30 years. The discount rate was 3%.

Step 4: Set up the ROI calculation

The gain from investment is the NPV of cost-savings from the vaccination (discounted for the years to market) multiplied by the number of persons vaccinated that year (i.e. target population coverage). This generates the cost impact of rolling out the vaccine. The following formula summarizes the ROI calculation:

$$\begin{aligned}\text{ROI} = \Big[&(\text{Target Population} \times \text{Target Coverage} \times \text{Vaccine Compliance} \\ &\times \text{Cost-Savings per Vaccinated Case}) \\ &/(1 + \text{Discount Rate})^{\text{Years to Market}}\Big] \\ &- \text{Vaccine Development Cost}\end{aligned}$$

This study assumed 6 million infected persons eligible for vaccination and a compliance rate of 100% and determined the ROI for both the desired and minimally acceptable vaccine target (changing the cost-savings per vaccinated case).

Step 5: Conduct sensitivity analyses

As many values for the ROI are unknown, sensitivity analyses varying these parameters are essential for determining their impact on the ROI and elucidating the return under various conditions. These variables include: years to market, vaccine development costs, vaccine characteristics and the target population coverage. As already described, this study ranged the years to market from 15 to 30 years and vaccine development costs from USD 100 to USD 200 million. In this example, the target population coverage is ranged from 1% to 7%, while the vaccine characteristics varied from the desired target vaccine to the minimally acceptable vaccine target. Experiments including evaluating the ROI with these various combinations of parameter values to determine the ROI and the thresholds.

Table 9 shows the results for the ROI analysis for the desired TPP vaccine. Assuming 6 million people would be eligible for vaccination and the vaccine reached the market in 15 years, the additional return on investment costs would be USD 2–USD 18 million if $\geq$3% of the target population was vaccinated in the first year and development costs were USD 100 million, or 4% if development costs were USD 150 million, 5.5% if development costs were USD 200 million and 11% if development costs were USD 400 million. Even a vaccine with a desired target TPP could take 30 years to reach the market and still generate a positive ROI. However, a vaccine meeting the minimally acceptable TPP would have difficultly yielding a positive ROI, regardless of time to market and vaccine coverage values analyzed. The analysis examines only the potential return after one year of vaccination, therefore the ROI may be underestimated (Lee et al., 2012a).

Table 9 Minimum number of vaccinated individuals and minimum target population coverage needed in the first year of vaccination to achieve a positive return-on-investment for a desired target product profiles therapeutic Chagas vaccine in Mexico

Years to market	Number vaccinated	Target population coverage (%)
Total development cost: USD 100 million		
15	180,000	3.0
20	210,000	3.5
30	270,000	4.5
Total development cost: USD 150 million		
15	240,000	4.0
20	300,000	5.0
30	420,000	7.0
Total development cost: USD 200 million		
15	330,000	5.5
20	420,000	7.0
30	540,000	9.0
Total development cost: USD 400 million		
15	660,000	11.0
20	780,000	13.0
30	1,020,000	17.0

5.2 Internal rate of return

While the ROI shows what someone will get back by investing a certain amount, the IRR calculates what return is needed to break even or earn back what was initially invested. The IRR sets the NPV of a programme or intervention to zero – allowing us to calculate the 'break even' point (i.e. the rate of return that is equivalent to spending zero present-day dollars). Thus, a lower rate will lose money and a higher rate will make money. The IRR intrinsically includes discounted cash flow analysis this means that it calculates the discount rate (independent of inflation) that is required to make the NPV of costs (negative cash flows) equal to the NPV of benefits (positive cash flows) – thus it is an intrinsic calculation of the potential costs and benefits of a programme or intervention.

The following formula calculates the IRR:

$$0 = P_0 + P_n/(1 + IRR)^n$$

where P_0 represents the investment as a negative value and P_n represents the generated gains (cash flows) or savings each period (n) for the investment made. In other words, P_n is the expected return each period (e.g. year) for the given investment. This formula determines the IRR for which the total return is equal to the investment, or 0.

Decision-makers can use this information to efficiently allocate resources among different programme and intervention options, as well as to other, more general allocation possibilities – such as prevailing interest rates, for example. This allows decision-makers to understand the potential costs and benefits of certain programmes in the larger economic context. For example, if USD 50,000 worth of bed nets were purchased and distributed for the prevention of visceral leishmaniasis, we can calculate the IRR. Let us assume that these bed nets have a lifetime of 3 years and avert USD 18,000 per year in health care costs. Our formula would be

$$0 = -\text{USD } 50,000 + (\text{USD } 18,000/(1 + \text{IRR})) + \left(\text{USD } 18,000/(1 + \text{IRR})^2\right) + \left(\text{USD } 18,000/(1 + \text{IRR})^3\right)$$

Solving, we find that the IRR is approximately 3.9%.

5.3 Rate of return studies in the scientific literature

Table 10 summarizes the ROI and IRR literature for parasitic NTDs. As can be seen, these types of analyses in regards to NTDs are severely lacking – there are only two published studies. We have already described the ROI of a therapeutic Chagas vaccine in detail. The remaining study evaluated

Table 10 Summary of literature of return-on-investment (ROI) and internal rate of return (IRR) studies for neglected tropical disease

Disease	Setting	Intervention	ROI or IRR (in USD)
Chagas	Mexico	Therapeutic vaccine	9.1–29.6 million (Lee et al., 2012a)
Leishmaniasis	Nepal	Early diagnosis, prompt and complete treatment, early detection and containment of epidemics, indoor residual spraying, community health education, regular supply of drugs, health personnel and field research	873 million (65 million Nepalese Rupees); (35% IRR) (Adhikari and Supakankunti, 2010)

the IRR for leishmaniasis control in Nepal over a 13-year period (2003–2015) (Adhikari and Supakankunti, 2010). This study included not only the costs of resources for the interventions (e.g. drugs, vehicles, personnel, equipment, etc.), but also costs for the undesirable impacts, such as environmental impacts of insect spraying and toxicity of drugs (among others). Direct and indirect costs for both providers and patients are included in this model and come from published and unpublished literature. The benefits of leishmaniasis control included productivity gains and costs saved due to averted cases and reduced morbidity and mortality. The IRR was calculated by estimating the social discount rate when the NPV was set to zero. The total discounted net benefits were high, approximately 65 million Nepalese Rupees (Table 8) with a 35% IRR. This gives a ratio of 71:1 suggesting that for every one Rupee invested, the intervention will yield 71 Rupees in return. The authors note that this may be a conservative estimate, as other economic benefits of the intervention (poverty reduction, improved economic welfare, etc.) were not included.

Although there is only one study that directly calculates IRR for NTD interventions, it suggests that these interventions may have extremely high rates of internal return. There is a substantial gap in the literature for these types of economic analyses, gaps that if filled could be very beneficial to future NTD funding, research and prevention and control.

6. FINANCING NTD PREVENTION AND CONTROL

The terms funding and financing may seem synonymous and interchangeable but subtle differences exist. **Financing** is the mechanism, structure or procedures of providing (or trying to provide) funding (i.e. resources) for an activity. For example, how do you fund a vector control programme? You find money to pay for the insecticide, bed nets, personnel and other needs. How do you finance a vector control programme? This requires details on how specifically you would get and apply those funds. Who would provide the funds? If necessary, how would they be remunerated? What procedures and programmes need to be established or altered to help these funds to get to where they are needed? Who else would be involved?

Financing of a country's NTD control programme depends heavily on that country's government's available **fiscal space**. Fiscal space is the existing capacity in a government's (or other relevant organization's) budget to provide resources without jeopardizes its stability or the surrounding economy. Governments are unlikely to fund an activity when it does not have adequate fiscal space. Governments can increase their fiscal space by either reducing existing costs, borrowing or procuring more resources (e.g. from raising taxes or identifying donors). Calculating a government's current fiscal space gives stakeholders an idea of a government's ability to implement an NTD control programme and the funding gap that exists. Budget impact analyses tabulate how a particular programme or intervention may affect stakeholder's budget. That stakeholder can be a government, business, another type of organization or a person. For example, if MDA will cost a government over a million dollars and there is only budget for USD 200,000, implementing MDA will not be feasible without increasing the available budget. When not enough fiscal space is available, funding for an NTD programme in a country can come from within the country via **domestic resource mobilization** (DRM), i.e. shifting/utilizing resources from within the country (e.g. local businesses making investments). DRM is preferable to procuring resources from outside the country because it keeps ownership of the NTD control programmes within the country rather than increasing the country's dependence on other countries. A government's fiscal policy, i.e. taxation and spending, heavily determines the degree and nature of DRM that may occur. For example, tax breaks could favour activities that promote or facilitate NTD control and taxation can discourage activities that impede NTD control. Available resources (i.e. **capital**) can come in many different forms, such as buildings, vehicles and other equipment (real capital), currency or the equivalent (finance capital), and people (human capital). An important characteristic of a resource is its liquidity, i.e. how readily the resource can be converted into other flexible resources such as cash. A country can have considerable capital but if this capital is relatively illiquid (i.e. not easily convertible into other resources), the resources may not be useful for particular purposes. For instance, a country that has a lot of its resources invested into weapons may not be able to divert these resources towards NTD control unless the weapons can be easily and quickly sold for cash.

When a country cannot muster enough resources to fund its NTD control programmes, it often relies on foreign entities via donations, borrowing, trade or **foreign direct investments** (FDI). These approaches have many disadvantages. Donations are rarely sustainable as the donors do not have unlimited resources. Borrowing increases a country's debt to others. Many resource-poor countries may even default on loans due to their ability to pay back their debts. Trade presumes that the country has something of value to exchange for funds. Even when a country has valuable enough assets to trade, they rarely do so to bolster NTD control programmes, having many other priorities. An FDI occurs when a foreign company, government or organization purchases something (e.g. land, equipment or buildings) in a country, i.e. resulting in a controlling interest (ownership) in that country. FDIs can be predatory, as the entity making the FDI is seeking to establish some type of position within or taking advantage of the country. Therefore, getting foreign entities involved in a country's NTD control can be financially quite complicated.

Marginal analyses can help an entity to determine the effect of a change in a policy or activity. The marginal cost of a change is the resources needed to make a one unit change (e.g. the marginal cost of immunizing one more child may be USD 5). The marginal benefit is the gain of that one unit change (e.g. the marginal benefit of immunizing one more child may be USD 20). A marginal analysis would then be net of the benefits minus the costs for that one unit change (e.g. USD 20−USD 5 = USD 15). This would suggests that it would be favourable to immunize another child. The marginal analysis can vary depending on the number of children already immunized. Immunizing the first few children may have tremendous benefits because no one at the time is protected. However, once enough of the population is immunized and a high level of herd immunity is achieved, affording nonimmunized children protection, the marginal benefits of immunizing an additional child may be lower.

Financing NTD prevention and control is challenging given their tendency to affect resource-starved environments and the comparative lack of attention from the global community. Businesses may not see an obvious profit-making opportunity that can convince them to participate. This makes it especially important to identify funding sources and develop creative financing mechanisms to secure funding. However, the scientific literature is currently rather short on studies that serve these purposes. It is not clear whether this shortage is due to lack of attention or the complicated nature of the situation or some combination of both. Some studies, commentaries and reports in the scientific literature help to profile and assess the current sources and amount of funding for NTDs and the amount that people and organizations are willing to pay to control NTDs. These range from discussions on the structure of NTD control financing to calls for NTD awareness and advocacy for control measures. Currently, the funding sources for NTD control are relatively limited, compared with funding sources for more prominent diseases. The total funding available for all NTDs combined pales in comparison with other, more prominent diseases such as HIV/AIDs and malaria. A survey conducted in 2007 sought to find the global investment into research and development (R&D) of new pharmaceuticals, products to prevent, manage or cure diseases of the developing world (Moran et al., 2009). In 2007, 90% of funding for NTDs came from public and philanthropic donors (totalling USD 2.3 billion). Funding for STHs totalled ~USD 51.6 million (2.02% of total funding). Compared with HIV/AIDS (42.30% of total funding) and malaria (18.30% of total funding), the amount invested in NTD R&D is substantially less. Since many NTD programmes operate in resource-poor environments, outside organizations may be required to maintain coverage and continue sustainability (Garba et al., 2009, 2006; Guyatt, 2003). Outside organizations include pharmaceutical and medical device manufacturers, foreign governments, and nongovernmental organizations donating money, medications, diagnostics, medical equipment and personnel. To make programmes sustainable, commitments need to be fairly long term (Zhang et al., 2010). Organizations can also play important roles by advocating for particular disease control programmes and funding and highlight funding gaps (Fenwick et al., 2009).

As there can be considerable cross-over among NTD control programmes and also between NTD control programmes and other disease control programmes, some studies have evaluated the feasibility of coordinating different disease programmes, which in turn could gain efficacies and help to leverage the same set of resources across multiple diseases (Linehan et al., 2011). One strategy may be pairing NTD control programmes with related disease control programmes that can garner more attention and resources. For example, pool of available funds for malaria control may be greater than that for an NTD control. Malaria control-designated funds could support vector control activities that may additionally help the NTD control. Another strategy is to find creative ways to leverage existing community assets for NTD programmes, such as using vehicles, buildings and personnel that already serve other purposes (Vazquez-Prokopec et al., 2009; Boelee and Laamrani, 2004).

Economic studies can help to determine the financial feasibility to achieve desired NTD control goals and prioritize efforts. Even if some goals may be biologically or epidemiologically possible, they may be prohibitively expensive. An example is a study evaluated the feasibility of control and elimination of lymphatic filariasis, onchocerciasis, schistosomiasis, trachoma and STH in 14 Latin American and Caribbean countries (Bitran et al., 2009). This study simulated NTD interventions including MDA, illness level control (e.g. surgery for trachoma), vector control, social mobilization and education. Results estimated the total NPV of control and elimination at USD 169 million (excluding safe water provision and sanitation). Lymphatic filariasis elimination made up the largest portion of total costs. This study suggested that lymphatic filariasis, onchocerciasis and trachoma can feasibly and affordably be eliminated by 2020 at a cost of USD 128 million. NTD control cost less than USD 0.51 per capita and represented less than 1% of government expenditures on health for all countries except Guyana and Haiti. Another study evaluated a methodology using multiple criteria to (e.g. cost-effectiveness programmes, disease severity, the loss of productivity and quality of evidence for intervention effectiveness) to rank the priority of NTDs. The authors simultaneously took into account several relevant criteria for NTDs and suggested a multi-criteria methodology to identify NTD targets for prevention and control (Stromme et al., 2013).

Willingness-to-pay (WTP) studies can provide insight into how people value an intervention. They also can help to determine how subgroups in a community may be affected, if cost recovery strategies were introduced. A few WTP studies evaluated various programmes. In Haiti, the mean WTP was USD 491 per year for lymphatic filariasis treatment for one person and USD 5.57 per month per household to prevent disease transmission. Additionally, 39% of households were not willing to pay for treatment and 7% were not willing to pay for prevention (Rheingans et al., 2004). A community's WTP for school-based chemotherapy programme for schistosomiasis and STHs in rural Tanzania ranged from USD 0 to USD 25 per child per year for a household and USD 1.25 for the community (i.e. median household WTP). Additionally, the community was willing to pay an extra USD 0.125 per year to cover costs for schoolchildren whose parents could not afford or were unable to pay (Lwambo et al., 2005). In a region of China, only 30.9% of households were willing to pay for schistosomiasis screening, diagnosis and treatment (Yu et al., 2001). In a similar type of study, authors evaluated the affordability of community-directed treatment with ivermectin for households in onchocerciasis endemic Nigeria (Onwujekwe et al., 2001). This study concluded that households could afford the programme (consuming 0.01–0.05% of the average annual household income, depending on programme cost), however the final decision of affordability lies with the household.

A noticeable gap in the literature is new financial mechanisms to fund NTD control efforts. For example, **advanced market commitment** (AMC) helped to transform immunization in low- and middle-income countries (LMICs). Prior to AMCs, many vaccine manufacturers were reluctant to develop and manufacture vaccines for LMICs as they were concerned about low profit margins and uncertain demand. AMC changed this dynamic by getting donors to commit to paying a predetermined price for vaccines in the future thereby guaranteeing vaccine manufacturers revenue. It seems that NTD control programmes could benefit from similar arrangements or other types of creative mechanisms. More dialogue within the scientific literature could help to germinate such new approaches.

7. SUMMARY

Economic and financing studies are particularly important for decision making when resources are considerably limited, which is the case for NTDs. In fact, the definition of NTDs is an economic one. The shortage of resources for NTD control may be due in large part to the fact that the burden of NTDs and economic value of control measures have not been fully characterized. A number of study methodologies are available. Cost of illness can quantify the extent, magnitude and change of a problem. Cost of intervention studies can profile the feasibility and guide the design of a policy or intervention. Cost–benefit, cost-effectiveness and return-on-investment studies can determine the potential value of different policies and interventions.

NTDs have unique characteristics that necessitate special attention in economic analyses (highlighted in Box). While the existing literature has made important findings, there is a need for substantially more work. Many of the NTDs and interventions and policies require more evaluation. Approaches used for other diseases may need adjustments for NTDs to capture to full impact of NTDs. New approaches may emerge. With increasing literature in this area, NTDs may not be as 'neglected' in the future as they are now.

- NTDs affect the poorest of the poor and have a strong association with poverty, which makes it challenging to disentangle cause and effect. Impoverished conditions can promote transmission, exacerbate symptoms, impede monitoring and delay diagnosis and proper treatment. At the same time, NTDs can also lead to poverty, as and their chronic and cognitive outcomes can impair a population's ability to overcome poverty.
- NTD infection can lead to a wide spectrum of outcomes, many of which are chronic, atypical and have subtle effects. It can take years to develop symptoms after initial infection and these outcomes can have a long time horizon. Some outcomes and their long-term ramifications can be hard to measure, as is the case with cognitive delay. Thus, NTDs are not easily quantifiable as other diseases.
- The true disease burden can be difficult to determine. In addition to evading detection, under-reporting and misdiagnosis, not all infected individuals seek care. Additionally, misdiagnosis can occur as some disease outcomes can be attributed to other causes, so disentangling true cause can be extremely difficult.

REFERENCES

AbouZahr, C., Vaughan, J.P., 2000. Assessing the burden of sexual and reproductive ill-health: questions regarding the use of disability-adjusted life years. Bull. World Health Organ. 78 (5), 655–666.

Abuhab, A., Trindade, E., Aulicino, G.B., Fujii, S., Bocchi, E.A., Bacal, F., 2013. Chagas' cardiomyopathy: the economic burden of an expensive and neglected disease. Int. J. Cardiol. 168 (3), 2375–2380.

Alvar, J., Yactayo, S., Bern, C., December 2006. Leishmaniasis and poverty. Trends Parasitol. 22 (12), 552–557.

Adhikari, S.R., Supakankunti, S., Khan, M.M., 2010. Kala azar in Nepal: estimating the effects of socioeconomic factors on disease incidence. Kathmandu Univ. Med. J. 8 (29), 73–79.

Albonico, M., Montresor, A., Crompton, D.W., Savioli, L., 2006. Intervention for the control of soil-transmitted helminthiasis in the community. Adv. Parasitol. 61, 311–348.

Anderson, R.M., Truscott, J.E., Pullan, R.L., Brooker, S.J., Hollingsworth, T.D., 2013. How effective is school-based deworming for the community-wide control of soil-transmitted helminths? PLoS Negl. Trop. Dis. 7 (2), e2027.

Agapova, M., Busch, M.P., Custer, B., 2010. Cost-effectiveness of screening the US blood supply for *Trypanosoma cruzi*. Transfusion 50 (10), 2220–2232.

Adhikari, S.R., Supakankunti, S., 2010. A cost benefit analysis of elimination of kala-azar in Indian subcontinent: an example of Nepal. J. Vector Borne Dis. 47 (3), 127–139.

Basta, S.S., Soekirman, Karyadi, D., Scrimshaw, N.S., 1979. Iron deficiency anemia and the productivity of adult males in Indonesia. Am. J. Clin. Nutr. 32 (4), 916–925.

Benner, C., Carabin, H., Sanchez-Serrano, L.P., Budke, C.M., Carmena, D., 2010. Analysis of the economic impact of cystic echinococcosis in Spain. Bull. World Health Organ. 88 (1), 49–57.

Budke, C.M., Deplazes, P., Torgerson, P.R., 2006. Global socioeconomic impact of cystic echinococcosis. Emerg. Infect. Dis. 12 (2), 296–303.

Brooker, S., Kabatereine, N.B., Gyapong, J.O., Stothard, J.R., Utzinger, J., 2009. Rapid mapping of schistosomiasis and other neglected tropical diseases in the context of integrated control programmes in Africa. Parasitology 136 (13), 1707–1718.

Brady, M.A., Hooper, P.J., Ottesen, E.A., 2006. Projected benefits from integrating NTD programs in sub-Saharan Africa. Trends Parasitol. 22 (7), 285–291.

Bacon, K.M., Hotez, P.J., Kruchten, S.D., Kamhawi, S., Bottazzi, M.E., Valenzuela, J.G., et al., 2013. The potential economic value of a cutaneous leishmaniasis vaccine in seven endemic countries in the Americas. Vaccine 31 (3), 480–486.

Boelee, E., Laamrani, H., September 2004. Environmental control of schistosomiasis through community participation in a Moroccan oasis. Trop. Med. Int. 9 (9), 997–1004.

Bitran, R., Martorell, B., Escobar, L., Munoz, R., Glassman, A., 2009. Controlling and eliminating neglected diseases in Latin America and the Caribbean. Health Aff. 28 (6), 1707–1719.

Castillo-Riquelme, M., Guhl, F., Turriago, B., Pinto, N., Rosas, F., Martinez, M.F., et al., 2008. The costs of preventing and treating chagas disease in Colombia. PLoS Negl. Trop. Dis. 2 (11), e336.

Colston, J., Saboya, M., 2013. Soil-transmitted helminthiasis in Latin America and the Caribbean: modelling the determinants, prevalence, population at risk and costs of control at sub-national level. Geospatial Health 7 (2), 321–340.

Engels, D., Savioli, L., 2006. Reconsidering the underestimated burden caused by neglected tropical diseases. Trends Parasitol. 22 (8), 363–366.

Das, M., Banjara, M., Chowdhury, R., Kumar, V., Rijal, S., Joshi, A., et al., 2008. Visceral leishmaniasis on the Indian sub-continent: a multi-centre study of the costs of three interventions for the control of the sandfly vector, *Phlebotomus argentipes*. Ann. Trop. Med. Parasitol. 102 (8), 729–741.

Evans, D., McFarland, D., Adamani, W., Eigege, A., Miri, E., Schulz, J., et al., 2011. Cost-effectiveness of triple drug administration (TDA) with praziquantel, ivermectin and albendazole for the prevention of neglected tropical diseases in Nigeria. Ann. Trop. Med. Parasitol. 105 (8), 537–547.

Fincham, J.E., Markus, M.B., Adams, V.J., 2003. Could control of soil-transmitted helminthic infection influence the HIV/AIDS pandemic. Acta Trop. 86 (2–3), 315–333.

Fevre, E.M., Odiit, M., Coleman, P.G., Woolhouse, M.E., Welburn, S.C., 2008. Estimating the burden of rhodesiense sleeping sickness during an outbreak in Serere, eastern Uganda. BMC Public Health 8, 96.

Fasihi Harandi, M., Budke, C.M., Rostami, S., 2012. The monetary burden of cystic echinococcosis in Iran. PLoS Negl. Trop. Dis. 6 (11), e1915.

Fenwick, A., Molyneux, D., Nantulya, V., 2005. Achieving the millennium development goals. Lancet 365 (9464), 1029–1030.

Fatiregun, A.A., Osungbade, K.O., Olumide, A.E., January 2009. Cost-effectiveness of screening methods for urinary schistosomiasis in a school-based control programme in Ibadan, Nigeria. Health Policy 89 (1), 72–77.

Fenwick, A., Webster, J.P., Bosque-Oliva, E., Blair, L., Fleming, F.M., Zhang, Y., et al., 2009. The schistosomiasis control initiative (SCI): rationale, development and implementation from 2002-2008. Parasitology 136 (13), 1719–1730.

Global Tuberculosis Control: Surveillance, Planning, Financing, 2007. World Health Organization.

Gold, M.R., Stevenson, D., Fryback, D.G., 2002. HALYs and QALYs and DALYs, oh my: similarities and differences in summary measures of population health. Annu. Rev. Public Health 23, 115–134.

Goldman, A.S., Guisinger, V.H., Aikins, M., Amarillo, M.L., Belizario, V.Y., Garshong, B., et al., 2007. National mass drug administration costs for lymphatic filariasis elimination. PLoS Negl. Trop. Dis. 1 (1), e67.
Gutman, J., Richards Jr., F.O., Eigege, A., Umaru, J., Alphonsus, K., Miri, E.S., 2009. The presumptive treatment of all school-aged children is the least costly strategy for schistosomiasis control in Plateau and Nasarawa states, Nigeria. Ann. Trop. Med. Parasitol. 103 (6), 501–511.
Gabrielli, A.F., Toure, S., Sellin, B., Sellin, E., Ky, C., Ouedraogo, H., et al., 2006. A combined school- and community-based campaign targeting all school-age children of Burkina Faso against schistosomiasis and soil-transmitted helminthiasis: performance, financial costs and implications for sustainability. Acta Trop. 99 (2–3), 234–242.
Guyatt, H.L., Brooker, S., Kihamia, C.M., Hall, A., Bundy, D.A., 2001. Evaluation of efficacy of school-based anthelmintic treatments against anaemia in children in the United Republic of Tanzania. Bull. World Health Organ. 79 (8), 695–703.
Garba, A., Toure, S., Dembele, R., Boisier, P., Tohon, Z., Bosque-Oliva, E., et al., 2009. Present and future schistosomiasis control activities with support from the schistosomiasis control initiative in West Africa. Parasitology 136 (13), 1731–1737.
Garba, A., Toure, S., Dembele, R., Bosque-Oliva, E., Fenwick, A., July 2006. Implementation of national schistosomiasis control programmes in West Africa. Trends Parasitol. 22 (7), 322–326.
Guyatt, H., 2003. The cost of delivering and sustaining a control programme for schistosomiasis and soil-transmitted helminthiasis. Acta Trop. 86 (2–3), 267–274.
Hotez, P.J., Ottesen, E., Fenwick, A., Molyneux, D., 2006. The neglected tropical diseases: the ancient afflictions of stigma and poverty and the prospects for their control and elimination. Adv. Exp. Med. Biol. 582, 23–33.
Hotez, P.J., Ferris, M.T., 2006. The antipoverty vaccines. Vaccine 24 (31–32), 5787–5799.
Hashimoto, K., Alvarez, H., Nakagawa, J., Juarez, J., Monroy, C., Cordon-Rosales, C., et al., 2012. Vector control intervention towards interruption of transmission of Chagas disease by *Rhodnius prolixus*, main vector in Guatemala. Memorias do Inst. Oswaldo Cruz 107 (7), 877–887.
Hirve, S., Singh, S.P., Kumar, N., Banjara, M.R., Das, P., Sundar, S., et al., 2010. Effectiveness and feasibility of active and passive case detection in the visceral leishmaniasis elimination initiative in India, Bangladesh, and Nepal. Am. J. Trop. Med. Hyg. 83 (3), 507–511.
Huda, M.M., Hirve, S., Siddiqui, N.A., Malaviya, P., Banjara, M.R., Das, P., et al., 2012. Active case detection in national visceral leishmaniasis elimination programs in Bangladesh, India, and Nepal: feasibility, performance and costs. BMC Public Health 12, 1001.
Hu, G.H., Hu, J., Song, K.Y., Lin, D.D., Zhang, J., Cao, C.L., et al., 2005. The role of health education and health promotion in the control of schistosomiasis: experiences from a 12-year intervention study in the Poyang Lake area. Acta Trop. 96 (2–3), 232–241.
Jia, T.W., Utzinger, J., Deng, Y., Yang, K., Li, Y.Y., Zhu, J.H., et al., 2011. Quantifying quality of life and disability of patients with advanced schistosomiasis japonica. PLoS Negl. Trop. Dis. 5 (2), e966.
Jia, T.W., Zhou, X.N., Wang, X.H., Utzinger, J., Steinmann, P., Wu, X.H., 2007. Assessment of the age-specific disability weight of chronic schistosomiasis japonica. Bull. World Health Organ. 85 (6), 458–465.
Jimenez, S., Perez, A., Gil, H., Schantz, P., Ramalle, E., Juste, R., 2002. Progress in control of cystic echinococcosis in La Rioja, Spain: decline in infection prevalences in human and animal hosts and economic costs and benefits. Acta Trop. 83 (3), 213–221.

King, C.H., 2010. Parasites and poverty: the case of schistosomiasis. Acta Trop. 113 (2), 95–104.
King, C.H., Dickman, K., Tisch, D.J., 2005. Reassessment of the cost of chronic helmintic infection: a meta-analysis of disability-related outcomes in endemic schistosomiasis. Lancet 365 (9470), 1561–1569.
King, C.H., Bertino, A.M., 2008. Asymmetries of poverty: why global burden of disease valuations underestimate the burden of neglected tropical diseases. PLoS Negl. Trop. Dis. 2 (3), e209.
Kruchten, S.D., Bacon, K.M., Lee, B.Y., 2014. The impact of human immunodeficiency virus (HIV) co-infection on the economic Burden of cutaneous leishmaniasis (CL) in Brazil and potential value of new CL drug treatments. Am. J. Trop. Med. Hyg. 91 (3), 520–527.
Keating, J., Yukich, J.O., Mollenkopf, S., Tediosi, F., 2014. Lymphatic filariasis and onchocerciasis prevention, treatment, and control costs across diverse settings: a systematic review. Acta Trop. 135, 86–95.
Kreimer, P., Romero, L.A., Bilder, P., 2010. [To treat or not to treat? Scientific controversy about the treatment for chronic Chagas' disease patients.]. Asclepio; Archivo Iberoamericano de Historia de la Medicina y Antropologia Medica 62 (2), 405–428 (in Spanish).
Kabatereine, N.B., Tukahebwa, E.M., Kazibwe, F., Twa-Twa, J.M., Barenzi, J.F., Zaramba, S., et al., 2005. Soil-transmitted helminthiasis in Uganda: epidemiology and cost of control. Trop. Med. Int. Health 10 (11), 1187–1189.
Lozano, R., Naghavi, M., Foreman, K., Lim, S., Shibuya, K., Aboyans, V., et al., 2012. Global and regional mortality from 235 causes of death for 20 age groups in 1990 and 2010: a systematic analysis for the Global Burden of Disease Study 2010. Lancet 380 (9859), 2095–2128.
Leslie, J., Garba, A., Oliva, E.B., Barkire, A., Tinni, A.A., Djibo, A., et al., 2011. Schistosomiasis and soil-transmitted helminth control in Niger: cost effectiveness of school based and community distributed mass drug administration [corrected]. PLoS Negl. Trop. Dis. 5 (10), e1326.
Leslie, J., Garba, A., Boubacar, K., Yaye, Y., Sebongou, H., Barkire, A., et al., 2013. Neglected tropical diseases: comparison of the costs of integrated and vertical preventive chemotherapy treatment in Niger. Int. Health 5 (1), 78–84.
Lee, B.Y., Bacon, K.M., Bottazzi, M.E., Hotez, P.J., 2013. Global economic burden of Chagas disease: a computational simulation model. Lancet Infect. Dis. 13 (4), 342–348.
Lee, B.Y., Bacon, K.M., Bailey, R.R., Wiringa, A.E., Smith, K.J., 2011. The potential economic value of a hookworm vaccine. Vaccine 29, 1201–1210.
Lee, B.Y., Bacon, K.M., Connor, D.L., Willig, A.M., Bailey, R.R., 2010. The potential economic value of a *Trypanosoma cruzi* (Chagas disease) vaccine in Latin America. PLoS Negl. Trop. Dis. 4 (12), e916.
Lee, B.Y., Bacon, K.M., Wateska, A.R., Bottazzi, M.E., Dumonteil, E., Hotez, P.J., 2012. Modeling the economic value of a Chagas' disease therapeutic vaccine. Hum. Vaccines Immunother. 8 (9), 1293–1301.
Lee, B.Y., Bacon, K.M., Shah, M., Kitchen, S.B., Connor, D.L., Slayton, R.B., 2012. The economic value of a visceral leishmaniasis vaccine in Bihar state, India. Am. J. Trop. Med. Hyg. 86 (3), 417–425.
Linehan, M., Hanson, C., Weaver, A., Baker, M., Kabore, A., Zoerhoff, K.L., et al., 2011. Integrated implementation of programs targeting neglected tropical diseases through preventive chemotherapy: proving the feasibility at national scale. Am. J. Trop. Med. Hyg. 84 (1), 5–14.
Lwambo, N.J., Siza, J.E., Mwenda, G.C., 2005. Community's willingness to pay for a school-based chemotherapy programme to control morbidity due to schistosomiasis

and soil-transmitted helminthiasis in children in rural Tanzania. Tanzan. Health Res. Bull. 7 (3), 149–153.
Malaria Finding & Resource Utilization: The First Decade of Roll Back Malaria, 2010. World Health Organization & Roll Back Malaria Partnership Secretariat, Geneva.
Murray, C.J., Vos, T., Lozano, R., Naghavi, M., Flaxman, A.D., Michaud, C., et al., 2012. Disability-adjusted life years (DALYs) for 291 diseases and injuries in 21 regions, 1990-2010: a systematic analysis for the Global Burden of Disease Study 2010. Lancet 380 (9859), 2197–2223.
Majorowski, M.M., Carabin, H., Kilani, M., Bensalah, A., 2005. Echinococcosis in Tunisia: a cost analysis. Trans. R. Soc. Trop. Med. Hyg. 99 (4), 268–278.
Moro, P.L., Budke, C.M., Schantz, P.M., Vasquez, J., Santivanez, S.J., Villavicencio, J., 2011. Economic impact of cystic echinococcosis in Peru. PLoS Negl. Trop. Dis. 5 (5), e1179.
Mannocci, A., La Torre, G., Chiaradia, G., De Waure, C., Mainelli, M.T., Cernigliaro, A., et al., 2007. Epidemiology and direct medical costs of human leishmaniasis in Italy. J. Prev. Med. Hyg. 48 (1), 27–36.
Meheus, F., Abuzaid, A.A., Baltussen, R., Younis, B.M., Balasegaram, M., Khalil, E.A., et al., 2013. The economic burden of visceral leishmaniasis in Sudan: an assessment of provider and household costs. Am. J. Trop. Med. Hyg. 89 (6), 1146–1153.
Modabber, F., 2010. Leishmaniasis vaccines: past, present and future. Int. J. Antimicrob. Agents 36 (Suppl. 1), S58–S61.
Montresor, A., Zin, T.T., Padmasiri, E., Allen, H., Savioli, L., 2004. Soil-transmitted helminthiasis in Myanmar and approximate costs for countrywide control. Trop. Med. Int. Health 9 (9), 1012–1015.
Montresor, A., Cong, D.T., Le Anh, T., Ehrhardt, A., Mondadori, E., Thi, T.D., et al., 2007. Cost containment in a school deworming programme targeting over 2.7 million children in Vietnam. Trans. R. Soc. Trop. Med. Hyg. 101 (5), 461–464.
Molyneux, D.H., Nantulya, V.M., 2004. Linking disease control programmes in rural Africa: a pro-poor strategy to reach Abuja targets and millennium development goals. BMJ 328 (7448), 1129–1132.
Malaviya, P., Singh, R.P., Singh, S.P., Hasker, E., Ostyn, B., Shankar, R., et al., 2011. Monitoring drug effectiveness in kala-azar in Bihar, India: cost and feasibility of periodic random surveys vs. a health service-based reporting system. Trop. Med. Int. Health 16 (9), 1159–1166.
Moran, M., Guzman, J., Ropars, A.L., McDonald, A., Jameson, N., Omune, B., et al., 2009. Neglected disease research and development: how much are we really spending? PLoS Med. 6 (2), e30.
Ndeffo Mbah, M.L., Poolman, E.M., Atkins, K.E., Orenstein, E.W., Meyers, L.A., Townsend, J.P., et al., 2013. Potential cost-effectiveness of schistosomiasis treatment for reducing HIV transmission in Africa—the case of Zimbabwean women. PLoS Negl. Trop. Dis. 7 (8), e2346.
Ndeffo Mbah, M.L., Kjetland, E.F., Atkins, K.E., Poolman, E.M., Orenstein, E.W., Meyers, L.A., et al., 2013. Cost-effectiveness of a community-based intervention for reducing the transmission of *Schistosoma haematobium* and HIV in Africa. Proc. Natl. Acad. Sci. USA 110 (19), 7952–7957.
Onwujekwe, O., Shu, E., Onwuameze, O., Ndum, C., Okonkwo, P., 2001. Onchocerciasis control in Nigeria: will households be able to afford community-directed treatment with ivermectin? Acta Trop. 80 (3), 277–281.
Report on the Global AIDS Epidemic, 2012. Joint United Nations Programme on HIV/AIDS, Washington, D.C.
Reidpath, D.D., Allotey, P.A., Kouame, A., Cummins, R.A., 2003. Measuring health in a vacuum: examining the disability weight of the DALY. Health Policy Plan. 18 (4), 351–356.

Ramsey, J.M., Elizondo-Cano, M., Sanchez-Gonzalez, G., Pena-Nieves, A., Figueroa-Lara, A., 2014. Opportunity cost for early treatment of Chagas disease in Mexico. PLoS Negl. Trop. Dis. 8 (4), e2776.

Ritmeijer, K., Davies, C., van Zorge, R., Wang, S.J., Schorscher, J., Dongu'du, S.I., et al., 2007. Evaluation of a mass distribution programme for fine-mesh impregnated bednets against visceral leishmaniasis in eastern Sudan. Trop. Med. Int. Health 12 (3), 404–414.

Robays, J., Raguenaud, M.E., Josenando, T., Boelaert, M., 2008. Eflornithine is a cost-effective alternative to melarsoprol for the treatment of second-stage human West African trypanosomiasis in Caxito, Angola. Trop. Med. Int. Health 13 (2), 265–271.

Reithinger, R., Coleman, P.G., 2007. Treating cutaneous leishmaniasis patients in Kabul, Afghanistan: cost-effectiveness of an operational program in a complex emergency setting. BMC Infect. Dis. 7, 3.

Rheingans, R.D., Haddix, A.C., Messonnier, M.L., Meltzer, M., Mayard, G., Addiss, D.G., 2004. Willingness to pay for prevention and treatment of lymphatic filariasis in Leogane, Haiti. Filaria J. 3, 2.

Spiegel, A., Tall, A., Raphenon, G., Trape, J.F., Druilhe, P., 2003. Increased frequency of malaria attacks in subjects co-infected by intestinal worms and *Plasmodium falciparum* malaria. Trans. R. Soc. Trop. Med. Hyg. 97 (2), 198–199.

Sonnenberg, F.A., Beck, J.R., 1993. Markov models in medical decision making: a practical guide. Med. Decis. Mak. 13 (4), 322–338.

Singh, B.B., Dhand, N.K., Ghatak, S., Gill, J.P., 2014. Economic losses due to cystic echinococcosis in India: need for urgent action to control the disease. Prev. Vet. Med. 113 (1), 1–12.

Sarnoff, R., Desai, J., Desjeux, P., Mittal, A., Topno, R., Siddiqui, N.A., et al., 2010. The economic impact of visceral leishmaniasis on rural households in one endemic district of Bihar, India. Trop. Med. Int. Health 15 (Suppl. 2), 42–49.

Sundar, S., Arora, R., Singh, S.P., Boelaert, M., Varghese, B., 2010. Household cost-of-illness of visceral leishmaniasis in Bihar, India. Trop. Med. Int. Health 15 (Suppl. 2), 50–54.

Singh, S.P., Hirve, S., Huda, M.M., Banjara, M.R., Kumar, N., Mondal, D., et al., 2011. Options for active case detection of visceral leishmaniasis in endemic districts of India, Nepal and Bangladesh, comparing yield, feasibility and costs. PLoS Negl. Trop. Dis. 5 (2), e960.

Sinuon, M., Tsuyuoka, R., Socheat, D., Montresor, A., Palmer, K.L., 2005. Financial costs of deworming children in all primary schools in Cambodia. Trans. R. Soc. Trop. Med. Hyg. 99, 664–668.

Shastry, G.K., Weil, D.N., 2003. How much of cross-country income variation is explained by health? J. Eur. Econ. Assoc. 1 (2/3), 387–396.

Sicuri, E., Munoz, J., Pinazo, M.J., Posada, E., Sanchez, J., Alonso, P.L., et al., 2011. Economic evaluation of Chagas disease screening of pregnant Latin American women and of their infants in a non endemic area. Acta Trop. 118 (2), 110–117.

Stromme, E.M., Baeroe, K., Norheim, O.F., 2013. Disease control priorities for neglected tropical diseases: lessons from priority ranking based on the quality of evidence, cost effectiveness, severity of disease, catastrophic health expenditures, and loss of productivity. Dev. World Bioeth. 14 (3), 132–141.

Turner, H.C., Osei-Atweneboana, M.Y., Walker, M., Tettevi, E.J., Churcher, T.S., Asiedu, O., et al., 2013. The cost of annual versus biannual community-directed treatment of onchocerciasis with ivermectin: Ghana as a case study. PLoS Negl. Trop. Dis. 7 (9), e2452.

Uniting to Combat Neglected Tropical Diseases, 2014. Delivering on Promises and Driving Progress - Second Progress Report.

Uranw, S., Meheus, F., Baltussen, R., Rijal, S., Boelaert, M., 2013. The household costs of visceral leishmaniasis care in south-eastern Nepal. PLoS Negl. Trop. Dis. 7 (2), e2062.

Vazquez-Prokopec, G.M., Spillmann, C., Zaidenberg, M., Kitron, U., Gurtler, R.E., 2009. Cost-effectiveness of chagas disease vector control strategies in Northwestern Argentina. PLoS Negl. Trop. Dis. 3 (1), e363.

Vanlerberghe, V., Diap, G., Guerin, P.J., Meheus, F., Gerstl, S., Van der Stuyft, P., et al., 2007. Drug policy for visceral leishmaniasis: a cost-effectiveness analysis. Trop. Med. Int. Health 12 (2), 274–283.

van der Werf, M.J., de Vlas, S.J., Brooker, S., Looman, C.W., Nagelkerke, N.J., Habbema, J.D., et al., May 2003. Quantification of clinical morbidity associated with schistosome infection in sub-Saharan Africa. Acta Trop. 86 (2–3), 125–139.

Wang, L., Wen, H., Feng, X., Jiang, X., Duan, X., 2012. Analysis of economic burden for patients with cystic echinococcosis in five hospitals in northwest China. Trans. R. Soc. Trop. Med. Hyg. 106 (12), 743–748.

van der Werf, M.J., Bosompem, K.M., de Vlas, S.J., 2003. Schistosomiasis control in Ghana: case management and means for diagnosis and treatment within the health system. Trans. R. Soc. Trop. Med. Hyg. 97 (2), 146–152.

WHO, 2006. Neglected Tropical Diseases: Hidden Successes, Emerging Opportunities. Geneva.

Wilson, L.S., Strosberg, A.M., Barrio, K., 2005. Cost-effectiveness of Chagas disease interventions in Latin america and the Caribbean: Markov models. Am. J. Trop. Med. Hyg. 73 (5), 901–910.

Yang, Y.R., Williams, G.M., Craig, P.S., Sun, T., Yang, S.K., Cheng, L., et al., 2006. Hospital and community surveys reveal the severe public health problem and socio-economic impact of human echinococcosis in Ningxia Hui Autonomous Region, China. Trop. Med. Int. Health 11 (6), 880–888.

Yu, D., Manderson, L., Yuan, L., Wei, W., He, H., Chen, Y., 2001. Is equity being sacrificed? Willingness and ability to pay for schistosomiasis control in China. Health Policy Plan. 16 (3), 292–301.

Zhou, X.N., Wang, L.Y., Chen, M.G., Wang, T.P., Guo, J.G., Wu, X.H., et al., 2005. An economic evaluation of the national schistosomiasis control programme in China from 1992 to 2000. Acta Trop. 96 (2–3), 255–265.

Zhang, Y., MacArthur, C., Mubila, L., Baker, S., 2010. Control of neglected tropical diseases needs a long-term commitment. BMC Med. 8, 67.

INDEX

Note: Page numbers followed by "f" and "t" indicate figures and tables respectively

E

F

G

H

N

O

P

Q

R

S

W

Y

CONTENTS OF VOLUMES IN THIS SERIES

Volume 46

Volume 47

Volume 48

Volume 49

Volume 50

Volume 51

Volume 52

Volume 53

Volume 54

Volume 55

Volume 61

Volume 62

Volume 63

Volume 64

Volume 65

Volume 66

Volume 67

Volume 68

Volume 69

Volume 70

Volume 71

Volume 72

Volume 73

Volume 74

Volume 75

Volume 76

Volume 77

Volume 78

Volume 79

Volume 80

Volume 81

Volume 82

Volume 83

Volume 84

Volume 85

Volume 86

Kolibri 360 GmbH
Ferdinand-Jühlke-Straße 7
99095 Erfurt, Deutschland
produktsicherheit@kolibri360.de